# INSTRUMENTATION
IN ELEMENTARY
PARTICLE PHYSICS

# Related Titles from AIP Conference Proceedings and the Subseries on High Energy Physics

**533**   Next Generation Nucleon Decay and Neutrino Detector: NNN99
Edited by Milind V. Diwan and Chang Kee Jung, August 2000, 1-56396-956-4

**531**   Particles and Fields: Seventh Mexican Workshop
Edited by Alejandro Ayala, Guillermo Contreras, and Gerardo Herrera, July 2000, 1-56396-954-8

**530**   Colliders and Collider Physics at the Highest Energies: Muon Colliders at 10 TeV to 100 TeV, HEMC'99 Workshop
Edited by Bruce J. King, August 2000, 1-56396-953-X

**512**   Nuclear Physics at Storage Rings: Fourth International Conference: STORI99
Edited by Hans-Otto Meyer and Peter Schwandt, June 2000, 1-56396-928-9

**508**   Hadron Physics: Effective Theories of Low Energy QCD
Edited by A. H. Blin, B. Hiller, M. C. Ruivo, C. A. Sousa, and E. van Beveren, March 2000, 1-56396-927-0

**496**   Workshop on Instabilities of High Intensity Hadron Beams in Rings
Edited by T. Roser and S. Y. Zhang, December 1999, 1-56396-910-6

**494**   New Directions in Quantum Chromodynamics
Edited by Chueng-Ryong Ji and Dong-Pil Min, November 1999, 1-56396-908-4

**490**   Particles and Fields: Eighth Mexican School
Edited by Juan Carlos D'Olivo, Gabriel López Castro, and Myriam Mondragón, November 1999, 1-56396-895-9

**488**   High Energy Physics at the Millennium: MRST'99
Edited by Pat Kalyniak, Stephen Godfrey, and B. Kamal, October 1999, 1-56396-902-5

**482**   RHIC Physics and Beyond: Kay Kay Gee Day
Edited by Berndt Müller and Robert Pisarski, July 1999, 1-56396-878-9

**459**   Heavy Quarks at Fixed Target
Edited by Harry W. K. Cheung and Joel N. Butler, February 1999, 1-56396-864-9

**453**   Particles, Fields, and Gravitation
Edited by Jakub Rembieliński, December 1998, 1-56396-837-1

**422**   Instrumentation in Elementary Particle Physics: The VII ICFA School
Edited by G. Herrera Corral and M. Sosa Aquino, April 1998, 1-56396-763-4

To learn more about these titles, or the AIP Conference Proceedings Series, please visit the webpage **http://www.aip.org/catalog/aboutconf.html**

# INSTRUMENTATION IN ELEMENTARY PARTICLE PHYSICS

VIII ICFA School

*Istanbul, Turkey    28 June–10 July 1999*

*EDITOR*
Sehban Kartal
*University of Istanbul*

Melville, New York, 2000
AIP CONFERENCE PROCEEDINGS ■ VOLUME 536

**Editor:**

Sehban Kartal
Department of Physics
Science Faculty
University of Istanbul
34459 Vezneciler, Istanbul
TURKEY

E-mail: sehban@istanbul.edu.tr

Authorization to photocopy items for internal or personal use, beyond the free copying permitted under the 1978 U.S. Copyright Law (see statement below), is granted by the American Institute of Physics for users registered with the Copyright Clearance Center (CCC) Transactional Reporting Service, provided that the base fee of $17.00 per copy is paid directly to CCC, 222 Rosewood Drive, Danvers, MA 01923. For those organizations that have been granted a photocopy license by CCC, a separate system of payment has been arranged. The fee code for users of the Transactional Reporting Service is: 1-56396-960-2/00/$17.00.

© 2000 American Institute of Physics

Individual readers of this volume and nonprofit libraries, acting for them, are permitted to make fair use of the material in it, such as copying an article for use in teaching or research. Permission is granted to quote from this volume in scientific work with the customary acknowledgment of the source. To reprint a figure, table, or other excerpt requires the consent of one of the original authors and notification to AIP. Republication or systematic or multiple reproduction of any material in this volume is permitted only under license from AIP. Address inquiries to Office of Rights and Permissions, Suite 1NO1, 2 Huntington Quadrangle, Melville, N.Y. 11747-4502; phone: 516-576-2268; fax: 516-576-2450; e-mail: rights@aip.org.

L.C. Catalog Card No. 00-106869
ISBN 1-56396-960-2
ISSN 0094-243X
Printed in the United States of America

## CONTENTS

Preface .................................................................. vii
Organization ............................................................ viii
Acknowledgments ......................................................... ix
Group Photograph ........................................................ xi

### LECTURE COURSES

Physics of Particle Detection ............................................. 3
    C. Grupen
Silicon Detectors ........................................................ 35
    P. Giubellino, E. Crescio, R. Hernandez, M. Idzik, D. Nouais,
    and A. Rivetti
RICH Detectors .......................................................... 60
    E. Nappi
Detectors for Particle Identification Time-of-Flight, dE/dx,
and Transition Radiation ................................................. 87
    M. Sheaff
Future Particle Detector Systems ........................................ 115
    A. G. Clark

### REVIEW TALKS

Perspectives in High-Energy Physics ..................................... 165
    C. Quigg
Confronting New Technological Challenges in HEP ......................... 192
    A. Savoy-Navarro
Advanced Photodetectors for High Energy Physics Particle
Astrophysics and Medical Imaging ........................................ 227
    M. Ataç
Visible Light Photon Counters for High Rate Tracking Medical
Imaging and Particle Astrophysics ....................................... 240
    M. Ataç
Detecting Dark Matter ................................................... 253
    R. L. Dixon
Review of HERA Experiments .............................................. 267
    G. Herrera Corral

### LABORATORY SESSIONS

Lifetime of Cosmic Ray Muons ............................................ 291
    J. Anderson, M. Ataç, S. Cihangir, N. Erduran, S. Kartal,
    and S. Erturk

**Measurement of Attenuation Length of Photons and Determination
of Photon Yields in Plastic Scintillators** .................................... 294
    M. Ataç, S. Cihangir, N. Erduran, S. Kartal, and S. Erturk
**Muon Lifetime Measurement** ............................................ 301
    S. A. Çetin, T. Çonka-Nurdan, and A. Mailov
**Principle of Operation of Micro Pattern Detectors** ......................... 324
    A. Cattai and R. Malina
**Use of De-Randomizing Buffers in a Data Acquisition System** ............... 329
    M. Johnson and M. Sheaff
**Tests of a Position Sensitive Photomultiplier and Measurement
of Diffraction Pattern by Counting Single Photons** ......................... 340
    S. Korpar, P. Križan, A. Gorišek, and A. Stanovnik
**Laboratory Course on Silicon Sensors** ................................... 349
    R. Brenner, S. Roe, and A. Rudge
**GaAs Pixel Detector with Integrated Readout for Digital Mammography** ..... 381
    S. R. Amendolia, M. G. Bisogni, U. Bottigli, P. Delogu,
    G. Dipasquale, M. E. Fantacci, V. Marzulli, P. Maestro,
    E. Pernigotti, V. Rosso, A. Stefanini, S. Stumbo, and O. Venier

**List of Participants** ...................................................... 401
**Author Index** ........................................................... 405

# PREFACE

The "1999 ICFA School VIII on Instrumentation in Elementary Particle Physics" was held at the University of Istanbul, Istanbul, Turkey, from June 28 to July 10, 1999. There were approximately 100 participants, including graduate students, postdocs, and faculty from all continents except Australia and Antarctica. Some 42 distinguished physicists from the U.S.A., Europe, Japan, and Mexico contributed to the program, which was organised by the ICFA Panel on Instrumentation, Innovation and Future Development. In addition to lectures and "hands-on" laboratory courses on the state-of-the-art instrumentation techniques for high energy physics, there were a number of special talks on future experiments, collaboration with industry, and instrumentation for medical imaging and particle physics and particle astrophysics.

The main organisers of the school were M. N. Erduran, B. Akkus, and S. Kartal from Istanbul University and M. Atac from Fermilab. The major contributors in financing the school were CERN, DOE/Fermilab, NSF, DESY, ICTP, as well as the Universities of Istanbul and Bogazici. The students were given a Certificate of Accomplishment following the school in recognition of their very enthusiastic participation in the school program. As in the past, the school was very successful. At a wrap-up session, the students indicated satisfaction with the program presented. They all seemed to agree that the experience would be valuable to them in their future careers. The social program was also appreciated and everyone went back home happy. The University of Istanbul provided a delightful one day excursion to the Bosphorus and nearby islands, which was very much enjoyed by all.

*Sehban Kartal*
Coordinator of the School


"As the physics shapes life and life shapes the physics, man's greatest struggle in the new millennium will be to shape himself. While man is engaged in this struggle, as a result of questioning himself and his activities, a new kind of civilized man will be created in the new millennium. Greetings to all those who are engaged in this struggle and those who will be engaged in this struggle."

S. Kartal, 28.12.1999

# ORGANIZATION

## ICFA Panel of Instrumentation Innovation and Development

### International Organizing Committee

| | |
|---|---|
| Ariella Cattai | CERN (Switzerland) |
| Boris Dolgoshein | MPEI (Russia) |
| Tord Ekelöf | Uppsala University (Sweden) |
| Igor A. Golutvin | JINP (Russia) |
| Atul Gurtu | RIFR (India) |
| Gerardo Herrera | CINVESTAV (Mexico) |
| Takakiko Kondo | KEK (Japan) |
| Stan Majewski | TJNAF (USA) |
| Simon W.L. Kwan | FNAL (USA) |
| H. Okuno | INS (Japan) |
| Aurore Savoy-Navarro | LPNHE- Paris 6 et 7 (France) |
| Marleigh Sheaff | Wisconsin (USA)/ CINVESTAV (Mexico) |
| Veniamin Sidorov | INP (Russia) |
| Jaroslav Va'vra | SLAC (USA) |
| A. P. Vorobiev | IHEP (Russia) |
| Heinrich Walenta | Siegen (Germany) |

### Local Organizing Committee

| | |
|---|---|
| M. Nizamettin Erduran | (Univ. of Istanbul) - *Chairman* |
| Sehban Kartal | (Univ. of Istanbul) - *Coordinator* |
| Baki Akkus | (Univ. of Istanbul) - *Coordinator* |
| Engin Arik | (Bogazici Univ.) |
| Metin Arik | (Bogazici Univ.) |
| Muzaffer Ataç | (FNAL / UCLA) |
| Melih Bostan | (Univ. of Istanbul) |
| Ayla Çelikel | (Ankara Univ.) |
| Serkan Çetin | (Bogazici Univ.) |
| Tuba Çonka | (Bogazici Univ.) |
| Kıvanç Cankoçak | (Bogazici Univ.) |
| Arif Mailov | (Bogazici Univ.) |
| Saleh Sultansoy | (Ankara Univ.) |

# ACKNOWLEDGEMENTS

The 1999 ICFA Instrumentation School took place from June 28 - July 10 at University of Istanbul. It was organised by the International Committee for Future Accelerators, Instrumentation Panel, and the University of Istanbul. On behalf of the Local Organizing Committee, we would like to thank the International Organizing Committee for its role in the selection of the scientific program and the Laboratory and Institutions for providing the necessary equipment for the laboratory courses, which is an important feature of the school. The support of CERN, DESY, FNAL, INFN, Josef Stefan Institute, Siegen University, are gratefully acknowledged. The school was sponsored by CERN, DESY, FNAL, DOE, NSF, ICTP, IN2P3, RAL, and the Turkish Physical Society.

The ICFA Instrumentation School also received grants from the University of Istanbul and Bogazici University Centre for Turkish Balkan Physics and Applications (CTBP). We would like, in particular, to give our warmest thanks to Prof. Dr. Kemal Alemdaroglu, the Rector of the University of Istanbul, Prof. Dr. Dincer Gulen, the Dean of the Science Faculty, and Prof. Dr. Metin Arik, Director of CTBP, and Prof. Dr. Marleigh Sheaff, Wisconsin (USA)/ CINVESTAV (Mexico).

Thanks are due to members of the Local Organizing Committee, particularly to Azmi Barut, Tuba Conka, Serkant Cetin, Yesim Cetin, Isa Dumanoglu, Sefa Erturk, Gulhan Gurdal, Mehtap Yalcınkaya whose energy and commitment helped participants to solve all kinds of practical problems throughout the school. Also, special thanks to Mr. Tabib Derinbay, the Director of the University Publication Office and Mr. Namik Kaya, Director of the University Staff Club.

Prof. Dr. Muzaffer Atac played a major role not only in organizing the school here in Istanbul but also in providing guidance and advice on difficult matters which arose from time to time and stimulating discussions on important issues from transporting equipment to school lecture programming. We would like to convey our deepest appreciation for his endless efforts. Many thanks Muzaffer.

Finally, we would like to thank all lecturers, instructors, and students who participated in the school, whose collective interest, enthusiasm, and efforts together made this school a success.

*Sehban Kartal*
Coordinator of the School

*M. Nizamettin Erduran*
Director of the School

# LECTURE COURSES

# Physics of Particle Detection

Claus Grupen

*Department of Physics, University of Siegen*
*D-57068 Siegen, Germany*
*e-mail: grupen@siux00.physik.uni-siegen.de*

**Abstract.** In this review the basic interaction mechanisms of charged and neutral particles are presented. The ionization energy loss of charged particles is fundamental to most particle detectors and is therefore described in more detail. The production of electromagnetic radiation in various spectral ranges leads to the detection of charged particles in scintillation, Cherenkov and transition radiation counters. Photons are measured via the photoelectric effect, Compton scattering or pair production, and neutrons through their nuclear interactions.

A combination of the various detector methods helps to identify elementary particles and nuclei. At high energies absorption techniques in calorimeters provide additional particle identification and an accurate energy measurement.

## INTRODUCTION

The detection and identification of elementary particles and nuclei is of particular importance in high energy, cosmic ray and nuclear physics [1-6]. Identification means that the mass of the particle and its charge is determined. In elementary particle physics most particles have unit charge. But in the study e.g. of the chemical composition of primary cosmic rays different charges must be distinguished.

Every effect of particles or radiation can be used as a working principle for a particle detector.

The deflection of a charged particle in a magnetic field determines its momentum $p$; the radius of curvature $\rho$ is given by

$$\rho \propto \frac{p}{z} = \frac{\gamma m_0 \beta c}{z} \qquad (1)$$

where $z$ is the particle's charge, $m_0$ its rest mass and $\beta = \frac{v}{c}$ its velocity. The particle velocity can be determined e.g. by a time-of-flight method yielding

$$\beta \propto \frac{1}{\tau} \, , \qquad (2)$$

where $\tau$ is the flight time. A calorimetric measurement provides a determination of the kinetic energy

$$E^{\text{kin}} = (\gamma - 1) m_0 c^2 \qquad (3)$$

where $\gamma = \frac{1}{\sqrt{1-\beta^2}}$ is the Lorentz factor.

From these measurements the ratio of $m_0/z$ can be inferred, i.e. for singly charged particles we have already identified the particle. To determine the charge one needs another $z$-sensitive effect, e.g. the ionization energy loss

$$\frac{dE}{dx} \propto \frac{z^2}{\beta^2} \ln(a\beta\gamma) \qquad (4)$$

($a$ is a material dependent constant.)

Now we know $m_0$ and $z$ separately. In this way even different isotopes of elements can be distinguished.

The basic principle of particle detection is that every physics effect can be used as an idea to build a detector. In the following we distinguish between the interaction of charged and neutral particles. In most cases the observed signature of a particle is its ionization, where the liberated charge can be collected and amplified, or its production of electromagnetic radiation which can be converted into a detectable signal. In this sense neutral particles are only detected indirectly, because they must first produce in some kind of interaction a charged particle which is then measured in the usual way.

# INTERACTION OF CHARGED PARTICLES

## Kinematics

Four-momentum conservation allows to calculate the maximum energy transfer of a particle of mass $m_0$ and velocity $v = \beta c$ to an electron initially at rest to be [2]

$$E^{\text{max}}_{\text{kin}} = \frac{2 m_e c^2 \beta^2 \gamma^2}{1 + 2\gamma \frac{m_e}{m_0} + \left(\frac{m_e}{m_0}\right)^2} = \frac{2 m_e p^2}{m_0^2 + m_e^2 + 2 m_e E/c^2} \quad , \qquad (5)$$

here $\gamma = \frac{E}{m_0 c^2}$ is the Lorentz factor, $E$ the total energy and $p$ the momentum of the particle.

For low energy particles heavier than the electron ($2\gamma \frac{m_e}{m_0} \ll 1$; $\frac{m_e}{m_0} \ll 1$) eq. 5 reduces to

$$E^{\text{max}}_{\text{kin}} = 2 m_e c^2 \beta^2 \gamma^2 \quad . \qquad (6)$$

For relativistic particles ($E_{\text{kin}} \approx E$; $pc \approx E$) one gets

$$E^{\max} = \frac{E^2}{E + m_0^2 c^2/2m_e} \quad . \tag{7}$$

For example, in a $\mu$-$e$ collision the maximum transferable energy is

$$E^{\max} = \frac{E^2}{E + 11} \qquad E \text{ in GeV} \tag{8}$$

showing that in the extreme relativistic case the complete energy can be transferred to the electron.

If $m_0 = m_e$, eq. 5 is modified to

$$E_{\text{kin}}^{\max} = \frac{p^2}{m_e + E/c^2} = \frac{E^2 - m_e^2 c^4}{E + m_e c^2} = E - m_e c^2 \quad . \tag{9}$$

## Scattering

### Rutherford Scattering

The scattering of a particle of charge $z$ on a target of nuclear charge $Z$ is mediated by the electromagnetic interaction (figure 1).

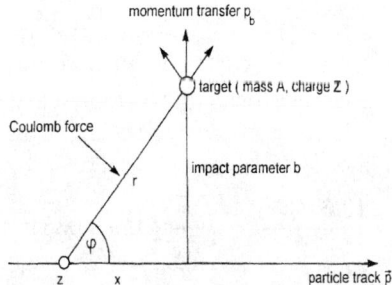

**FIGURE 1.** Kinematics of Coulomb scattering of a particle of charge $z$ on a target of charge $Z$

The Coulomb force between the incoming particle and the target is written as

$$\vec{F} = \frac{z \cdot e \cdot Z \cdot e}{r^2} \frac{\vec{r}}{r} \quad . \tag{10}$$

For symmetry reasons the net momentum transfer is only perpendicular to $\vec{p}$ along the impact parameter $b$

$$p_b = \int_{-\infty}^{+\infty} F_b \, dt = \int_{-\infty}^{+\infty} \frac{z \cdot Z \cdot e^2}{r^2} \cdot \frac{b}{r} \cdot \frac{dx}{\beta c} \quad , \qquad (11)$$

with $b = r \sin \varphi$, $dt = dx/v = dx/\beta c$, and $F_b$ force perpendicular to $p$.

$$p_b = \frac{z \cdot Z \cdot e^2}{\beta c} \int_{-\infty}^{+\infty} \frac{b \, dx}{(\sqrt{x^2 + b^2})^3} = \frac{z \cdot Z \cdot e^2}{\beta c b} \underbrace{\int_{-\infty}^{+\infty} \frac{d(x/b)}{\left(\sqrt{1 + \left(\frac{x}{b}\right)^2}\right)^3}}_{=2} \qquad (12)$$

$$p_b = \frac{2z \cdot Z \cdot e^2}{\beta c b} = \frac{2 r_e m_e c}{b \beta} z \cdot Z \quad , \qquad (13)$$

where $r_e$ is the classical electron radius. This consideration leads to a scattering angle

$$\Theta = \frac{p_b}{p} = \frac{2z \cdot Z \cdot e^2}{\beta c b} \cdot \frac{1}{p} \quad . \qquad (14)$$

The cross section for this process is given by the well-known Rutherford formula

$$\frac{d\sigma}{d\Omega} = \frac{z^2 Z^2 r_e^2}{4} \left(\frac{m_e c}{\beta p}\right)^2 \frac{1}{\sin^4 \Theta/2} \quad . \qquad (15)$$

## Multiple Scattering

From eq. 15 one can see that the average scattering angle $\langle \Theta \rangle$ is zero. To characterize the different degrees of scattering when a particle passes through an absorber one normally uses the so-called "average scattering angle" $\sqrt{\langle \Theta^2 \rangle}$. The projected angular distribution of scattering angles in this sense leads to an average scattering angle of [6]

$$\sqrt{\langle \Theta^2 \rangle} = \Theta_{\text{plane}} = \frac{13.6 \, \text{MeV}}{\beta c p} z \cdot \sqrt{\frac{x}{X_0}} \left\{ 1 + 0.038 \ln \left(\frac{x}{X_0}\right) \right\} \qquad (16)$$

with $p$ in MeV/c and $x$ the thickness of the scattering medium measured in radiation lengths $X_0$ (see **Bremsstrahlung**). The average scattering angle in three dimensions is

$$\Theta_{\text{space}} = \sqrt{2} \, \Theta_{\text{plane}} = \sqrt{2} \, \Theta_0 \quad . \qquad (17)$$

The projected angular distribution of scattering angles can approximately be represented by a Gaussian

$$P(\Theta) d\Theta = \frac{1}{\sqrt{2\pi} \Theta_0} \exp \left\{ -\frac{\Theta^2}{2 \Theta_0^2} \right\} d\Theta \quad . \qquad (18)$$

# Energy Loss of Charged Particles

Charged particles interact with a medium via electromagnetic interactions by the exchange of photons. If the range of photons is short, the absorption of virtual photons constituting the field of the charged particle gives rise to ionization of the material. If the medium is transparent Cherenkov radiation can be emitted above a certain threshold. But also sub-threshold emission of electromagnetic radiation can occur, if discontinuities of the dielectric constant of the material are present (transition radiation) [7]. The emission of real photons by decelerating a charged particle in a Coulomb field also constitutes an important energy loss (bremsstrahlung).

## Ionization Energy-Loss

### *Bethe-Bloch Formula*

This energy-loss mechanism represents the scattering of charged particles off atomic electrons, e.g.

$$\mu^+ + \text{atom} \rightarrow \mu^+ + \text{atom}^+ + e^- \quad . \tag{19}$$

The momentum transfer to the electron is (see eq. 13)

$$p_b = \frac{2 r_e m_e c}{b \beta} z \quad ,$$

and the energy transfer in the classical approximation

$$\varepsilon = \frac{p_b^2}{2 m_e} = \frac{2 r_e^2 m_e c^2}{b^2 \beta^2} z^2 \quad . \tag{20}$$

The interaction probability per (g/cm$^2$), given the atomic cross-section $\sigma$, is

$$\phi(\text{g}^{-1}\text{cm}^2) = \frac{N}{A} \sigma [\text{cm}^2/\text{atom}] \tag{21}$$

where $N$ is Avogadro's constant.

The differential probability to hit an electron in the area of an annulus with radii $b$ and $b + db$ (see figure 2) with an energy transfer between $\varepsilon$ and $\varepsilon + d\varepsilon$ is

$$\phi(\varepsilon) d\varepsilon = \frac{N}{A} 2\pi b \, db \, Z \quad , \tag{22}$$

because there are $Z$ electrons per target atom.

Inserting $b$ from eq. 20 into eq. 22 gives

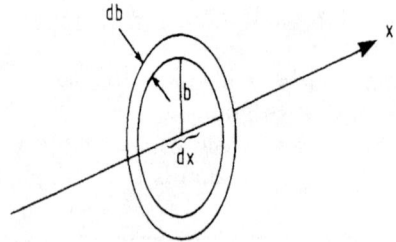

**FIGURE 2.** Sketch explaining the differential collision probability

$$b^2 = \frac{2r_e^2 m_e c^2}{\beta^2} z^2 \cdot \frac{1}{\varepsilon}$$

$$2|b\,db| = \frac{2r_e^2 m_e c^2}{\beta^2} z^2 \cdot \frac{d\varepsilon}{\varepsilon^2}$$

$$\phi(\varepsilon) d\varepsilon = \frac{N}{A} \pi \frac{2r_e^2 m_e c^2}{\beta^2} z^2 \cdot Z \cdot \frac{d\varepsilon}{\varepsilon^2}$$

$$= \frac{2\pi r_e^2 m_e c^2 N}{\beta^2} \cdot \frac{Z}{A} \cdot z^2 \cdot \frac{d\varepsilon}{\varepsilon^2}, \quad (23)$$

showing that the energy spectrum of δ-electrons or knock-on electrons follows an $1/\varepsilon^2$ dependence (figure 3, [8]).

The energy loss is now computed from eq. 22 by integrating over all possible impact parameters [5]

$$-dE = \int_0^\infty \phi(\varepsilon) \cdot \varepsilon \cdot dx$$

$$= \int_0^\infty \frac{N}{A} 2\pi b \cdot db \cdot Z \cdot \varepsilon \cdot dx$$

$$-\frac{dE}{dx} = \frac{2\pi N}{A} \cdot Z \int_0^\infty \varepsilon \cdot b \cdot db$$

$$= 2\pi \frac{Z \cdot N}{A} \cdot \frac{2r_e^2 m_e c^2}{\beta^2} z^2 \int_0^\infty \frac{db}{b}. \quad (24)$$

This classical calculation yields an integral which diverges for $b = 0$ as well as for $b = \infty$. This is not a surprise because one would not expect that our approximations hold for these extremes.

a) The $b = 0$ case: Let us approximate the "size" of the target electron seen from the rest frame of the incident particle by half the de Broglie wavelength. This gives a minimum impact parameter of

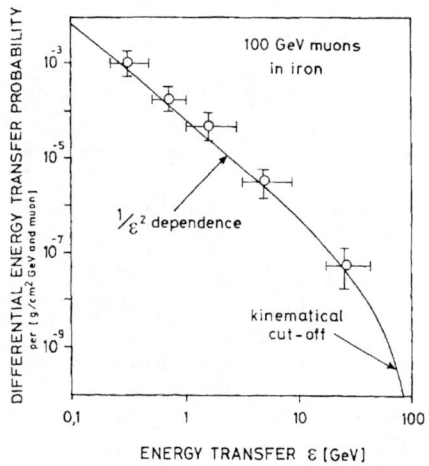

**FIGURE 3.** $1/\varepsilon^2$-dependence of the knock-on electron production probability [8]

$$b_{\min} = \frac{h}{2p} = \frac{h}{2\gamma m_e \beta c} \quad . \qquad (25)$$

b) The $b = \infty$ case: If the revolution time $\tau_R$ of the electron in the target atom becomes smaller than the interaction time $\tau_i$, the incident particle "sees" a more or less neutral atom

$$\tau_i = \frac{b_{\max}}{v}\sqrt{1-\beta^2} \quad . \qquad (26)$$

The factor $\sqrt{1-\beta^2}$ takes into account that the field at high velocities is Lorentz-contracted. Hence the interaction time is shorter. For the revolution time we have

$$\tau_R = \frac{1}{\nu_Z \cdot Z} = \frac{h}{I} \quad , \qquad (27)$$

where $I$ is the mean excitation energy of the target material, which can be approximated by

$$I = 10eV \cdot Z \qquad (28)$$

for elements heavier than sulphur.

The condition to see the target as neutral now leads to

$$\tau_R = \tau_i \quad \Rightarrow \quad \frac{b_{\max}}{v}\sqrt{1-\beta^2} = \frac{h}{I}$$

$$b_{\max} = \frac{\gamma \hbar \beta c}{I} \quad . \tag{29}$$

With the help of eq. 25 and 29 we can solve the integral in eq. 24

$$-\frac{dE}{dx} = 2\pi \cdot \frac{Z}{A} N \cdot \frac{2r_e^2 m_e c^2}{\beta^2} z^2 \cdot \ln \frac{2\gamma^2 \beta^2 m_e c^2}{I} \quad . \tag{30}$$

Since for long-distance interactions the Coulomb field is screened by the intervening matter one has

$$-\frac{dE}{dx} = \kappa z^2 \cdot \frac{Z}{A} \frac{1}{\beta^2} \left[ \ln \frac{2\gamma^2 \beta^2 m_e c^2}{I} - \eta \right] \quad , \tag{31}$$

where $\eta$ is a screening parameter (density parameter) and

$$\kappa = 4\pi N r_e^2 m_e c^2 \quad .$$

The exact treatment of the ionization energy loss of heavy particles leads to [6]

$$-\frac{dE}{dx} = \kappa z^2 \cdot \frac{Z}{A} \cdot \frac{1}{\beta^2} \left[ \frac{1}{2} \ln \frac{2 m_e c^2 \gamma^2 \beta^2}{I^2} E_{\text{kin}}^{\max} - \beta^2 - \frac{\delta}{2} \right] \tag{32}$$

which reduces to eq. 31 for $\gamma m_e/m_0 \ll 1$ and $\beta^2 - \frac{\delta}{2} = \eta$.

The energy-loss rate of muons in iron is shown in Figure 4 [6]. It exhibits a $\frac{1}{\beta^2}$-decrease until a minimum of ionization is obtained for $3 \leq \beta\gamma \leq 4$.

Due to the $\ln \gamma$-term the energy loss increases again (relativistic rise, logarithmic rise) until a plateau is reached (density effect, Fermi plateau).

The energy loss is usually expressed in terms of the area density $ds = \rho dx$ with $\rho$-density of the absorber. It varies with the target material like $Z/A$ ($\leq 0.5$ for most elements). Minimum ionizing particles lose 1.94 MeV/(g/cm$^2$) in helium decreasing to 1.08 MeV/(g/cm$^2$) in uranium. The energy loss of minimum ionizing particles in hydrogen is exceptionally large, because here $Z/A = 1$.

The relativistic rise saturates at high energies because the medium becomes polarized, effectively reducing the influence of distant collisions. The density correction $\delta/2$ can be described by

$$\frac{\delta}{2} = \ln \frac{\hbar \omega_p}{I} + \ln \beta\gamma - \frac{1}{2} \tag{33}$$

where

$$\hbar \omega_p = \sqrt{4\pi N_e r_e^3 m_e c^2 / \alpha} \tag{34}$$

is the plasma energy and $N_e$ the electron density of the absorbing material.

For gases the Fermi-plateau, which saturates the relativistic rise, is about 60% higher compared to the minimum of ionization. Figure 5 shows the measured energy-loss rates of electrons, muons, pions, kaons, protons and deuterons in the PEP4/9-TPC (185 $dE/dx$ measurements at 8.5 atm in Ar-CH$_4$ = 80 : 20) [9].

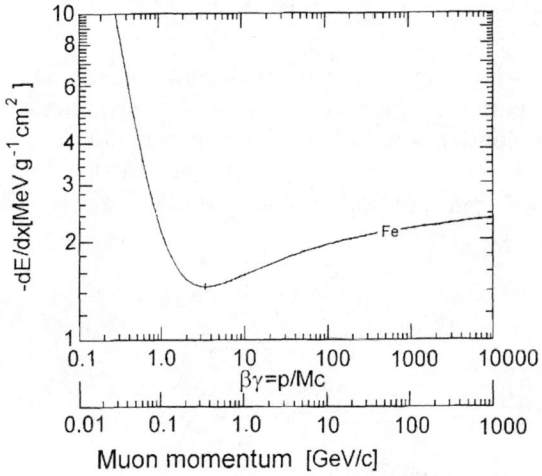

**FIGURE 4.** Energy loss of muons in iron [6]

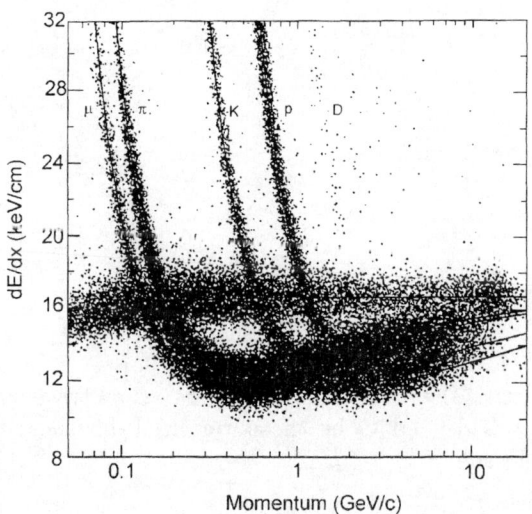

**FIGURE 5.** Measured ionization energy loss of electrons, muons, pions, kaons, protons and deuterons in the PEP4/9-TPC [9]

## Landau Distributions

The Bethe-Bloch formula describes the average energy loss of charged particles. The fluctuation of the energy loss around the mean is described by an asymmetric distribution, the Landau distribution [10,11].

The probability $\phi(\varepsilon)d\varepsilon$ that a singly charged particle loses an energy between $\varepsilon$ and $\varepsilon + d\varepsilon$ per unit length of an absorber was (eq. 23)

$$\phi(\varepsilon) = \frac{2\pi N e^4}{m_e v^2} \frac{Z}{A} \cdot \frac{1}{\varepsilon^2} \quad . \tag{35}$$

Let us define

$$\xi = \frac{2\pi N e^4}{m_e v^2} \cdot \frac{Z}{A} x \quad , \tag{36}$$

where $x$ is the area density of the absorber:

$$\phi(\varepsilon) = \xi(x) \frac{1}{x\varepsilon^2} \quad . \tag{37}$$

Numerically one can write

$$\xi = \frac{0.1536}{\beta^2} \frac{Z}{A} \cdot x \quad [\text{keV}] \quad , \tag{38}$$

where $x$ is measured in mg/cm$^2$.

For an absorber of 1 cm Ar we have for $\beta = 1$

$$\xi = 0.123 \text{ keV} \quad .$$

We define now

$$f(x, \Delta) = \frac{1}{\xi} \omega(\lambda) \tag{39}$$

as the probability that the particle loses an energy $\Delta$ on traversing an absorber of thickness $x$. $\lambda$ is defined to be the normalized deviation from the most probable energy loss $\Delta^{\text{m.p.}}$

$$\lambda = \frac{\Delta - \Delta^{\text{m.p.}}}{\xi} \quad . \tag{40}$$

The most probable energy loss is calculated to be [10,12]

$$\Delta^{\text{m.p.}} = \xi \left\{ \ln \frac{2m_e c^2 \beta^2 \gamma^2 \xi}{I^2} - \beta^2 + 1 - \gamma_E \right\} \quad , \tag{41}$$

where $\gamma_E = 0.577\ldots$ is Euler's constant.

Landau's treatment of $f(x, \Delta)$ yields

$$\omega(\lambda) = \frac{1}{\pi} \int_0^\infty e^{-u \ln u - \lambda u} \sin \pi u \, du \quad , \tag{42}$$

which can be approximated by [12]

$$\Omega(\lambda) = \frac{1}{\sqrt{2\pi}} \exp\left\{-\frac{1}{2}(\lambda + e^{-\lambda})\right\} \quad . \tag{43}$$

Figure 6 shows the energy loss distribution of 3 GeV electrons in an Ar/CH$_4$ (80:20) filled drift chamber of 0.5 cm thickness [13]. According to equation 35 the $\delta$-ray contribution to the energy loss falls inversely proportional to the energy transfer squared, producing a long tail, called Landau tail, in the energy-loss distribution up to the kinematical limit (see also figure 3).

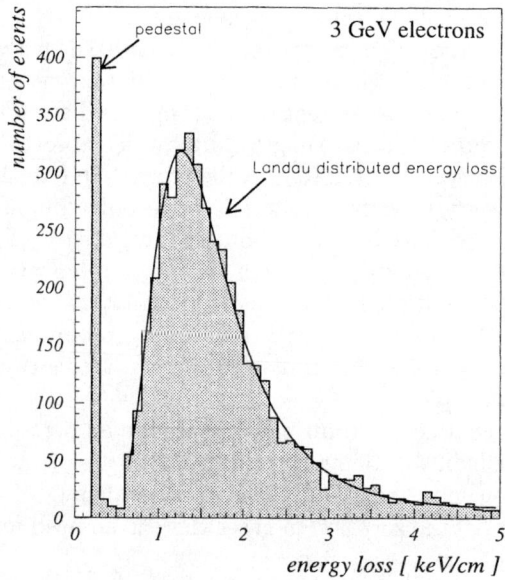

**FIGURE 6.** Energy-loss distribution of 3 GeV electrons in a thin-gap multiwire drift chamber [13]

The asymmetric property of the energy-loss distribution becomes obvious for thin absorbers. For larger absorber thicknesses or truncation techniques applied to thin absorbers the Landau distribution gets more symmetric.

# Scintillation in Materials

Scintillator materials can be inorganic crystals, organic liquids or plastics and gases. The scintillation mechanism in organic crystals is an effect of the lattice. Incident particles can transfer energy to the lattice by creating electron-hole pairs or taking electrons to higher energy levels below the conduction band. Recombination of electron-hole pairs may lead to the emission of light. Also electron-hole bound states (excitons) moving through the lattice can emit light when hitting an activator center and transferring their binding energy to activator levels, which subsequently deexcite. In thallium doped NaI-crystals about 25 eV are required to produce one scintillation photon. The decay time in inorganic scintillators can be quite long (1 $\mu$s in CsI (Tl); 0.62 $\mu$s in BaF$_2$).

In organic substances the scintillation mechanism is different. Certain types of molecules will release a small fraction ($\approx$ 3%) of the absorbed energy as optical photons. This process is expecially marked in organic substances which contain aromatic rings, such as polystyrene, polyvinyltoluene, and naphtalene. Liquids which scintillate include toluene or xylene [6].

This primary scintillation light is preferentially emitted in the UV-range. The absorption length for UV-photons in the scintillation material is rather short: the scintillator is not transparent for its own scintillation light. Therefore, this light is transferred to a wavelength shifter which absorbs the UV-light and reemits it at longer wavelengths (e.g. in the green). Due to the lower concentration of the wavelength shifter material the reemitted light can get out of the scintillator and be detected by a photosensitive device. The technique of wavelength shifting is also used to match the emitted light to the spectral sensitivity of the photomultiplier. For plastic scintillators the primary scintillator and wavelength shifter are mixed with an organic material to form a polymerizing structure. In liquid scintillators the two active components are mixed with an organic base [2].

About 100 eV are required to produce one photon in an organic scintillator. The decay time of the light signal in plastic scintillators is substantially shorter compared to inorganic substances (e.g. 30 ns in naphtalene).

Because of the low light absorption in gases there is no need for wavelength shifting in gas scintillators.

Plastic scintillators do not respond linearly to the energy-loss density. The number of photons produced by charged particles is described by Birk's semi-empirical formula [6,14,15]

$$N = N_0 \frac{dE/dx}{1 + k_B\, dE/dx} \quad , \tag{44}$$

where $N_0$ is the photon yield at low specific ionization density, and $k_B$ is Birk's density parameter. For 100 MeV protons in plastic scintillators one has

$dE/dx \approx 10\,\text{MeV}/(\text{g/cm}^2)$ and $k_B \approx 5\,\text{mg}/(\text{cm}^2\text{MeV})$, yielding a saturation effect of $\sim 5\%$ [4].

For low energy losses eq. 44 leads to a linear dependence

$$N = N_0 \cdot dE/dx \quad, \tag{45}$$

while for very high $dE/dx$ saturation occurs at

$$N = N_0/k_B \quad. \tag{46}$$

There exists a correlation between the energy loss of a particle that goes into the creation of electron-ion pairs or the production of scintillation light because electron-ion pairs can recombine thus reducing the $dE/dx|_\text{ion}$-signal. On the other hand the scintillation light signal is enhanced because recombination frequently leads to excited states which deexcite yielding scintillation light.

## Cherenkov Radiation

A charged particle traversing a medium with refractive index $n$ with a velocity $v$ exceeding the velocity of light $c/n$ in that medium, emits Cherenkov radiation. The threshold condition is given by

$$\beta_\text{thres} = \frac{v_\text{thres}}{c} \geq \frac{1}{n} \quad. \tag{47}$$

The angle of emission increases with the velocity reaching a maximum value for $\beta = 1$, namely

$$\Theta_o^\text{max} = \arccos \frac{1}{n} \quad. \tag{48}$$

The threshold velocity translates into a threshold energy

$$E_\text{thres} = \gamma_\text{thres} m_0 c^2 \tag{49}$$

yielding

$$\gamma_\text{thres} = \frac{1}{\sqrt{1 - \beta_\text{thres}^2}} = \frac{n}{\sqrt{n^2 - 1}} \quad. \tag{50}$$

The number of Cherenkov photons emitted per unit path length $dx$ is

$$\frac{dN}{dx} = 2\pi \alpha z^2 \int \left(1 - \frac{1}{n^2 \beta^2}\right) \frac{d\lambda}{\lambda^2} \tag{51}$$

for $n(\lambda) > 1$, $z$ – electric charge of the incident particle, $\lambda$ – wavelength, and $\alpha$ – fine structure constant. The yield of Cherenkov radiation photons is proportional to $1/\lambda^2$, but only for those wavelengths where the refractive index is larger than unity. Since $n(\lambda) \approx 1$ in the X-ray region, there is no X-ray Cherenkov emission. Integrating eq. 51 over the visible spectrum ($\lambda_1 = 400\,\text{nm}$, $\lambda_2 = 700\,\text{nm}$) gives

$$\begin{aligned}\frac{\mathrm{d}N}{\mathrm{d}x} &= 2\pi\alpha z^2 \frac{\lambda_2 - \lambda_1}{\lambda_1 \lambda_2} \sin^2 \Theta_c \\ &= 490 \cdot z^2 \cdot \sin^2 \Theta_c\,[\text{cm}^{-1}] \quad .\end{aligned} \qquad (52)$$

The Cherenkov effect can be used to identify particles of fixed momentum by means of threshold Cherenkov counters. More information can be obtained, if the Cherenkov angle is measured by DIRC-counters (Detection of Internally Reflected Cherenkov light). In these devices some fraction of the Cherenkov light produced by a charged particle is kept inside the radiator by total internal reflection. The direction of the photons remains unchanged and the Cherenkov angle is conserved during the transport. When exiting the radiator the photons produce a Cherenkov ring on a planar detector (figure 7).

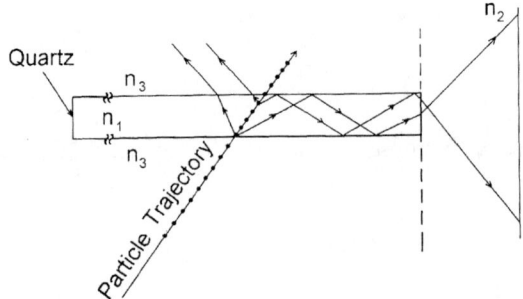

**FIGURE 7.** Imaging principle of a DIRC-counter [16]

The pion/proton separation achieved with such a system is shown in figure 8 [16].

Ring-imaging Cherenkov-counters (RICH-counters) have become extraordinary useful in the field of elementary particles and astrophysics. Figure 9 shows the Cherenkov ring radii of electrons, muons, pions and kaons in a $C_4F_{10}$-Ar (75:25) filled Rich-counter read out by a 100-channel photomultipiler of $10 * 10\,\text{cm}^2$ active area [17].

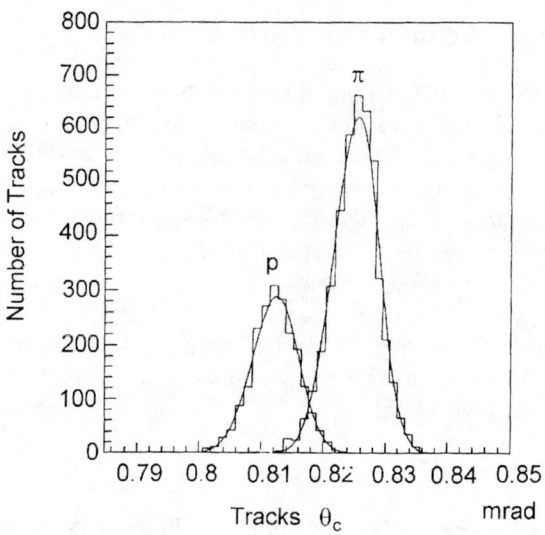

**FIGURE 8.** Cherenkov-angle distribution for pions and protons of 5.4 GeV/c in a DIRC-counter [16].

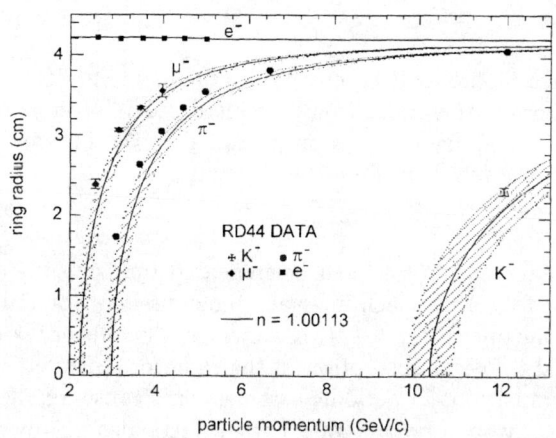

**FIGURE 9.** Cherenkov-ring radii of $e, \mu, \pi, K$ in a $C_4F_{10}$-Ar (75:25) RICH-counter. The solid curves show the expected radii for an index of refraction of n = 1.00113. The shaded regions represent a 5% uncertainty in the absolute momentum scale [17]

# Transition Radiation

Transition radiation is emitted when a charged particle traverses a medium with discontinuous dielectric constant. A charged particle moving towards a boundary, where the dielectric constant changes, can be considered to form together with its mirror charge an electric dipole whose field strength varies in time. The time dependent dipole field causes the emission of electromagnetic radiation. This emission can be understood in such a way that although the dielectric displacement $\vec{D} = \varepsilon\varepsilon_0\vec{E}$ varies continuously in passing through a boundary, the electric field does not.

The energy radiated from a single boundary (transition from vacuum to a medium with dielectric constant $\varepsilon$) is proportional to the Lorentz-factor of the incident charged particle [6,14,18]:

$$S = \frac{1}{3}\alpha z^2 \hbar \omega_p \gamma \quad , \tag{53}$$

where $\hbar\omega_p$ is the plasma energy (see equation 34). For commonly used plastic radiators (styrene or similar materials) one has

$$\hbar\omega_p \approx 20\,\text{eV} \quad . \tag{54}$$

The typical emission angle of transition radiation is proportional to $1/\gamma$. The radiation yield drops sharply for frequencies

$$\omega > \gamma\omega_p \quad . \tag{55}$$

The $\gamma$-dependence of the emitted energy originates mainly from the hardening of the spectrum rather than from the increased photon yield. Since the radiated photons also have energies proportional to the Lorentz factor of the incident particle, the number of emitted transition radiation photons is

$$N \propto \alpha z^2 \quad . \tag{56}$$

The number of emitted photons can be increased by using many transitions (stack of foils, or foam). At each interface the emission probability for an X-ray photon is of the order of $\alpha = 1/137$. However, the foils or foams have to be of low $Z$ material to avoid absorption in the radiator. Interference effects for radiation from transitions in periodic arrangements cause an effective threshold behaviour at a value of $\gamma \approx 1000$. These effects also produce a frequency dependent photon yield. The foil thickness must be comparable to or larger than the formation zone

$$D = \gamma c/\omega_p \tag{57}$$

which in practical situations ($\hbar\omega_p = 20\,\text{eV}$; $\gamma = 5 \cdot 10^3$) is about $50\,\mu$m. Transition radiation detectors are mainly used for $e/\pi$-setaration. In cosmic ray experiments transition radiation emission can also be employed to measure the energy of muons in the TeV-range.

# Bremsstrahlung

If a charged particle is decelerated in the Coulomb field of a nucleus a fraction of its kinetic energy will be emitted in form of real photons (bremsstrahlung). The energy loss by bremsstrahlung for high energies can be described by [2]

$$-\frac{dE}{dx} = 4\alpha N_A \frac{Z^2}{A} \cdot z^2 r^2 E \ln \frac{183}{Z^{1/3}} \quad , \tag{58}$$

where $r = \frac{1}{4\pi\varepsilon_0} \cdot \frac{e^2}{mc^2}$. Bremsstrahlung is mainly produced by electrons because

$$r_e \propto \frac{1}{m_e} \quad . \tag{59}$$

Equation 58 can be rewritten for electrons

$$-\frac{dE}{dx} = \frac{E}{X_0} \quad , \tag{60}$$

where

$$X_0 = \frac{A}{4\alpha N_A Z(Z+1) r_e^2 \ln(183\, Z^{-1/3})} \tag{61}$$

is the radiation length of the absorber in which bremsstrahlung is produced. Here we have included also radiation from electrons ($\sim Z$, because there are $Z$ electrons per nucleus). If screening effects are taken into account $X_0$ can be more accurately described by [6]

$$X_0 = \frac{716.4\, A}{Z(Z+1) \ln(287/\sqrt{Z})} \quad [\text{g/cm}^2] \quad . \tag{62}$$

The important point about bremsstrahlung is that the energy loss is proportional to the energy. The energy where the losses due to ionization and bremsstrahlung for electrons are the same is called critical energy

$$\left.\frac{dE_c}{dx}\right|_{\text{ion}} = \left.\frac{dE_c}{dx}\right|_{\text{brems}} \quad . \tag{63}$$

For solid or liquid absorbers the critical energy can be approximated by [6]

$$E_c = \frac{610\,\text{MeV}}{Z + 1.24} \quad , \tag{64}$$

while for gases one has [6]

$$E_c = \frac{710 \text{ MeV}}{Z + 0.92} \quad . \tag{65}$$

The difference between gases on the one hand and solids and liquids on the other hand comes about because the density corrections are different in these substances, and this modifies $\frac{dE}{dx}\big|_{\text{ion}}$.

The energy spectrum of bremsstrahlung photons is $\sim E_\gamma^{-1}$, where $E_\gamma$ is the photon energy.

At high energies also radiation from heavier particles becomes important and consequently a critical energy for these particles can be defined. Since

$$\frac{dE}{dx}\bigg|_{\text{brems}} \propto \frac{1}{m^2} \tag{66}$$

the critical energy e.g. for muons in iron is

$$E_c = \frac{610 \text{ MeV}}{Z + 1.24} \cdot \left(\frac{m_\mu}{m_e}\right)^2 = 960 \text{ GeV} \quad . \tag{67}$$

## Direct Electron Pair Production

Direct electron pair production in the Coulomb field of a nucleus via virtual photons ("tridents") is a dominant energy loss mechanism at high energies. The energy loss for singly charged particles due to this process can be represented by

$$-\frac{dE}{dx}\bigg|_{\text{pair}} = b(Z, A, E) \cdot E \quad . \tag{68}$$

It is essentially - like bremsstrahlung - also proportional to the particle's energy. Because bremsstrahlung and direct pair production dominate at high energies this offers an attractive possibility to build also muon calorimeters [2]. The average rate of muon energy losses can be parametrized as

$$\frac{dE}{dx} = a(E) + b(E) \cdot E \tag{69}$$

where $a(E)$ represents the ionization energy loss and $b(E)$ is the sum of direct elektron pair production, bremsstrahlung and photonuclear interactions.

The various contributions to the energy loss of muons in standard rock (Z = 11; A = 22; $\rho = 3 g/cm^3$) are shown in figure 10.

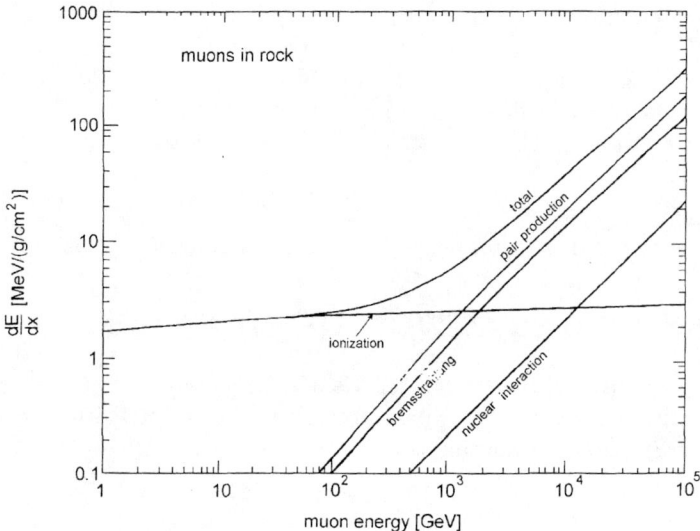

**FIGURE 10.** Contributions to the energy loss of muons in standard rock ($Z = 11$; $A = 22$; $\rho = 3 g/cm^3$).

## Nuclear Interactions

Nuclear interactions play an important role in the detection of neutral particles other than photons. They are also responsible for the development of hadronic cascades. The total cross section for nucleons is of the order of 50 mbarn and varies slightly with energy. It has an elastic ($\sigma_{el}$) and inelastic part ($\sigma_{inel}$). The inelastic cross section has a material dependence

$$\sigma_{inel} \approx \sigma_0 A^\alpha \tag{70}$$

with $\alpha = 0.71$. The corresponding absorption length $\lambda_a$ is [2]

$$\lambda_a = \frac{A}{N_A \cdot \rho \cdot \sigma_{inel}} [cm] \tag{71}$$

($A$ in g/mol, $N_A$ in mol$^{-1}$, $\rho$ in g/cm$^3$, and $\sigma_{inel}$ in cm$^2$).
This quantity has to be distinguished from the nuclear interaction length $\lambda_w$, which is related to the total cross section

$$\lambda_w = \frac{A}{N_A \cdot \rho \cdot \sigma_{total}} [cm] \quad . \tag{72}$$

Since $\sigma_{total} > \sigma_{inel}$, $\lambda_w < \lambda_a$ holds.

Strong interactions have a multiplicity which grows logarithmically with energy. The particles are produced in a narrow cone around the forward direction with an average transverse momentum of $p_T = 350\,\text{MeV}/c$, which is responsible for the lateral spread of hadronic cascades.

A useful relation for the calculation of interaction rates per $(\text{g}/\text{cm}^2)$ is

$$\phi((\text{g}/\text{cm}^2)^{-1}) = \sigma_N \cdot N_A \tag{73}$$

where $\sigma_N$ is the cross section per nucleon and $N_A$ Avogadro's number.

## INTERACTION OF PHOTONS

Photons are attenuated in matter via the processes of the photoelectric effect, Compton scattering and pair production. The intensity of a photon beam varies in matter according to

$$I = I_0\, e^{-\mu x} \quad , \tag{74}$$

where $\mu$ is mass attenuation coefficient. $\mu$ is related to the photon cross sections $\sigma_i$ by

$$\mu = \frac{N_A}{A} \sum_{i=1}^{3} \sigma_i \quad . \tag{75}$$

### Photoelectric Effect

Atomic electrons can absorb the energy of a photon completely

$$\gamma + \text{atom} \to \text{atom}^+ + e^- \quad . \tag{76}$$

The cross section for absorption of a photon of energy $E_\gamma$ is particularly large in the $K$-shell (80% of the total cross section). The total cross section for photon absorption in the $K$-shell is

$$\sigma^K_{\text{Photo}} = \left(\frac{32}{\varepsilon^7}\right)^{1/2} \alpha^4 Z^5 \sigma_{\text{Thomson}}[\text{cm}^2/\text{atom}] \quad , \tag{77}$$

where $\varepsilon = E_\gamma/m_e c^2$, and $\sigma_{\text{Thomson}} = \frac{8}{3}\pi r_e^2 = 665\,\text{mbarn}$ is the cross section for Thomson scattering. For high energies the energy dependence becomes softer

$$\sigma^K_{\text{Photo}} = 4\pi r_e^2 Z^5 \alpha^4 \cdot \frac{1}{\varepsilon} \quad . \tag{78}$$

The photoelectric cross section has sharp discontinuities when $E_\gamma$ coincides with the binding energy of atomic shells. As a consequence of a photoabsorption in the $K$-shell characteristic X-rays or Auger electrons are emitted [2].

# Compton Scattering

The Compton effect describes the scattering of photons off quasi-free atomic electrons

$$\gamma + e \to \gamma' + e' \quad . \tag{79}$$

The cross section for this process, given by the Klein-Nishina formula, can be approximated at high energies by

$$\sigma_c \propto \frac{\ln \varepsilon}{\varepsilon} \cdot Z \tag{80}$$

where $Z$ is the number of electrons in the target atom. From energy and momentum conservation one can derive the ratio of scattered ($E'_\gamma$) to incident photon energy ($E_\gamma$)

$$\frac{E'_\gamma}{E_\gamma} = \frac{1}{1 + \varepsilon(1 - \cos \Theta_\gamma)} \quad , \tag{81}$$

where $\Theta_\gamma$ is the scattering angle of the photon with respect to its original direction.

For backscattering ($\Theta_\gamma = \pi$) the energy transfer to the electron $E_{\mathrm{kin}}$ reaches a maximum value

$$E_{\mathrm{kin}}^{\mathrm{max}} = \frac{2\varepsilon^2}{1 + 2\varepsilon} m_e c^2 \quad , \tag{82}$$

which, in the extreme case ($\varepsilon \gg 1$), equals $E_\gamma$.

In Compton scattering only a fraction of the photon energy is transferred to the electron. Therefore, one defines an energy scattering cross section

$$\sigma_{cs} = \frac{E'_\gamma}{E_\gamma} \sigma_c \tag{83}$$

and an energy absorption cross section

$$\sigma_{ca} = \sigma_c - \sigma_{cs} = \sigma_c \frac{E_{\mathrm{kin}}}{E_\gamma} \quad . \tag{84}$$

At accelerators and in astrophysics also the process of inverse Compton scattering is of importance [2].

# Pair Production

The production of an electron-positron pair in the Coulomb field of a nucleus requires a certain minimum energy

$$E_\gamma \geq 2m_e c^2 + \frac{2m_e^2 c^2}{m_{\text{nucleus}}} \quad . \tag{85}$$

Since for all practical cases $m_{\text{nucleus}} \gg m_e$, one has effectively $E_\gamma \geq 2m_e c^2$.

The total cross section in the case of complete screening $\left(\varepsilon \gg \frac{1}{\alpha Z^{1/3}}\right)$; i.e. at reasonably high energies ($E_\gamma \gg 20\,\text{MeV}$), is

$$\sigma_{\text{pair}} = 4\alpha r_e^2 Z^2 \left(\frac{7}{9} \ln \frac{183}{Z^{1/3}} - \frac{1}{54}\right) \quad [\text{cm}^2/\text{atom}] \quad . \tag{86}$$

Neglecting the small additive term $1/54$ in eq. 86 one can rewrite, using eq. 58 and eq. 61,

$$\sigma_{\text{pair}} = \frac{7}{9} \frac{A}{N_A} \cdot \frac{1}{X_0} \quad . \tag{87}$$

The partition of the energy to the electron and positron is symmetric at low energies ($E_\gamma \ll 50\,\text{MeV}$) and increasingly asymmetric at high energies ($E_\gamma > 1\,\text{GeV}$) [2].

Figure 11 shows the photoproduction of an electron-positron pair in the Coulomb-field of an electron ($\gamma + e^- \rightarrow e^+ + e^- + e^-$) and also a pair-production in the field of a nucleus ($\gamma + \text{nucleus} \rightarrow e^+ + e^- + \text{nucleus}'$) [19].

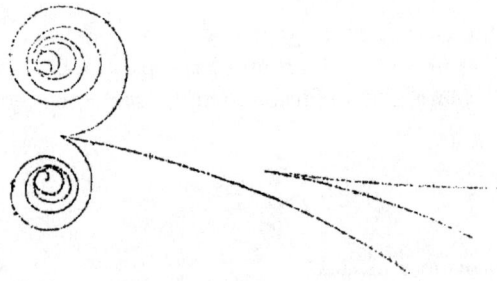

**FIGURE 11.** Photoproduction in the Coulomb-field of an electron ($\gamma + e^- \rightarrow e^+ + e^- + e^-$) and on a nucleus ($\gamma + \text{nucleus} \rightarrow e^+ + e^- + \text{nucleus}'$) [19]

## Mass-Attenuation Coefficients

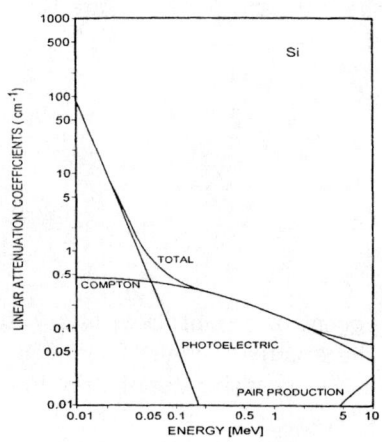

FIGURE 12. Mass attenuation coefficients for photon interactions in silicon [20]

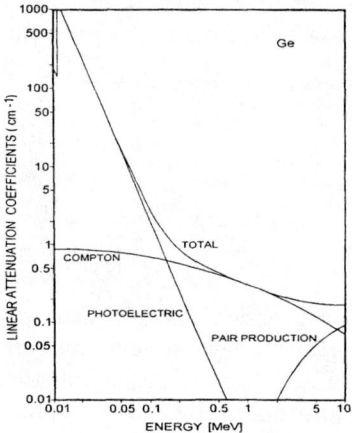

FIGURE 13. Mass attenuation coefficients for photon interactions in germanium [20]

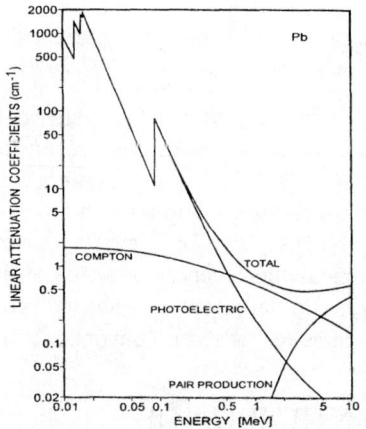

FIGURE 14. Mass attenuation coefficients for photon interactions in lead [20]

The mass-attenuation coefficients for photon interactions are shown in figures 12-14 for silicon, germanium and lead [20]. The photoelectric effect dominates at low energies ($E_\gamma < 100\,\text{keV}$). Superimposed on the continuous photoelectric attenuation coefficient are absorption edges characteristic of the absorber material. Pair production dominates at high energies ($> 10\,\text{MeV}$). In the intermediate region Compton scattering prevails.

# INTERACTION OF NEUTRONS

In the same way as photons are detected via their interactions also neutrons have to be measured indirectly. Depending on the neutron energy various reactions can be considered which produce charged particles which are then detected via their ionization or scintillation [2].

a) Low energies ($< 20\,\mathrm{MeV}$)

$$\begin{aligned} n + {}^6\mathrm{Li} &\to \alpha + {}^3\mathrm{H} \\ n + {}^{10}\mathrm{B} &\to \alpha + {}^7\mathrm{Li} \\ n + {}^3\mathrm{He} &\to p + {}^3\mathrm{H} \\ n + p &\to n + p \end{aligned} \qquad (88)$$

The conversion material can be a component of a scintillator (e.g. LiI (Tl)), a thin layer of material in front of the sensitive volume of a gaseous detector (boron layer), or an admixture to the counting gas of a proportional counter ($BF_3$, $^3$He, or protons in $CH_4$).

b) Medium energies ($20\,\mathrm{MeV} \leq E_{\mathrm{kin}} \leq 1\,\mathrm{GeV}$)

The $(n,p)$-recoil reaction can be used for neutron detection in detectors which contain many quasi-free protons in their sensitive volume (e.g. hydrocarbons).

c) High energies ($E > 1\,\mathrm{GeV}$)

Neutrons of high energy initiate hadron cascades in inelastic interactions which are easy to identify in hadron calorimeters.

Neutrons are detected with relatively high efficiency at very low energies. Therefore, it is often useful to slow down neutrons with substances containing many protons, because neutrons can transfer a large amount of energy to collision partners of the same mass. In some fields of application, like in radiation protection at nuclear reactors, it is of importance to know the energy of fission neutrons, because the relative biological effectiveness depends on it. This can e.g. be achieved with a stack of plastic detectors interleaved with foils of materials with different threshold energies for neutron conversion [21].

# INTERACTIONS OF NEUTRINOS

Neutrinos are very difficult to detect. Depending on the neutrino flavor the following inverse beta decay like interactions can be considered:

$$\begin{aligned} \nu_e + n &\to p + e^- \\ \bar{\nu}_e + p &\to n + e^+ \\ \nu_\mu + n &\to p + \mu^- \end{aligned}$$

$$\bar{\nu}_\mu + p \to n + \mu^+ \tag{89}$$
$$\nu_\tau + n \to p + \tau^-$$
$$\bar{\nu}_\tau + p \to n + \tau^+$$

The cross section for $\nu_e$-detection in the MeV-range can be estimated as [22]

$$\sigma(\nu_e N) = \frac{4}{\pi} \cdot 10^{-10} \left(\frac{\hbar p}{(m_p c)^2}\right)^2$$
$$= 6.4 \cdot 10^{-44} \text{cm}^2 \text{ for } 1\,\text{MeV} \quad . \tag{90}$$

This means that the interaction probability of e.g. solar neutrinos in a water Cherenkov counter of $d = 100$ meter thickness is only

$$\phi = \sigma \cdot N_A \cdot d = 3.8 \cdot 10^{-16} \quad . \tag{91}$$

Since the coupling constant of weak interactions has a dimension of $1/\text{GeV}^2$, the neutrino cross section must rise at high energies like the square of the center-of-mass energy. For fixed target experiments we can parametrize

$$\sigma(\nu_\mu N) = 0.67 \cdot 10^{-38} E_\nu[\text{GeV}] \quad \text{cm}^2/\text{nucleon}$$
$$\sigma(\bar{\nu}_\mu N) = 0.34 \cdot 10^{-38} E_\nu[\text{GeV}] \quad \text{cm}^2/\text{nucleon} \tag{92}$$

This shows that even at 100 GeV the neutrino cross section is lower by 11 orders of magnitude compared to the total proton-proton cross section.

## ELECTROMAGNETIC CASCADES

The development of cascades induced by electrons, positrons or photons is governed by bremsstrahlung of electrons and pair production of photons. Secondary particle production continues until photons fall below the pair production threshold, and energy losses of electrons other than bremsstrahlung start to dominate: the number of shower particles decays exponentially.

Already a very simple model can describe the main features of particle multiplication in electromagnetic cascades: A photon of energy $E_0$ starts the cascade by producing an $e^+e^-$-pair after one radiation length. Assuming that the energy is shared symmetrically between the particles at each multiplication step, one gets at the depth $t$

$$N(t) = 2^t \tag{93}$$

particles with energy

$$E(t) = E_0 \cdot 2^{-t} \quad . \tag{94}$$

The multiplication continues until the electrons fall below the critical energy $E_c$

$$E_c = E_0 \cdot 2^{-t_{\max}} \quad . \tag{95}$$

From then on $(t > t_{\max})$ the shower particles are only absorbed. The position of the shower maximum is obtained from eq. 95

$$t_{\max} = \frac{\ln E_0/E_c}{\ln 2} \propto \ln E_0 \quad . \tag{96}$$

The total number of shower particles is

$$S = \sum_{t=0}^{t_{\max}} N(t) = \sum 2^t = 2^{t_{\max}+1} - 1 \approx 2^{t_{\max}+1}$$

$$S = 2 \cdot 2^{t_{\max}} = 2 \cdot \frac{E_0}{E_c} \propto E_0 \quad . \tag{97}$$

If the shower particles are sampled in steps $t$ measured in units of $X_0$, the total track length is obtained as

$$S^* = \frac{S}{t} = 2\frac{E_0}{E_c} \cdot \frac{1}{t} \quad , \tag{98}$$

which leads to an energy resolution of

$$\frac{\sigma}{E_0} = \frac{\sqrt{S^*}}{S^*} = \frac{\sqrt{t}}{\sqrt{2E_0/E_c}} \propto \frac{\sqrt{t}}{\sqrt{E_0}} \quad . \tag{99}$$

In a more realistic description the longitudinal development of the electron shower can be approximated by [6]

$$\frac{dE}{dt} = \mathrm{const} \cdot t^a \cdot e^{-bt} \quad , \tag{100}$$

where $a, b$ are fit parameters.

Figure 15 shows muon induced electromagnetic cascades in a multi-plate cloud chamber [23].

The lateral spread of an electromagnetic shower is mainly caused by multiple scattering. It is described by the Molière radius

$$R_m = \frac{21\,\mathrm{MeV}}{E_c} X_0 \,[\mathrm{g/cm}^2] \quad . \tag{101}$$

95% of the shower energy in a homogeneous calorimeter is contained in a cylinder of radius $2R_m$ around the shower axis.

Figure 16 demonstrates the interplay of the longitudinal and lateral development of an electromagnetic shower [2].

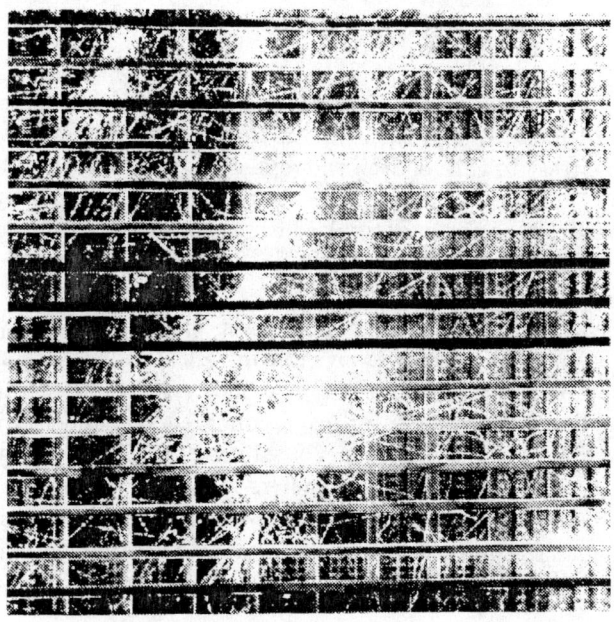

**FIGURE 15.** Some muon induced electromagnetic cascades in a multi-plate cloud chamber operated in a concrete shielded air shower array [23]

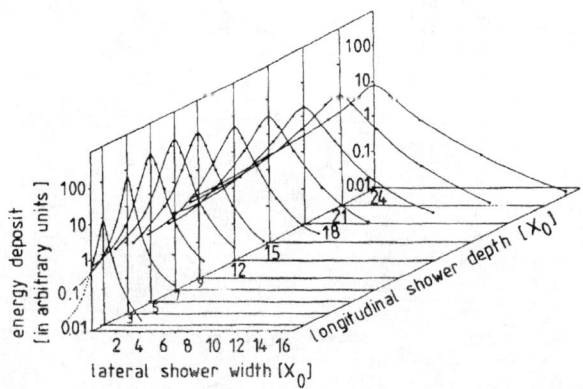

**FIGURE 16.** Sketch of the longitudinal and lateral development of an electromagnetic cascade in a homogeneous absorber [2]

# HADRON CASCADES

The longitudinal development of electromagnetic cascades is characterized by the radiation length $X_0$ and their lateral width is determined by multiple scattering. In contrast to this, hadron showers are governed in their longitudinal structure by the nuclear interaction length $\lambda$ and by transverse momenta of secondary particles as far as lateral width is concerned. Since for most materials $\lambda \gg X_0$, and $\langle p_T^{\text{interaction}} \rangle \gg \langle p_T^{\text{multiple scattering}} \rangle$ hadron showers are longer and wider.

Part of the energy of the incident hadron is spent to break up nuclear bonds. This fraction of the energy is invisible in hadron calorimeters. Further energy is lost by escaping particles like neutrinos and muons as a result of hadron decays. Since the fraction of lost binding energy and escaping particles fluctuates considerably, the energy resolution of hadron calorimeters is systematically inferior to electron calorimeters.

The longitudinal development of pion induced hadron cascades is plotted in figure 17. Figure 18 shows a comparison between proton, iron, and photon induced cascades in the atmosphere [24].

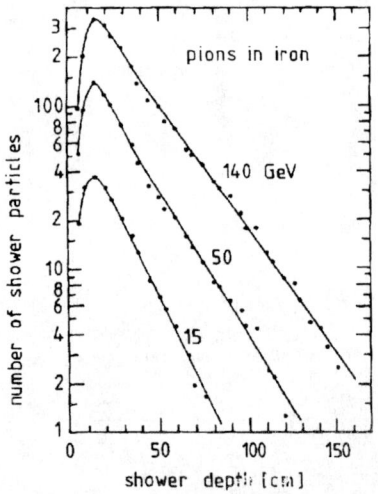

**FIGURE 17.** Longitudinal development of pion induced hadron cascades. Reprinted from [25], copyright 1978, with permission from Elsevier Science.

The different response of calorimeters to electrons and hadrons is an undesirable feature for the energy measurement of jets of unknown particle composition. By appropriate compensation techniques, however, the electron to hadron response can be equalized.

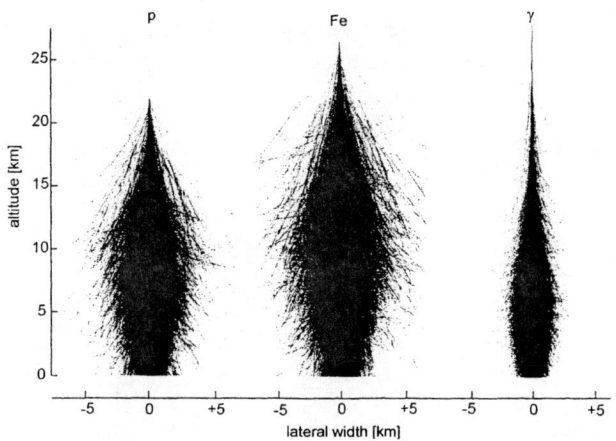

**FIGURE 18.** Comparison between proton, iron, and photon induced cascades in the atmosphere. The primary energy in each case is $10^{14}$eV [24].

# PARTICLE IDENTIFICATION

Particle identification is based on measurements which are sensitive to the particle velocity, its charge and its momentum. Figure 19 sketches the different possibilities to separate photons, electrons, positrons, muons, charged pions, protons, neutrons and neutrinos in a mixed particle beam using a general purpose detector.

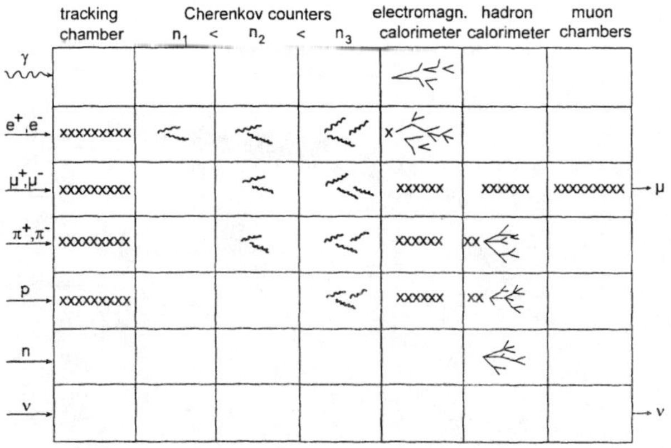

**FIGURE 19.** Particle identification using a detector consisting of a tracking chamber, Cherenkov counters, calorimetry and muon chambers.

Figure 20 shows the particle separation power of a balloon borne experiment using momentum, time-of-flight, dE/dx and Cherenkov radiation measurements [26].

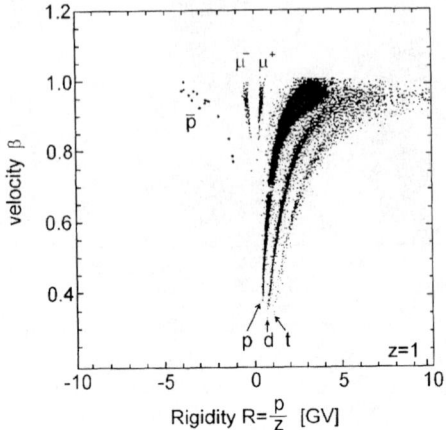

**FIGURE 20.** Particle identification in a balloon borne experiment using momentum, time-of-flight, dE/dx and Cherenkov radiation information [26].

Even the abundance of different helium isotopes can be determined from a velocity and momentum measurement (figure 21 [27]). This is feasable, because at fixed momentum the lighter isotope $^3$He is faster than the more abundant $^4$He.

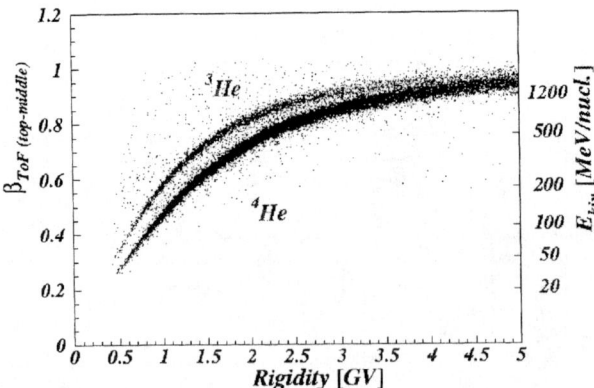

**FIGURE 21.** Isotopic abundance of energetic cosmic ray helium nuclei [27]

# CONCLUSION

Basic physical principles can be used to identify all kinds of elementary particles and nuclei. The precise measurement of the particle composition in high energy physics experiments at accelerators and in cosmic rays is essential for the insight into the underlying physics processes. This is an important ingredient for the progress in the fields of elementary particles and astrophysics aiming at the unification of forces and the understanding of the evolution of the universe.

# ACKNOWLEDGEMENTS

The author thanks Mrs. C. Hauke (figures) and Dipl. Phys. G. Prange (text and layout) for their help in preparing the manuscript.

# REFERENCES

1. K. Kleinknecht, *Detectors for Particle Radiation* , Cambridge University Press 1998
2. C. Grupen, *Particle Detectors* , Cambridge University Press 1996
3. R. Fernow, *Introduction to Experimental Particle Physics* , Cambridge University Press 1989
4. W.R. Leo, *Techniques for Nuclear and Particle Physics Experiments* , Springer, Berlin 1987
5. B. Rossi, *High Energy Particles* , Prentice-Hall (1952)
6. Particle Data Group, R.M. Barnett et al., *Phys.Rev.* **D54** (1996) 1; Eur. Phys. J. **C3** (1998) 1
7. W.W.M. Allison, P.R.S. Wright, *Oxford Univ. Preprint* **35/83** (1983) and W.W.M. Allison, J.H. Cobb, *Ann.Rev.Nucl.Sci.* **30** (1980) 253
8. C. Grupen, *Ph.D. Thesis*, University of Kiel 1970
9. D. R. Nygren, J. N. Marx, Physics Today **31** (1978) 46
10. D.H. Wilkinson, *Nucl.Instr.Meth.* **A 383** (1996) 513
11. L.D. Landau, *J.Exp.Phys. (USSR)* **8** (1944) 201
12. S. Behrens, A.C. Melissinos, *Univ. of Rochester Preprint* **UR-776** (1981)
13. K. Affholderbach et al., Nucl. Instr. Meth. **A 410** (1998) 166
14. R.C. Fernow, *Brookhaven Nat.Lab. Preprint* **BNL-42114** (1988)
15. J. Birks, *Theory and Practice of Scintillation Counting* , MacMillan 1964
16. I. Adam et al. SLAC-Pub-7706, Nov. 1997; and hep-ex/9712001
17. R. Debbe et al. hep-ex/9503006 (1995)
18. S. Paul, *CERN-PPE* **91-199** (1991)
19. F. Close, M. Marten, C. Sutton, *The Particle Explosion*, Oxford University Press 1987
20. Harshaw Chemical Company (H. Lentz, L. van Gelderen, Th. Courbois) 1969
21. E. Sauter, *Grundlagen des Strahlenschutzes* , Thiemig, München 1982

22. D.H. Perkins *Introduction to High Energy Physics*, Addison-Wesley, 1986
23. W. Wolter, private communication 1999
24. J. Knapp, D. Heck *Luftschauer-Simulationsrechnungen mit dem Corsika-Programm* Forschungszentrum Karlsruhe Nachrichten **30** (1998) 27
25. M. Holder et al., *Nucl.Instr.Meth.* **151** 69 (1978)
26. J. W. Mitchell et al. Phys. Rev. Lett. **76** (1996) 3057
27. M. Simon, H. Göbel, private communication 1999

# Silicon Detectors

P. Giubellino, E. Crescio, R. Hernandez, M. Idzik, D. Nouais,
A. Rivetti

*INFN, Torino, Italy*

**Abstract.** This paper is an elementary introduction to the use of silicon detectors in High-Energy Physics. The principles of operation are described together with an overview of the most commonly used detector types.

## INTRODUCTION

Silicon detectors are used in almost all High-Energy Physics experiments built in the last fifteen years, from large collider experiments to fixed-target ones, and also in many specialized detectors like spectrometers for space or detectors for medical diagnostics. This success has grown along with the progress of micro-electronics and interconnections, which has a direct effect on the quality of the fabrication technology of the detectors and on the possibility of reading out detectors of increasing complexity in a faster *and* more effective way. Some of the characteristics which are at the basis of the success of the silicon detectors, making them excellent devices for both energy and position measurements, are the following:

- Speed of the order of 10 ns

- Spatial resolution of the order of 10 $\mu$m

- Flexibility of design, with feature-size of the order of 10 $\mu$m

- Small amount of material (0.003 $X_0$ for a typical 300 $\mu$m thick detector)

- Excellent mechanical properties

- Linearity of the response vs. the deposited energy

- Good resolution in the deposited energy (3.6 eV of deposited energy are needed to create a pair of charges, vs. 30 eV in a gas detector)

- Tolerance to high radiation doses

An intuitive, non-rigorous description of a silicon detector can provide some grasp on its main aspects. In the following section these concepts will be reformulated in a more proper way. A silicon detector works pretty much like an ionization chamber: the impinging ionizing particles generate electron-hole pairs, which drift to the electrodes under the electric field present in the detector volume. The electron-hole current in the detector induces a signal at the electrodes on the detector faces. If the electrodes are segmented, the detector is position-sensitive. The amount of charge deposited in the typical 300 $\mu$m of thickness of a silicon detector is very small (25,000 electrons is the average value for a relativistic, singly-charged particle traversing the detector orthogonally to its surface), and therefore it would be masked by the fluctuations of the current which the applied electric field [1] makes flow even in high-resistivity, hyper-pure silicon. The solution which has been found is to implant on e.g. $n$-type silicon a thin layer of $p$-type dopants, so that the whole detector behaves as a $p$-$n$ diode. If we reverse-bias the diode, we will have the necessary electric field and only a very small current (the so-called leakage current of the diode, of the order of few to few tens of nanoamps per cm$^2$ at room temperature).

In a silicon crystal each silicon atom is bound to its neighbours by the four covalent bounds formed by its four valence electrons. If dopant atoms with five valence electrons are introduced, it also forms the same number of bonds with its neighbouring Si atoms. Therefore, the fifth electron is loosely bound, and even at room temperature it is essentially free of moving around the lattice, leaving the atom as a positively charged ion. Such a crystal will be called $n$-type because of the presence of the free negative charges, available for conduction, and the dopant is called a donor. In case of a trivalent dopant atom, one electron is missing to form the four covalent bonds, and this 'hole' can be filled by another electron of the lattice, which in turn will leave its position vacant. The hole can be treated as a free positive charge, and the trivalent atom, now with four covalent bonds, will remain as a negatively charged ion. In this case the dopant atom is called an acceptor, and the doped crystal will be called $p$-type because of the presence of free positive charge carriers. If we put in contact $p$-type and $n$-type silicon we obtain a $p$-$n$ junction, (in practice this is obtained by implanting a lot of acceptor atoms on the surface of a wafer of $n$-type silicon, which outnumber the donors and change sign to the doping in a shallow surface region). A very simple scheme of the formation of a $p$-$n$ junction is drawn in Fig. 1. The electrons are free in the $n$-type, the holes in the $p$-type (Fig. 1a).

When the two types are in contact (Fig. 1b), both types of free carriers diffuse, as in any system with nonzero temperature and no binding force. Therefore electrons will migrate to the $p$-side and holes to the $n$-side, and there will recombine with the free charges of the opposite sign. This process leaves behind a net positive charge on the $n$ side and a net negative charge on the $p$ side (the charged ions fixed in the lattice). This net charge generates an electric field which is opposed to the diffusion

---

[1] needed to collect the deposited charge, and thus generate the current

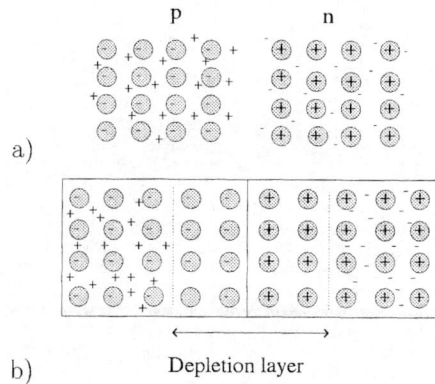

**FIGURE 1.** Simple scheme of the formation of a $p$-$n$ junction.

process, and eventually balances it, creating a stable state in which a thin layer at the boundary between the $n$ and the $p$ type silicon is left without free carriers, with a non zero charge density on the two sides and an electric field across it. The layer is called 'depletion region'. In fact this is already a functional detector: if we deposit charge in the depletion layer, the built-in potential will push the charges inducing a signal on the collection electrodes. If we connect an amplifier to a diode and we produce ionization at its surface, for example with an $\alpha$-source, we will see a signal even without applying any external potential. If we do apply a voltage difference, we can extend the depletion layer, up to the full thickness of the silicon wafer (a 300 $\mu$m thick wafer with a resistivity of 3 k$\Omega$·cm will need 100 V of reverse voltage between the surfaces to be fully depleted; the depletion voltage grows quadratically with the thickness and is inversely proportional to the resistivity). The depletion layer is the sensitive region of the detector, thanks to the presence of the electric field, while the non-depleted region is essentially field-free (the non-depleted silicon is a conductor, since the dopant atoms provide abundant charge carriers). The charge generated by the energy loss of a ionizing particle will be collected if in the depleted region, while in the undepleted region it will recombine. Therefore, the thicker the depleted layer, the more efficiently we collect the charge. So we will generally operate our detector applying a reverse bias voltage large enough to fully deplete it.

A schematic section of a Si detector is shown in Fig. 2. If, as in the figure, the junction is segmented, each section acts as an individual detector, with the boundary between two sections more or less halfway between the junctions. The potential to segment the detector is fantastic, and has allowed innumerable variations, some of which will be described in the following sections. The principal limit is the ability to implant very thin $p$-type regions, but since, as was said, it can be of just few microns, in practice the limit will derive from other constraints, such as the capability to connect many readout electronics channels. In addition, the charge generated

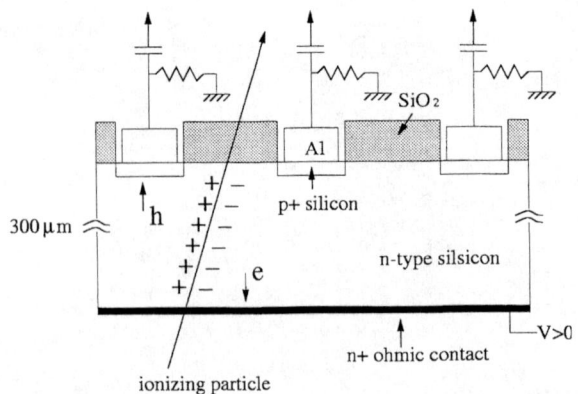
**FIGURE 2.** Scheme of a silicon particle detector.

by the crossing particles will diffuse during its movement to the electrodes to a size of about ≈10$\mu$m diameter (r.m.s.) and therefore there is no point in segmenting to a much smaller width. The size of the detector can be very large, in fact up to the largest size of silicon wafers which the industry can handle. At the moment detectors are produced mostly out of wafers of four, five or sometimes six inches of diameter. As an example of the freedom one has to design the shape of the electrodes, in Fig. 3 is shown one of detectors of the Multiplicity Detector of the NA50 experiment at CERN [1]. In this detector, the strips, each a $20^0$ section of a circle, have a pitch which grows from 50 $\mu$m for the innermost ones to 1.2 mm for the outermost ones, with a change in surface of two orders of magnitude! Since the fabrication technology of silicon detectors is analogous to the one of integrated electronics circuits, one can include in the detector additional features: very frequently resistors, capacitors and even transistors are integrated on the detector substrate.

## Energy response

Among the main reasons for the success of silicon detectors is the fact that they produce a signal that is proportional to the energy deposited in the detector volume, which is generally referred to as "being linear", and the fact that the energy needed to produce a charge pair is very small (3.6 eV in silicon), which translates in a very good ratio signal/energy deposit. These characteristics have very important implications on possible applications and on the performance of the detectors. One should remember, though, that the signal released by a traversing particle in a thin silicon detector is very small, and even smaller for X-rays, so the possibility to profit of these features will depend on the noise of the readout electronics connected to it.

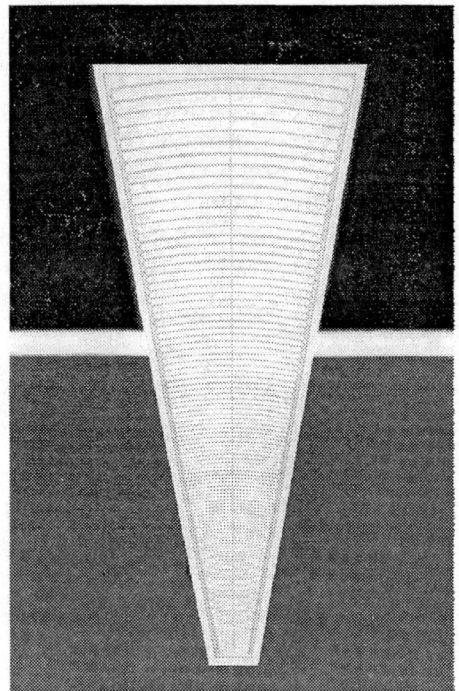

**FIGURE 3.** A silicon detector of the Multiplicity Detector of the Na50 experiment.

## Energy deposition and resolution

The energy resolution $\sigma^2$ which can be achieved is the quadratic sum of two factors: the noise of the readout chain (including detector-related effects such as the leakage current) and the statistical fluctuations in the generation of charge carriers. Two main cases should be considered: low-energy regime, in which essentially all of the particle's energy is deposited in the detector volume, and high-energy regime, in which the particle traverses the detector losing a small fraction of its energy.

*Low energy* The $\sigma$ of the generated charge is in this case simply the square root of the number of generated charges, i.e. $E/\delta$ where E is the energy of the detected particle and $\delta$ is the average energy needed for the creation of an electron-hole pair. This naive expression must be multiplied by a further factor F (between 0 and 1), called Fano factor, which depends on the specific material and takes into account the fact that the charges are not generated in a fully uncorrelated way.

For charged particles most of the energy loss is due to ionization. For low-energy photons (X-rays) most of the energy loss is due to the photoelectric effect, so if a photon interacts it loses its full energy, giving a gaussian-shaped spectrum. For higher energies (for silicon starting in the tens of KeV) the Compton effect becomes important. In a Compton scattering the photon loses part of its energy, and the secondary photon escapes detection, since the detector volume is small and the photon interaction probability is consequently also small. This gives rise to a shoulder of signals lower than the full absorption peak. At higher energies (about 1 MeV), electron-positron pair creation becomes the dominant process. Yet, the interaction probability of photons decreases rapidly with energy, and is very small already at few tens of keV, so thin Si detectors will almost never be used to detect photons outside of the region in which the photoelectric effect dominates.

*High Energy* The average energy loss of relativistic particles other than electrons (high-energy electrons lose most of their energy through bremsstrahlung) is described by the Bethe-Bloch formula:

$$-\frac{dE}{dx} = Kz^2\frac{Z}{A}\frac{1}{\beta^2}[\frac{1}{2}ln(\frac{2m_ec^2\beta^2\gamma^2T_{max}}{I^2}) - \beta^2 - \frac{\delta}{2}] \quad (1)$$

Where z is the charge of the incident particle, Z is the atomic number of the medium, A is the atomic mass of the medium, $\beta$ and $\gamma$ are the kinematic variables of the incident particle, $m_ec^2$ is the electron mass, I is the mean excitation energy, $T_{max}$ is the max. kinetic energy which can be transferred to a free electron in one collision, $\delta$ is a density correction factor. The resulting curve is shown in Fig. 4 (dashed curve). Its characteristic features are readily visible: for low energies

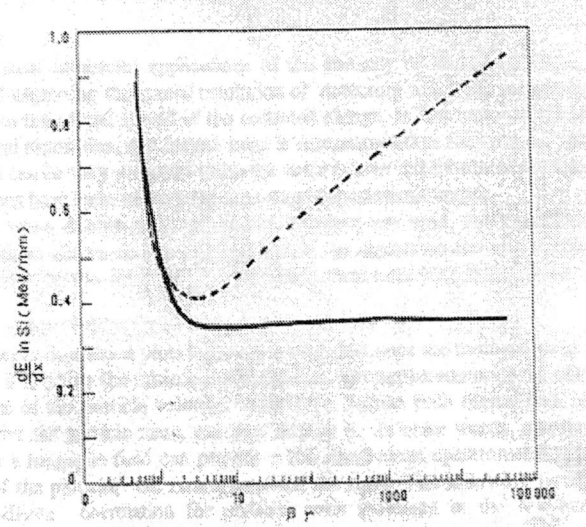

**FIGURE 4.** Energy loss (dashed curve) and restricted energy loss as a function of $\beta\gamma$.

the energy loss decreases as $\beta^2$, and after a minimum it grows slowly. Particles with energy above the so-called $1/\beta^2$ region are called minimum ionizing particles

(MIPs). A significant fraction of the energy loss goes to knock-on electrons of relatively high energy, which escape from a thin detector. Therefore the energy deposited is lower than the energy lost, and is expressed by the restricted energy loss, for which the largest energy transfers are excluded. In the plot the energy loss in silicon and the restricted one are compared: the rise at high energies, mostly due to the high energy transfers, is not visible with a thin silicon detector. For a MIP, the statistical fluctuations in the energy deposited in a thin silicon detector follow the distribution shown in Fig. 5, which is usually referred to as a Landau curve, but it is a bit wider, and more complex, than the one calculated from the Landau formula.

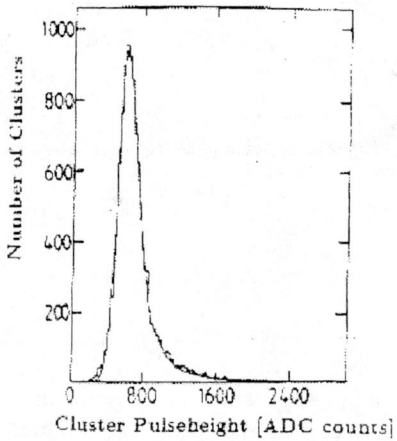

**FIGURE 5.** Pulseheight distribution for a MIP.

## Applications

*Tracking* One of the most important applications of the linearity of silicon detectors lies in the possibility of improving the spatial resolution of detectors which are segmented to a pitch comparable to the spatial spread of the collected charge. In this way, the charge is shared among several electrodes, and the impact point is reconstructed as the centroid of the signals [2]. This method can be very powerful: resolutions down to few microns have been achieved even in large experimental setups.

---

[2] A signal is induced by the movement of the charge also on the electrodes which do not collect it, but it is in general quite small.

*Particle identification* From the energy deposition plots, it is clear that once the momentum of a particle is known, and it is below the minimum of ionization, the measurement of the $dE/dx$ allows a measurement of the particle velocity. Therefore, known both momentum and velocity, one can derive the particle mass, and thus identify it. A system of silicon detectors in a magnetic field can provide a full stand-alone spectrometer, measuring the trajectory of the particle, the momentum and the mass. Figure 6 shows the momentum-$dE/dx$ correlation for particles with momenta in the few-hundred MeV region.

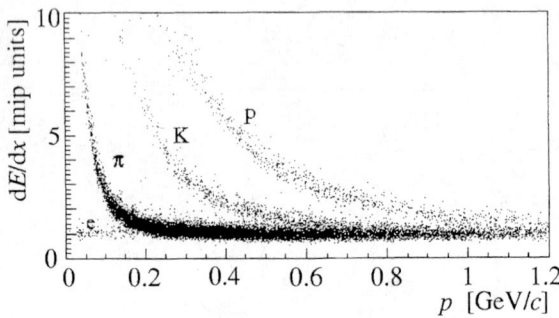

**FIGURE 6.** Momentum-$dE/dx$ correlation for particles with momenta in the few-hundred MeV region.

It is clear, though, that an individual energy deposition follows the Landau curve, so an individual measurement will fall anywhere on the curve, giving a very rough measurement of the average energy loss of a particle of that momentum. What is normally done is to use a few measurements (which come naturally if the silicon detectors form a tracking system, in which each particle crosses several detector layers) and perform a "truncated mean" of them, i.e. take the average of the lowest ones. This method improves the result very rapidly, and already with just three or four measurements one gets good resolution, around 10%. The energy deposition is proportional to the square of the charge of the crossing particle, so it is straightforward to identify particles with charge more than one (ions). This method is used, for example, for experimental setups which fly on satellites or will fly on the space shuttle. They use silicon detectors to study the composition of primary cosmic rays, using for example (NINA experiment) a simple telescope of silicon microstrip detectors to get the charge, the range (and thus the energy), and the angle of incidence of nuclei in the cosmic rays.

*Spectroscopy* When one has to measure an X-ray line, the quality factors are *resolution* and *statistics*, both essential to the determination of the peak. The resolution depends on the factors observed before: the noise of the readout system and the statistics of charge carriers. The second factor gives a small advantage to Germanium relative to silicon (2.96 eV/pair against 3.62), but the excellent low-noise electronics which has been developed for silicon detectors can offset this advantage.

The reason for which Ge, although rather cumbersome to use, since it needs to be cooled to 77 K, is still the material normally used comes from statistics: the cross section for the photoelectric effect is proportional to the charge of the medium nuclei to the power 4 to 5, so Ge is 40 times more efficient than silicon in detecting low-energy photons. In addition, it is easier to fabricate Ge detectors of large volume, again an important factor for efficiency. One could think of offsetting these problems by simply increasing the time of the measurement. This is not only annoying, but often impossible, since the measurement has to be performed in a time short compared with mean life of the source, which for cases of practical interest can be very short. In other cases, a low detection efficiency would result in the need to irradiate a lot more the sample to be studied, which might be impractical or unsafe. Therefore for spectroscopic measurements Ge detectors are normally the preferred solution. Yet, silicon detectors are nowadays considered for measurements of low-energy X-rays in several applications in which spatial resolution and simple installation are important factors, namely for imaging detectors for medical diagnostics.

## CHARACTERISTICS OF SILICON DETECTORS

The characteristics of semiconductors are due to the relatively low value of the energy gap $E_g$ between the valence band and the conduction band (1.12 eV at 300 K for silicon). In fact at nonzero temperatures the electrons in the valence band can acquire energy sufficient to reach the conduction band, thus creating holes in the valence one. The conductivity of these materials is then due to two types of charge carriers, electrons and holes, and can be written:

$$\sigma = en\mu_e + ep\mu_h$$

where $\mu_e$ and $\mu_h$ are the mobilities of electrons and holes respectively (for silicon at 300 K $\mu_e$=1350 cm$^2$/Vs and $\mu_h$=480 cm$^2$/Vs). In a pure (intrinsic) semiconductor the concentration of negatively charged electrons is equal to the concentration of positively charged holes:

$$p = n = n_i = (N_c N_v)^{1/2} e^{\frac{-E_g}{2k_B T}} \propto T^{3/2} e^{\frac{-E_g}{2k_B T}}$$

where $N_c$ and $N_v$ are the effective density of states in the conduction band and in the valence band. The concentrations of electrons and holes may be changed by doping, i.e. by introducing impurities into the semiconductor. The *n*-type doping supplies electrons to the conduction band. In this case the electrons become the majority carriers. In *p*-type silicon the majority carriers are the holes. The energy levels introduced by *n*-type dopants and *p*-type dopants are close respectively to the bottom of the conduction band and to the top of the valence band, i. e. the impurities introduced by doping are "shallow'. A very small energy (often less than a thermal energy $k_B T$) is needed to ionize a shallow impurity. Deep impurities (with

**TABLE 1.** Properties of silicon and germanium.

| | Si | Ge |
|---|---|---|
| Z | 14 | 32 |
| A | 28.09 | 72.60 |
| Density (300 K), g/cm$^3$ | 2.33 | 5.33 |
| Atoms/cm$^3$ | $4.96 \times 10^{22}$ | $4.41 \times 10^{22}$ |
| $\varepsilon_r$ | 12 | 16 |
| $E_g$ (300 K), eV | 1.115 | 0.665 |
| Intrinsic carriers density (300 K), cm$^{-3}$ | $1.5 \times 10^{10}$ | $2.4 \times 10^{13}$ |
| Intrinsic resistivity (300 K), $\Omega$cm | $2.3 \times 10^5$ | 47 |
| $\mu_e$ (300 K), cm$^2$/Vs | 1350 | 3900 |
| $\mu_h$ (300 K), cm$^2$/Vs | 480 | 1900 |

energy levels close to the center of the gap) affect the behavior of the semiconductor, acting as traps and facilitating capture and recombination processes. In table 1 the main properties of silicon and germanium are listed.

## The p-n junction

Let us consider a reverse biased *p-n* junction (Fig. 7), with uniform doping concentration $N_D$ in the *n* region and $N_A$ in the *p* region, with $N_A \gg N_D$ which is normally the case. We can treat this simple geometry as one dimensional. Without external potential there exists a built-in potential $\Psi_0$ in the junction, created by the thermal diffusion of electrons and holes:

$$\Psi_0 = V_T ln(\frac{N_A N_D}{n_i^2}) \qquad (2)$$

where:

$$V_T = \frac{k_B T}{q} \cong 26 mV \qquad (3)$$

at T=300 K. $n_i$ is the concentration of the charged carriers in the silicon ($1.5 \times 10^{10}$ cm$^{-3}$ at 300 K). As an effect, at the boundary between the *p* and the *n* regions there exists a layer without free charge carriers (depletion layer), but with fixed negative acceptor ions and positive donor ions which create the electric field in it. When an external bias $V_R$ is applied, the total voltage drop in the junction is $\Psi_0 + V_R$. If the depletion layer thickness is $W_A$ in the *p* region and $W_D$ in the *n* region then one can write the charge balance equation, which expresses the fact that our crystal remains electrically neutral, as:

$$W_A N_A = W_D N_D \qquad (4)$$

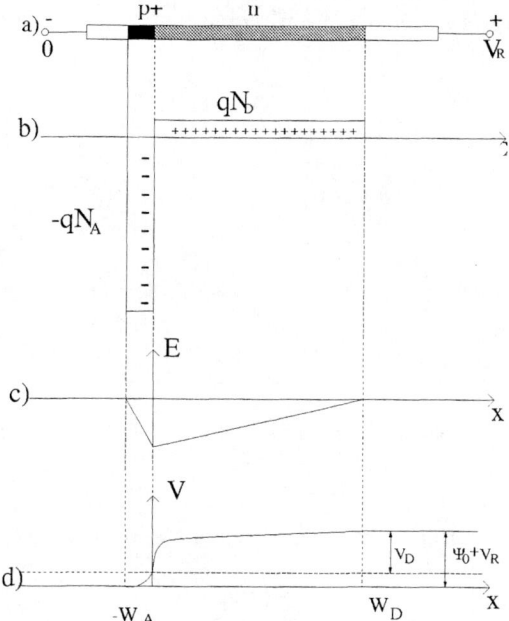

**FIGURE 7.** Here are shown 1-dimensional distributions of the following quantities in $p$-$n$ junctions: a) general junction scheme, b) concentration of ionized dopants, c) electric field, d) potential.

In the region $-W_A \leq x \leq 0$ we can write the one dimensional Poisson equation:

$$\frac{d^2V}{dx^2} = -\frac{\rho}{\varepsilon_r\varepsilon_0} = \frac{qN_A}{\varepsilon_r\varepsilon_0} \tag{5}$$

Setting the boundary condition $E(x) = 0$ for $x = -W_A$, from eq. 5 one can get (Fig. 7c):

$$E = -\frac{dV}{dx} = -\frac{qN_A}{\varepsilon_r\varepsilon_0}(x + W_A) \tag{6}$$

in this region. Integrating eq. 6 with the boundary condition $V(x) = 0$ for $x = -W_A$, one can get (Fig. 7d):

$$V = \frac{qN_A}{\varepsilon_r\varepsilon_0}\left(\frac{x^2}{2} + W_A x + \frac{W_A^2}{2}\right) \tag{7}$$

for $-W_A \leq x \leq 0$. Fixing the potential $V(0) = V_A$, for $x = 0$ we get from eq. 7:

$$V_A = \frac{qN_A}{\varepsilon_r\varepsilon_0}\frac{W_A^2}{2} \tag{8}$$

After similar considerations for the n region one can calculate the voltage drop between the points $x = 0$ and $x = W_D$ as:

$$V_D = \frac{qN_D}{\varepsilon_r \varepsilon_0} \frac{W_D^2}{2} \tag{9}$$

Summing the expressions equations 8 and 9 we get the total voltage drop for the junction:

$$\Psi_0 + V_R = V_A + V_D = \frac{q}{2\varepsilon_r\varepsilon_0}(N_A W_A^2 + N_D W_D^2) \tag{10}$$

which after replacing eq. 4 gives:

$$\Psi_0 + V_R = \frac{qW_A^2 N_A}{2\varepsilon_r\varepsilon_0}(1 + \frac{N_A}{N_D}) \tag{11}$$

From this expressions we can calculate the thickness of the depletion layer in the $p$ region as a function of the external voltage set and of the doping concentration:

$$W_A = \sqrt{\frac{2\varepsilon_r\varepsilon 0(\Psi_0 + V_R)}{qN_A(1 + \frac{N_A}{N_D})}} \tag{12}$$

In a similar way for the $n$ region we get:

$$W_D = \sqrt{\frac{2\varepsilon_r\varepsilon 0(\Psi_0 + V_R)}{qN_D(1 + \frac{N_D}{N_A})}} \tag{13}$$

From equations 12 and 13 it is seen that the thickness of the depletion layer decreases with growing doping concentration and it is proportional to the square root of the set voltage. In practice abrupt junctions are used, i.e. a very thin layer of $p^+$ region is produced (few microns) with very high doping concentration, on a thick $n$ layer (few hundred microns) with low doping concentration (few orders of magnitude less then $p^+$). So finally the total depletion thickness equals:

$$W = W_A + W_D \approx W_D$$

In order to get an ohmic contact between the $n$ region and the metal electrode, a thin $n^+$ layer is added.

## Capacitance of the junction

The depletion voltage is the value of the reverse detector bias voltage for which the whole detector volume is depleted, i.e. the $p$-$n$ junction or the electric field fills the whole detector volume. In the following we show that the measurement of the capacitance of the detector versus the reverse applied voltage allows the correct estimation of the depletion voltage. One can define the junction capacitance per unit area as:

$$C_j = \frac{dQ}{dV_R} = \frac{dQ}{dW}\frac{dW}{dV_R} \qquad (14)$$

Having in mind the charge balance equation 4 one can write:

$$dQ = -qN_A dW_A = qN_D dW_d \qquad (15)$$

taking the donor part and then differentiating eq. 13 one gets:

$$\frac{dW_D}{dV_R} = \sqrt{\frac{\varepsilon_r \varepsilon_0}{2qN_D(1+\frac{N_D}{N_A})(\Psi_0 + V_R)}} \qquad (16)$$

Substituting equations 15 and 16 into 14 we can finally obtain:

$$C_j = \sqrt{\frac{q\varepsilon_r \varepsilon_0 (N_A N_D)}{2(N_D + N_A)(\Psi_0 + V_R)}} \qquad (17)$$

The same result is obtained using the acceptor part of eq. 15 and subsequently eq. 12. For the usual case when $N_A \gg N_D$ one obtains:

$$C_j = \sqrt{\frac{q\varepsilon_r \varepsilon_0 N_D}{2(\Psi_0 + V_R)}} \qquad (18)$$

or taking into account equations 13 one can express capacitance as a function of the depletion layer thickness as:

$$C_j = \frac{\varepsilon_r \varepsilon_0}{W_D} \qquad (19)$$

It is seen from 18 and 19 that measuring the capacitance versus the applied voltage and making the plot of $1/C_j^2$ vs. $V_R$ one can easily find the depletion voltage of the detector (see Fig. 8). If the depletion layer is smaller then the thickness of the detector this relation should be linear, while when it reaches the full detector thickness the capacitance should reach the saturation value of equation 19.

## Leakage current

The leakage current of the reverse biased diode includes several components, among which the most important are: diffusion current from minority carriers diffusing from the neutral region outside the junction, generation current due to generation-recombination processes in the junction and surface current due to the surface charges and defects.

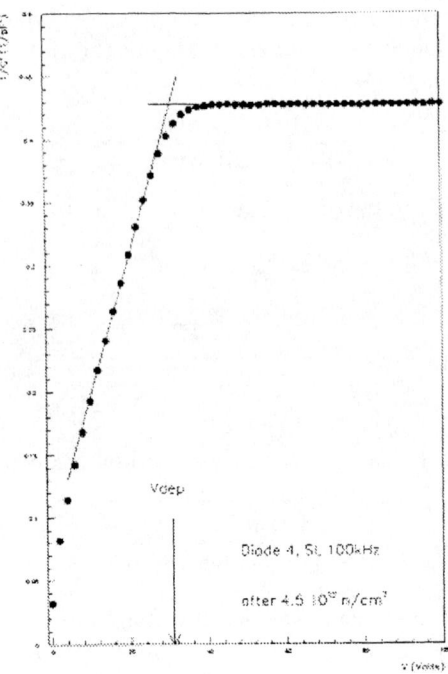

**FIGURE 8.** Typical $1/C_j^2$ vs. $V_R$ behaviour of a diode. The depletion voltage corresponds to the abscissa of the intersection point of the two lines.

*Diffusion current* Under reverse bias conditions there is limited probability of diffusion of electrons generated in the non depleted $p$-type region and holes generated in the $n$-type region to the junction layer. If it happens, then they drift to the appropriate electrodes under the effect of the electric field and so form the diffusion current. The diffusion current is expressed by the well known Shockley formula:

$$J_s = \frac{qD_p p_{n0}}{L_p} + \frac{qD_n n_{p0}}{L_n} \qquad (20)$$

where $D_x$ is the diffusion constant, $L_x$ is the diffusion length for electrons and holes respectively and $n_{p0}$, $p_{n0}$ are the equilibrium densities of electrons and holes on the $p$-side and $n$-side respectively. For the abrupt $p^+n$ junction only the left component (hole current) is significant. Taking also into account that $L_p = (D_p \tau_p)^{0.5}$ and $p_{n0} = n_i^2/N_D$, where $\tau$ is the lifetime of the minority carriers, one can write:

$$J_s = q\sqrt{\frac{D_p}{\tau_p}} \frac{n_i^2}{N_D} \qquad (21)$$

The diffusion component is the main one at room temperature for semiconductors with high $n_i$ like germanium ($n_i = 2.4 \times 10^{13}$ cm$^{-3}$) while for silicon ($n_i = 1.4 \times 10^{10}$ cm$^{-3}$) it is less important. For a fully depleted detector, and so without the regions with free carriers, the diffusion current is negligible. The main temperature dependence of the diffusion current is through $n_i^2$ and so one can approximately write:

$$J_s \approx e^{\frac{-E_g}{k_B T}} \qquad (22)$$

*Generation current* The current generated in the depletion layer may be expressed by the simple formula:

$$J_g = qgW \qquad (23)$$

where $q$ is the unit charge, $W$ is the thickness of the depletion layer and $g$ is the generation-recombination rate, which is derived from the Shockley-Read-Hall model. Having in mind that $W$ is proportional to the square root of the applied voltage, one could try to estimate the depletion voltage from the I-V curve (the current versus the applied voltage or rather its square root) for the case where the generation current is the most significant one. In a very rough approximation one can assume that the generation-recombination rate is proportional to $n_i$ and so:

$$J_g \approx e^{\frac{-E_g}{2k_B T}} \qquad (24)$$

which gives roughly an increase by a factor of two for a temperature increase of 8 K. Therefore any measurement of the leakage current should be coupled with the precise measurement of the temperature.

*Surface current* The currents generated in the region of the detector surface are of very complex origin. The main source of these currents are the ionic charges on and outside the semiconductor surface which affect the electric field distribution close to the surface. The absolute value of the surface leakage current is not analytically well determined and strongly depends on the technology and on the geometry of the detector. To limit the surface current in real detectors several protecting guard rings are usually added.

# DETECTOR TYPES

## Fabrication

Detectors can be fabricated both on $p$-type and $n$-type substrate, but since it is the most widely used, all our arguments will refer to detectors fabricated on $n$-type. The silicon used must be of high purity and high resistivity (which come together). A high resistivity, in the order of several k$\Omega$·cm, is needed to have a depletion voltage in the tens of volts (for the normal thickness of 300 $\mu$m). A special case are silicon drift detectors, for which also a high uniformity of the resistivity is necessary, to allow the drift of the charge over centimeters in a straight line. In this case special substrates doped with a uniform neutron flux are used (Neutron Transmutation Doped silicon). Based upon the different fabrication processes, silicon detectors can be classified in three main types, namely the "diffused detector", the "surface-barrier" detector and the ion-implanted detector.

Diffusion was the first technique available for detector fabrication and it is based on the diffusion of impurities through high temperature processing steps. The high temperature necessary for the processing, however, degrades the material resulting in higher leakage currents.

Surface barrier detectors, on the contrary, are made in a low temperature process, evaporating in vacuum a thin Au layer directly on the $n$-type crystal surface. In practice this metal-semiconductor interface is difficult to control and the leakage current of these devices is higher than that of the ones made by implantation.

Nowadays ion-implantation is the most commonly used technique. In Fig. 9 the main steps of the planar process are shown. First the $n$-type wafers are oxidized at 1030°C to have the whole surface passivated. Then, using photolithographic and

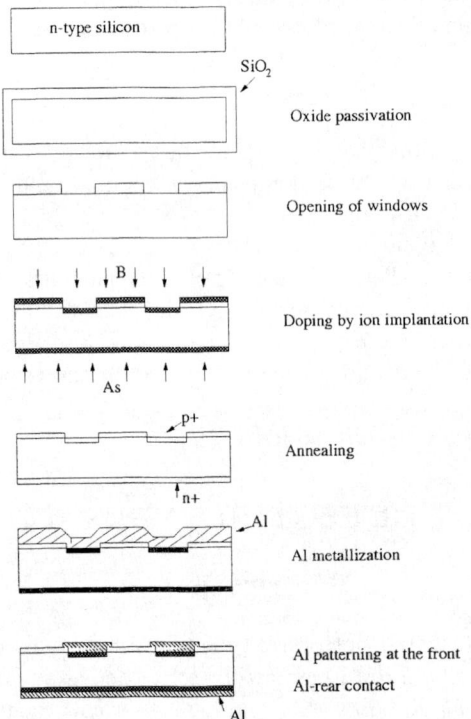

**FIGURE 9.** Processing steps of the planar fabrication of a silicon detector.

etching techniques, windows are made in the oxide to enable ion implantation in the desired areas. Different geometries of pads and strips can be achieved using appropriate masks. The next step is the doping of silicon by ion implantation, by which doping atoms are physically introduced into the lattice. Dopant ions are produced from a gaseous source by ionisation using high voltage. The ions are accelerated in an electric field to energies in the range of 10 keV to $\sim$ 100 keV and then the ion beam is directed to the windows made in the oxide. The $p^+$ strips are implanted with boron, while ions of phosphorus or arsenic are used for the $n^+$ contact. After the implantation an annealing process at 600°C allows partial recovery of the lattice from the damage caused by the irradiation. The next step is the metallisation process, which is required to make electrical contact to the silicon. During this process the surface and the back of the wafer are covered with aluminium. The desired pattern can be achieved using appropriate masks. Finally, last step before cutting is the passivation, which helps to maintain low leakage current and also protects the junction region from mechanical and ambient (humidity) degradation.

*Detector biasing* While the 'back' of the detector is connected to high positive voltage, the junction side is connected to ground. If the detector is segmented, the connection to ground has to go through a resistor (sometimes included in the readout circuit) to avoid shorting the sections together. These resistors, called *bias resistors* can be integrated in the detector: all the strips will be connected via the integrated resistor to a metal line deposited on the detector and running along the strips, which in turn will be connected to ground. These resistors have to be quite large (typically of the order of a M$\Omega$), and therefore they are not easy to fabricate. The value of a resistor is defined by the process used and by its length/width ratio, which therefore has to be quite large. In Fig. 10 is shown a photograph of a detail of a detector in which the strips are connected to a bias line via resistors, in this case made of polysilicon, which are fabricated in a thin, winding shape to allow a large resistance in a small space.

A very elegant method to realize large resistor values without introducing extra elements is based on the so called punch-through effect. The basis is the fact that an unbiased (floating) diode assumes a potential which is determined by the potential of the nearby ones. In this situation a steady current of holes flows between the implants close to the surface. The effective value of the bias resistor depends on the gap between the bias line, which runs orthogonal to the strips, and the strip inplant, and can reach very large values.

*AC coupled detectors* DC coupled detectors, i.e. detectors in which the readout electronics is connected directly to the strips, presents some difficulties. First of all, the first stage of the preamplifier sinks the leakage current, which can be large after a large radiation load, and therefore changes working conditions depending on its value. Unless the design compensates for this effect, one will have an undesirable bias in the response. In some cases DC coupling is quite complex, such as for double-sided detectors, in which at least one side sits at high voltage. Therefore, often the readout will go through a decoupling capacitor, which must

**FIGURE 10.** Detail of polysilicon biasing resistors.

be much larger than the capacitance to the neighbours to ensure good signal collection (typically over 100 pF). Such values are large, and usually a dedicated chip is used, connected on one side to the detector and on the other to the electronics. A more elegant solution consists in integrating a capacitor directly on the strips, using as plates the metal line and the inplant and a thin $SiO_2$ layer as dielectric. A schematic view of an AC detector is shown in in Fig. 11, to be compared with the DC one in Fig. 2.

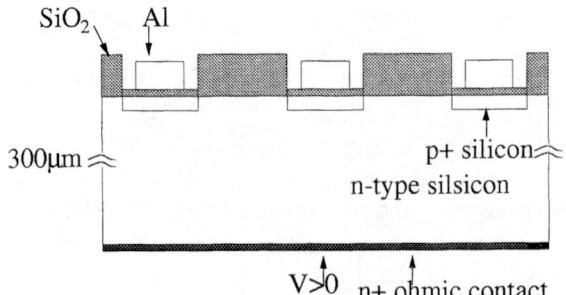
**FIGURE 11.** Cross section of a AC coupled silicon detector.

# Silicon Strip Detectors

A silicon detector segmented in long, narrow elements is called a micro-strip detector. Micro-strip detectors provide therefore the measurement of one coordinate of the particle's crossing point with high precision. Using very low noise readout electronics, the measurement of the centroid of the signal over more than one strip further improves the precision. In Fig. 12 the residual (the difference between the reconstructed and the effective crossing point) is plotted as a function of the distance between the particle impact point and the closest strip. The middle region, in which one can see a linear behaviour, corresponds to the absence of charge sharing between the two strips: the reconstructed impact point is the middle of the strip. When the impact point gets further from the central strip, the charge is shared among two neighbouring strips and the cluster is reconstructed as the centroid of two strips. In this case, the reconstructed point gets closer to the effective one and the residual improves when the crossing point is closer to the boundary between two strips. Precisions down to 1 $\mu$m for minimum ionizing particles have been

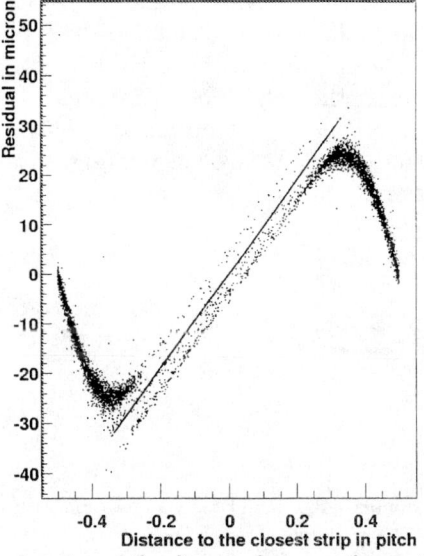

**FIGURE 12.** Residual as function of the distance between the particle impact point and the closest strip.

measured in test setups [2]. In complex experimental apparata, in which the conditions are less favourable and factors such as the precision of the alignment come into play, precisions of few microns have been reached. Clearly the precision of this procedure depends on the noise of the readout chain (including the quantization error introduced by the analog-to-digital converter, which is important when using small signals). If digital readout is used (strip hit or not hit), the resolution is simply $\sigma = \frac{pitch}{\sqrt{12}}$. In many applications the simplicity of digital readout, in which a simple threshold replaces the ADC, is considered more important than the gain in resolution.

The effect of charge sharing can be enhanced by reading out only every n-th strip, and leaving the intermediate one(s) floating. In this way, the signal of the intermediate strips is capacitively shared among the read-out ones and the resolution obtained is much better than what would be achieved, with the same number of electronics channels, using wider strips read out individually.

## Double sided micro-strip detectors

Since the micro-strip detector provides only one coordinate with good precision, the segmentation of the backplane is a natural way to provide a second coordinate and thus a space point without adding material on the trajectory of the particles. Yet, if one tries to simply subdivide the contacts on both sides of the detector, the presence of positive charge at the Si-SiO$_2$ interface induces in the n-type substrate an accumulation layer of electrons, resulting in a low (order of kOhms) resistance between the the strips on the back side. Therefore the signal charge spreads over many electrodes, making the subdivision ineffective.

To solve this problem, two methods have been invented. The first and most commonly used one [3] is to implant a p$^+$ blocking strip in between the n$^+$ ones, as shown in Fig. 13. The blocking strips are left floating, since their function is just to interrupt the conduction channel.

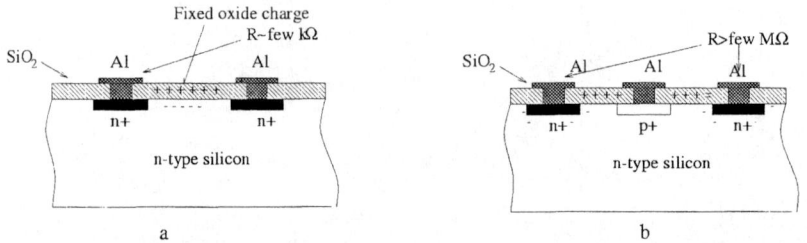

**FIGURE 13.** Problem of double side readout (a) and solution with the p$^+$ blocking strip (b).

The second one [4] uses an electric field to repel the electrons from the inter-strip region and establish insulation. The field is generated by applying a potential difference between the n$^+$ strips and a metal line which runs on top of them, separated by a thin insulator. The metal line is a few microns wider than the strip, and so the field generated effectively creates the insulation.

The use of double-sided micro-strip detectors allows the correlation of signals collected on the two sides, which apart from the readout electronics noise and response is the same, thus reducing multi-hit ambiguities.

# Pixel and Pad detectors

Producing a matrix of small diodes one can obtain in one detector true two-dimensional information. In general this is called a silicon pad detector, and it is connected to the readout electronics via a fanout circuit overlaid on the silicon wafer and wire bonded to the individual pads (such fanouts have been manufactured in ceramic, glass or Kapton). Sometimes, the space between diodes has been used to run traces which connect the pads to the edge of the detectors or, in even more sophisticated designs, a second layer of metal has been deposited on the detector, on top of a thin insulating layer, to create lines which connect the pads to the edge, where the traces are wire bonded to the electronics. It is clear, though, that silicon pad detectors cannot have too many detecting elements, or the problem of interconnections becomes unmanageable. A way out is to design the readout electronics in form of a matrix, with each channel occupying exactly the same surface as a detector element, and equip each channel of electronics and every element of the detector matrix with a bonding pad (an area of deposited metal with good surface quality not covered with passivation). Than, a tiny (few tens of microns of diameter) ball of solder (often an indium alloy) is deposited on the bonding pads, and the two chips are put in contact face-to-face. This device is called a *silicon pixel detector*, and a simple scheme is shown in Fig. 14.

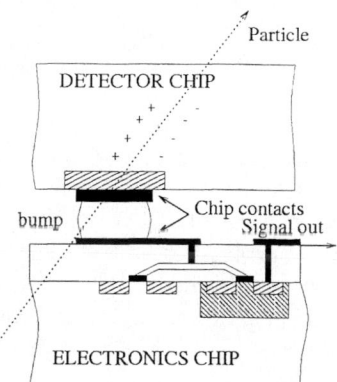

**FIGURE 14.** Scheme of the pixel detector operation principle.

Pixel detectors presently used in experiment by the NA57 collaboration [6] have cells 500 $\mu$m long and 50 $\mu$m wide. In Fig. 15 is shown a photograph of the matching electronics chip, with the solder bumps nicely visible (20 $\mu$m in diameter). Pixel detectors, combine the advantages of unambiguous two-dimensional readout with the geometrical precision, double-hit resolution, speed and simplicity of calibration characteristics of silicon micro-strip detectors. In addition, a high segmentation leads naturally to a low individual diode capacitance, resulting in an excellent signal-to-noise ratio at high speed. The price is a very large number of connections

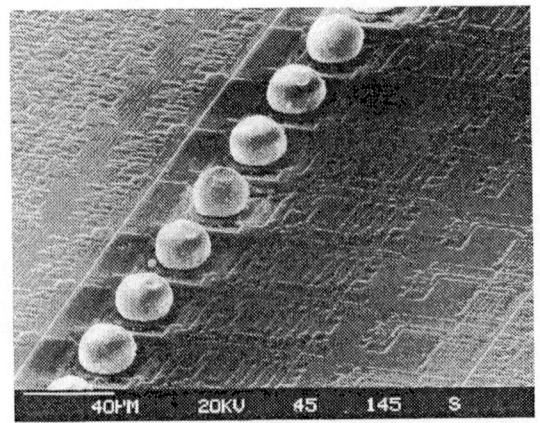

**FIGURE 15.** SEM photograph of solder bumps on the chip.

and of readout electronics channels. It is clear that to fit a full readout channel in the size of one of the detector cells is a formidable challenge! Only the progress in the component density achievable in CMOS micro-electronics chips has made it possible the production of practical silicon pixel devices. At the same time, fine-pitch surface packaging techniques of the type described before (flip-chip bonding) have become available in the electronics industry, making the assembly of very large numbers of channels possible with good yield. A detector like the one used in NA57, which includes several planes with several chips each, features over one million readout channels. Each includes a preamplifier, a discriminator, a delay line and control circuits. An additional advantage of pixel detectors, associated with the small size of the individual diode, is the very small leakage current per element, very important fact in applications involving very high radiation levels which cause a large increase in the leakage current per unit volume. Therefore, pixel detectors are ideally suited for the innermost layers of detectors at hadronic colliders, since they couple excellent granularity and spatial resolution in both coordinates with very good radiation tolerance.

## Silicon drift detectors

Silicon Drift Detectors (SDDs) [7]., like gaseous drift chambers, exploit the measurement of the transport time of the electrons generated by a traversing particle to determine one coordinate of its crossing point. A very simple scheme of the principle of a SDD is illustrated in Fig. 16. The electrons drift essentially in a direction parallel to the surface of the detector, under the effect of an applied electric field. The maximum drift path is typically of the order of few cm. The electrons are collected by an array of small size anodes (with a typical pitch of a

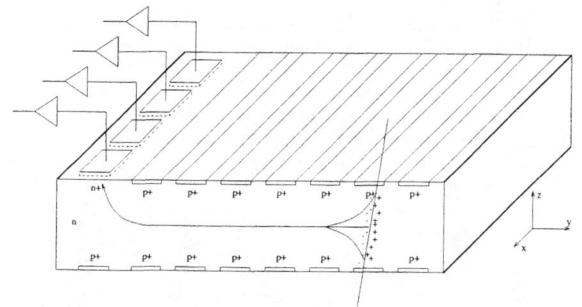

**FIGURE 16.** Schematic view of the principle of a Silicon Drift Detector.

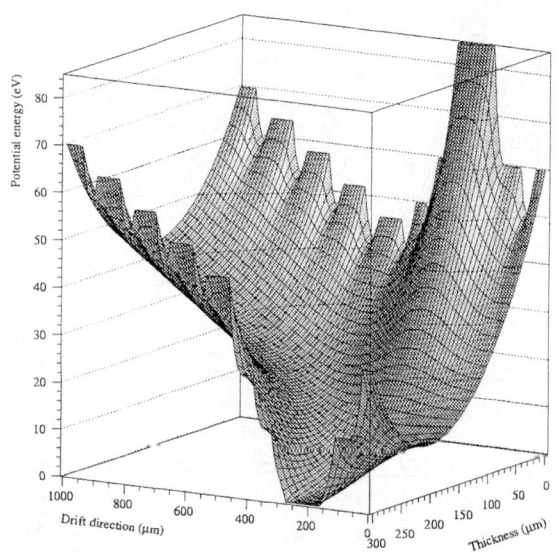

**FIGURE 17.** Simulated potential in a SDD

few hundreds of $\mu m$) placed at the end of the detector perpendicular to the drift direction. The second coordinate can be obtained from the centroid of the signals at the anodes, making the SDD a unambiguous 2D position sensitive detector. This type of detector is well suited to experiments with very high particle multiplicity with low event rates, since the read-out time is typically a few $\mu s$.

In practice, both the drift field and the depletion bias are produced by $p+$ parallel strips implanted in both faces of the detector. Each strip is polarized with a

**FIGURE 18.** Sketch of a SDD. (in order to see all structures, the number of drift strips is not respected.)

negative voltage proportional to its distance from the anode in order to produce the drift field. In this way, the $p^+n$–junctions are reverse polarized and can assure the depletion of the detector through a $n+$ ring connected to ground placed at the periphery of the detector (not seen in the scheme). The electrons generated by a crossing particle are focused into the middle plane of the detector due to the parabolic shape of the potential along the thickness axis, and drift towards collection region under the effect of the constant $\vec{E}$ field with a constant speed $v = \mu_e E$, where $\mu_e$ is the mobility parameter for the electrons. The holes are collected by the nearest $p^+$ electrode and do not contribute to the signal. The scale of potential is produced by a resistive voltage divider, which can be implanted directly on the silicon wafer. In the collection region, a quite complex polarization is needed to carry the electrons from the middle plane of the detector toward the anodes. A set of strips on both faces with adapted voltages is dedicated to this task. Figure 17 shows a numerical simulation of the potential in the drift detector, obtained when the detector is fully depleted for a typical drift field of $500\,V/cm$. The drift and the collection regions are clearly identified by the field shape.

In Fig. 18 a sketch of a drift detector is represented. The anodes are at the bottom surrounded by a grid dedicated to isolate them one from the other. The drift strips

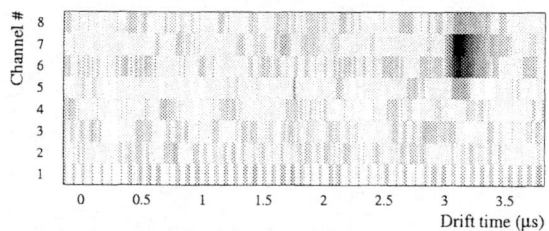

**FIGURE 19.** An example of SDD event with one cluster.

are connected to the resistive voltage divider, implanted on the top, through the side strips. These strips have also the essential function to reduce progressively the potential of the drift region (hundreds to a few thousand Volt) to ground. The ring surrounding all the detector is an $n+$ contact connected to ground, used to deplete the volume of the detector. SDD prototypes of very large sensitive area ($2 \times 35\,mm$ drift path $\times 75\,mm$ width) have been realized for the ALICE experiment [5].

The diffusion and Coulomb repulsion between electrons play a significant role in drift detectors since the drift time is of the order of a few $\mu s$. In the thickness direction, they are compensated by the parabolic potential, but generate an increase of the electron cloud size in both other directions. The electron cloud reaches the collection zone with a size increasing as a function of the total drift time. Thus the charge may be collected by more than one anode, so the coordinate along the direction perpendicular to the drift is determined by calculating the centroid of the charge deposited on the touched anodes. Typically, a $200\,\mu m$ pitch allows a precision of $30\mu m$. The signal measured on each anode is amplified and sampled with a typical frequency of a few tens of $MHz$, depending mainly on the drift velocity and the peaking time of the electronics. Fig. 19 shows an example of an event with one particle crossing a SDD where the signal is measured by four anodes. The coordinate along the drift direction is measured by calculating the elapsed time between an external trigger and the arrival of the charge.

## REFERENCES

1. B. Alessandro et al., *Nucl. Instr. and Meth.* **A409** 167 (1998).
2. J. Straver et al., *Nucl. Instr. and Meth.* **A348** 485 (1994).
3. L. Bosisio et al., *Nucl. Instr. and Meth.* **A273** 677 (1988).
4. V. Chabaud et al., *Nucl. Instr. and Meth.* **A368** 314 (1996).
5. ALICE Collaboration, CERN/LHCC, 99/12.
6. F. Antinori et al., *Nucl. Instr. and Meth.* **A360** 91 (1995).
7. E. Gatti and P. Rehak, *Nucl. Instr. and Meth.* **A225** 608-614 (1984).

# RICH detectors

E. Nappi

*CERN, Switzerland and INFN, Sez. Bari, Bari, Italy*

**Abstract.** Basic principles of the Ring Imaging CHerenkov (RICH) technique are given, the issues of designing devices based on such a technique and the factors which limit the particle discriminating power are discussed. The challenging designs adopted by current experiments are also reviewed.

## INTRODUCTION

The first time I was involved in the development of a RICH detector was many years ago, namely in 1985. In that period, I was participating in the NA35 experiment at CERN and after two years of data-taking, the Collaboration wished to upgrade the apparatus with a device having particle identification (PID) capability. When a RICH detector was proposed, frankly I must say that my first opinion on this technology was negative: a nasty photosensitive vapour called TMAE was addressed in its design and the pattern recognition seemed quite complicated with strange patterns to reconstruct. Moreover, the use of TMAE required a quite complicated detector geometry with blind electrodes and operation at high temperature. Despite my first judgment, by studying better the anticipated performance, it became clear to me that this technique was extremely powerful. The separation power reachable is unmatched among the current alternative technologies, therefore I decided to commit myself to the construction of a detector of that kind and after that many others followed. And now here I am, trying to fascinate you by showing how RICH detectors are impressively efficient in particle identification.

The RICH detector, conceived by A. Roberts [1], is the last born in the large family of devices based on the detection of Cherenkov radiation, named after the Russian physicist who discovered that charged particles moving through a medium with a velocity greater than the local phase velocity of light emanate a glow analogous to a sonic shock wave.

The spectrum and intensity of Cherenkov radiation can be calculated with high accuracy in terms of the classical theory of electromagnetism. The theoretical interpretation assumes that the atoms of the medium become polarized in the region along the charged particle track. Owing to the transient nature of this phenomenon, polarized atoms relax back to equilibrium by emitting a short electromagnetic pulse.

When the velocity of the particle does not exceed the local phase velocity of light, the emitted electromagnetic pulses interfere destructively because of the symmetrical distribution of the polarization field. Otherwise, a coherent wavefront, at a specific angle $\theta_c$ with respect to the particle direction, is produced when the particle moves faster than light since only in this case the polarization field is asymmetric along the particle track. The resulting radiation covers a band of frequencies corresponding to the different Fourier components of the electromagnetic pulses emitted by the medium dipoles. Although Cherenkov radiation is feeble, few hundred photons against the many thousands of photons emitted in scintillating material, the use of its directional properties is a powerful tool for the identification of charged particles.

This paper is based on two lectures given at the $8^{th}$ ICFA School on Instrumentation in Particle Physics (Istanbul, 1999) and consists of three parts. In the first part, the field is introduced by starting with an historical overview and later with the basic principles of the Cherenkov effect. The second part, after having outlined the different detector schemes, deals with the analysis of RICH performance with a detailed discussion on measurement errors. The third part describes the components of a RICH detector, and their main applications in current and future experiments are given.

The present paper is not meant to cover all aspects of the RICH technique: electronics and pattern recognition are two important issues that have been deliberately left out because of the limited time allocated to the lectures. There are a number of comprehensive textbooks for further reading, the most significant being listed in Ref. [2]. Recently proposed new ingenious detector schemes, like the DIRC counter, are also not mentioned in the present paper. Interested readers are referred to the Proceedings of the successful series of Workshops [3] devoted to Cherenkov light imaging since they represent an inexhaustible and updated source of information.

## HISTORICAL OVERVIEW

During the systematic work of reformulation of the Maxwell theory, anticipating by almost fifty years the experimental achievement of P. A. Cherenkov, Oliver Heaviside [4] showed that charged particles moving faster than light in vacuum emit an electromagnetic radiation whose wavefront propagates at a fixed angle with respect to the particle direction.

Although Heaviside made a wrong starting hypothesis since the condition of superluminality is not achievable in vacuum, his achievement is correct because the speed of charged particles moving in a dielectric medium with a refractive index larger than one can exceed the local phase velocity of light. Indeed, as early as 1919 M. Curie observed a faint blue light coming from concentrated solutions of radium in water. In 1934, the Russian P. A. Cherenkov, trying to understand the origin of the weak luminescience that salt solutions emit when struck by gamma rays, published a paper in which he proved that the light emission was caused by

Compton electrons moving quickly through the liquid and showed the relationship between the emission angle and the refractive index of the medium [5].

In 1937, Frank and Tamm formulated the theory of the Cherenkov effect and predicted the radiation spectrum by applying the principles of classical electrodynamics [6]. The quantum formulation of such a theory was elaborated by Ginsburg [7] a few years later.

In 1958, Cherenkov, Frank and Tamm were jointly awarded the Physics Nobel prize for the discovery and interpretation of 'Cherenkov radiation'.

The capability of using the Cherenkov radiation for PID, was already clear to its discoverer in 1937 [8]. In the early days, distilled water was used as radiator and photographic emulsions or the researcher eyes as photodetector.

A major breakthrough in this technique was provided in the 1940s by the availability of photomultipliers capable of detecting feeble light with high efficiency and fast response.

Since 1951, when Jelley developed the first device specifically for a physics experiment [9] employing photomultipliers, several other detectors have been designed and built for both nuclear physics and particle physics experiments and for astrophysics applications as well. Cherenkov detectors played a fundamental role in important high energy physics achievements, for example in the discovery of the antiproton [10].

The idea to discriminate particles by differentiating between different values of the emission angle $\theta$ was conceived by A. Roberts in 1960 [1] and proved to be an extremely powerful method for identifying particles, as T. Ypsilantis and J. Seguinot [11] practically demonstrated in 1977 by imaging Cherenkov photons directly in a gaseous photodetector. This technique was named RICH: an acronym for "Ring Imaging CHerenkov" coined by T. Ekelof as a good omen for the funding situation of the experimental group involved in the realization of such a counter [12].

In 1982, a RICH device was for the first time installed in a high-energy physics experiment (E605 at Fermilab [13]), many others have been built since then. More recently, considerable advances in the technologies associated with photodetectors have extended the potentiality of such devices. As a result a renewed development of RICH counters is taking place.

## PROPERTIES OF CHERENKOV RADIATION AND BASIC FORMULAE

The properties of Cherenkov radiation are contained in the following simple equation: [1]

---

[1] By taking into account the quantum theory of Cherenkov effect, the recoil of the charged particle of momentum $p$ slightly modifies the classical Cherenkov equation by providing an additive

$$\cos\theta_c = \frac{1}{n\beta} \tag{1}$$

which provides the emission angle $\theta_c$ of Cherenkov photons in terms of the velocity of the charged particle $\beta$, in units of the speed of light, and the refractive index $n$ of the medium. Since $|\cos\theta_c| \leq 1$, there exists a velocity threshold, expressed by the Lorentz factor

$$\gamma_t = (1 - 1/n^2)^{-\frac{1}{2}}. \tag{2}$$

When the velocity of the particle approaches the velocity of light in vacuum ($\beta \to 1$), $\theta_c$ takes the maximum value $\theta_{max} = \arccos 1/n$.

Eq. (1) alone, contains the two basic properties of the Cherenkov radiation that are exploited in practice, i.e. the existence of a threshold momentum and the peculiar direction of emission at an angle depending on the particle velocity.

For any practical purpose, the spectral dependence of the radiation must be taken into account. Frank and Tamm's equation describes the energy radiated per unit path length $dx$ by a particle of charge $Ze$:

$$\frac{d^2W}{dx\,d\omega} = \frac{Z^2e^2\omega}{c^2}\left(1 - \frac{1}{\beta^2 n^2(\omega)}\right), \tag{3}$$

where, due to the chromatic dispersion of the optical medium, $n$ is a function of the radiation frequency $\omega$. When integrated over the radiating path length L

$$\frac{dW}{d\omega} = \frac{LZ^2e^2\omega}{c^2}\left(1 - \frac{1}{\beta^2 n^2(\omega)}\right). \tag{4}$$

We deduce that the energy loss because of Cherenkov effect is much lower than the ionization energy loss. Actually, if we consider an electron that moves with $\beta \simeq 1$ across a 1 cm of water ($n = 1.33$), in the spectral range $\lambda = 400 - 700\ nm$ the electron loses about 500 eV by the Cherenkov effect, whilst its energy loss by ionization is 2 MeV.

The number $N$ of Cherenkov photons emitted with energy $\hbar\omega$ is a fundamental quantity for the detector design. It is easily deducible from Eq. (3):

$$N = \frac{LZ^2\alpha}{c}\int\left(1 - \frac{1}{\beta^2 n^2(\omega)}\right)d\omega \tag{5}$$

or

$$N = 2\pi LZ^2\alpha \int_{\beta n>1}\left[1 - \left(\frac{\beta_t(\lambda)}{\beta}\right)^2\right]\frac{d\lambda}{\lambda^2}, \tag{6}$$

---

term that is completely negligible for any practical application. The complete equation is: $\cos\theta_c = \frac{1}{n\beta} + \frac{\hbar}{\lambda p}\frac{n^2-1}{2n^2}$, where $\lambda$ is the Cherenkov photon wavelength.

where $\alpha$ is the fine structure constant.

It follows that in 1 cm of material with a refractive index n, the number of photons emitted in the spectral range of 1 eV by a particle of charge $Z$ moving with a $\beta \simeq 1$ is given by

$$N(cm^{-1}eV^{-1}) = 370Z^2 \left(1 - \frac{1}{n^2}\right). \tag{7}$$

The total number of photons emitted depends upon the wavelength integration, but in general, as Eq. (7) indicates, the number of photons emitted per unit length and per unit energy is a constant, where there are no absorption bands close to the interesting frequency range. This constant is just the mean of the Poisson distribution in the number of photons, since it is an inherently statistical process.
In summary:
a) energy loss for Cherenkov radiation is of the order of keV/cm;
b) the amount of Cherenkov radiation is proportional to the square of the particle charge and it is independent on the particle mass;
c) the photon yield per unit of wavelength interval $d\lambda$ is proportional to $d\lambda/\lambda^2$, consequently most of the photons are emitted in the UV region;
d) equal number of photons per unit path per unit frequency interval.

## PARTICLE IDENTIFICATION WITH CHERENKOV DETECTORS

A particle is univocally identified by its mass and electrical charge. The mass is provided by measuring at least two out of the three correlated quantities: momentum, kinetic energy and velocity. Practically, the choice is restricted to the momentum and velocity, in fact $p = mc\gamma\beta$.
The precision with which the mass is determined is given by

$$\left(\frac{dm}{m}\right)^2 = \left(\gamma^2 \frac{d\beta}{\beta}\right)^2 + \left(\frac{dp}{p}\right)^2. \tag{8}$$

If the momentum $p$ is relatively well measured, then the resolution of particles with masses $m_1$ and $m_2$ requires a velocity resolution, $\Delta\beta$, given by[2]

$$\frac{\Delta\beta}{\beta} \simeq \frac{m_1^2 - m_2^2}{2p^2}. \tag{9}$$

Actually, the deflection of the particle trajectory in a suitable static magnetic field provides the charge sign and the momentum value, whilst the velocity is achieved by means of one of the following methods: energy loss, time of flight (TOF), detection

---

2) from $m_1^2 - m_2^2 = p^2 \frac{\Delta\beta(\beta_1+\beta_2)}{(\beta_1\beta_2)^2}$

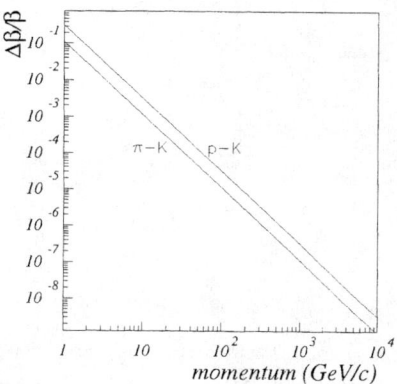

**FIGURE 1.** Resolution in velocity required to separate two particle species as a function of their momentum.

of Cherenkov radiation and detection of transition radiation.

As Fig. 1 shows, in order to separate kaons from pions already in the momentum range of a few GeV/c the velocity resolution must be better than a few percent. Such a precision can be achieved only by Cherenkov counters. Indeed, it can be easily proved that the PID capability of a 1 m long TOF system with an excellent time resolution of 50 ps is limited to momenta below 1.5 GeV/c when a $3\sigma$ mass separation is required.

## Threshold counters

By plotting the Cherenkov angle as a function of the particle velocity $\beta$, one realizes that the greatest sensitivity is provided by measuring the angle close to the threshold where $d\theta/d\beta$ is large (Fig. 2). However, the drawback is that the few photons emitted near the emission threshold cause the measurement of the Cherenkov angle to be affected by a large statistical error. Consequently the most effective way to exploit the threshold effect is achieved by counting the number $n_{p.e.}$ of detected photoelectrons. Near threshold $n_{p.e.} \simeq 0$, therefore the probability of not observing a signal is evaluated by means of Poisson distribution, $P(0) = e^{-n_{p.e.}}$, which gives an efficiency:

$$\epsilon = 1 - e^{-n_{p.e.}}. \tag{10}$$

This method is exploited in Cherenkov threshold detectors which employ a specific radiator medium whose refractive index $n$ is chosen in such a way that radiation is only emitted when particles move through it with a speed exceeding $c/n$, thus

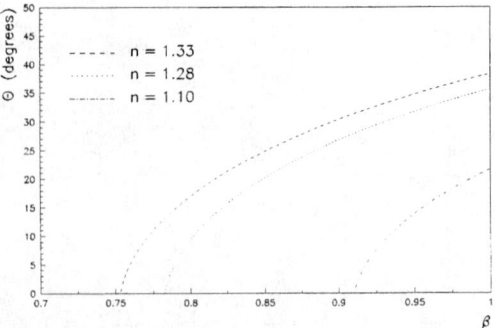

**FIGURE 2.** Variation of Cherenkov angle $\theta$ with particle velocity $\beta$ for three different refractive indices: n=1.33 (water), n=1.28 (liquid perfluorohexane) and n=1.1 (aerogel). Emission angle changes rapidly close to the velocity threshold, its variation flattens as particle velocity increases.

allowing them to be separated from slower particles ('below threshold')(Fig. 3). Equation (10) implies that to keep the detector inefficiency at the level of $10^{-2}$ at least 4.5 photoelectrons must be detected on average.

A 'modern' version of the Threshold Cherenkov detector was proposed by F. Piuz [14] in 1995 for performing the hadron identification in the 3-8 GeV/c momentum range in the CERN-NA44 heavy ion experiment. The device, called TIC (Threshold Imaging Cherenkov), exploits the property of a gaseous wire chamber equipped with a UV sensitive pad-segmented cathode to localize with high spatial accuracy Cherenkov photons. Oppositely to the traditional threshold counters, TIC

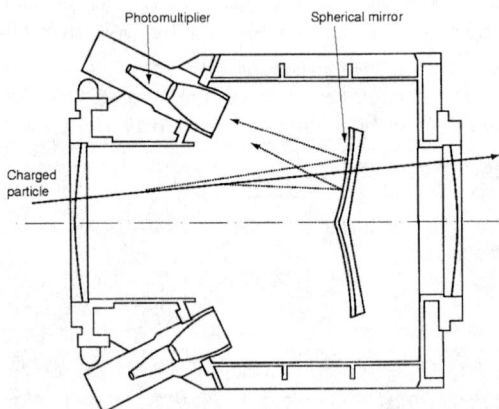

**FIGURE 3.** Schematic layout of a threshold Cherenkov detector.

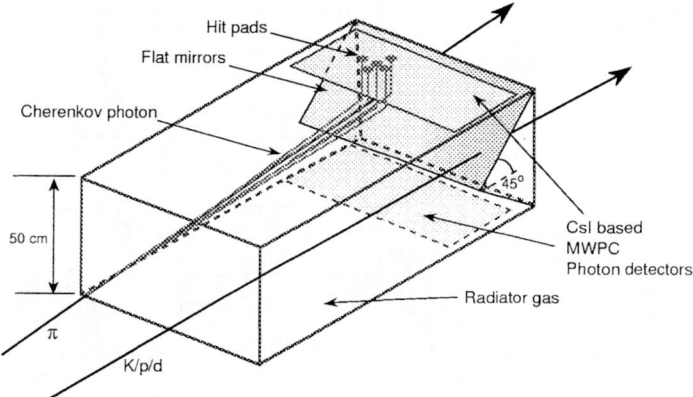

**FIGURE 4.** Schematic layout of a TIC detector. Two particles are traversing the gas radiator, one of them emits Cherenkov photons since it is above the Cherenkov threshold.

can be employed in experiments with several particles in the detector acceptance (Fig. 4).

## DISC counters

A significant step in the application of Cherenkov radiation to PID took place at the beginning of 1970s, when Litt and Menieur [15] invented the Differential Isochronous Self Collimating detector, named DISC. By taking into account the formidable accuracy achieved in the past, as good as $\Delta\beta/\beta = 10^{-7}$, DISC is still, so far, the most precise device ever built for measuring the speed of particles in primary beams.

A DISC counter is an improved version of Differential Cherenkov counters where photons are focused onto a matrix of photomultipliers placed behind an annular diaphragm by means of a spherical mirror. Consequently, the photomultipliers provide ring imaging only for those particles that emit Cherenkov light in the diaphragm aperture angle. The better angle resolution achieved by DISC counters is obtained by implementing a specifically designed optics just immediately in front of the diaphragm, with the aim of compensating the chromatic dispersion of the radiating medium (Fig. 5). PID is achieved by requiring the coincidence of several photomultipliers, therefore the efficiency $\epsilon_k$ for a k-fold coincidence is

$$\epsilon_k = (1 - e^{-n_{p.e.}/k})^k, \qquad (11)$$

where $n_{p.e.}$ is the number of detected photoelectrons.

Consequently, the overall efficiency rapidly decreases by requiring more photomultipliers in coincidence, it never reaches 90% even in the case of a large number

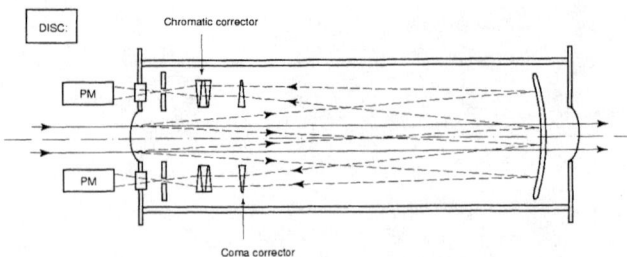

**FIGURE 5.** Schematic layout of a DISC counter. The spherical mirror shown on the right hand side focuses Cherenkov light onto a phototube matrix through a correcting optics and a ring diaphragm.

of detected photoelectrons, although the detector capability of rejecting unwanted particles is always very high as mentioned before. As for drawbacks, DISCs are quite complicated devices and are utilized only to tag the particles belonging to primary beams: therefore, although they have the best performance of all Cherenkov detectors, the limited phase-space acceptance makes them of no practical use for identifying secondary particles.

## RICH detectors

The small angular coverage of DISCs was overcome by the RICH counters which allow simultaneous measurement of the values of $\beta$ for several particles of different known momentum by determining the position of a certain number of Cherenkov photons. In a RICH detector, Cherenkov radiation, emitted from several particles in the same event, is transmitted through an optics, that could be either focusing with a spherical (or parabolic) mirror or not focusing (proximity-focusing), onto a photodetector that converts photons into photoelectrons with high spatial and time resolutions. Cherenkov footprints are visualized onto the plane of detection thus allowing the determination of the emission angle $\theta_c$ for each detected photon. The mass $m$ of the particle of known momentum $p$ is eventually given by

$$m = p\sqrt{n^2 \cos^2 \theta_c - 1} \; . \tag{12}$$

## DESIGN CRITERIA OF RICH DETECTORS

The great challenge of the RICH technique is the detection of signals of single electrons, i.e. analogous to the detector "noise". Therefore, a careful design is mandatory for achieving the best detector performance.

As previously mentioned, a RICH detector consists of two basic elements arranged in a focusing or in a proximity-focusing geometry: a transparent dielectric medium, called the radiator, whose refractive index is appropriate for the range of particle momentum being specifically studied (Eq. 2) and a photon detector. The latter provides information on the position of the photoelectron initiated by the conversion of the Cherenkov photons in a suitable photosensitive volume, or a conversion layer. The focusing arrangement (Fig. 6) is more suited for low refractive index radiators (mainly gas) due to the long length needed to provide a satisfactory number of detected photoelectrons per ring, whilst in the 'proximity-focusing' geometry (Fig. 7), a thin slab of radiator emits Cherenkov photons along a conical wavefront that enlarges in an inert gas volume between radiator and photodetector. The resolution of the Cherenkov rings is determined by the ratio of radiator thickness and photodetector distance. Quite compact designs are possible. In this configuration, the resolving power is worse than that for gaseous radiators, but it enables good PID in a momentum range where gaseous radiators are insensitive.

The design of a Cherenkov detector relies on Eq. (1), (3) and (7) and on the knowledge of the optical properties of the medium. Moreover, since Cherenkov radiation is linearly polarized with its electric vector lying in the plane defined by the particle direction and the photon direction, special care must be taken in evaluating the reflection losses at the medium interfaces and in choosing materials fully isotropic to polarized light.

In any practical case, the medium transparency and photon detector inefficiencies allow only a few Cherenkov photons to be detected. Indeed, for a photon detector with quantum efficiency $Q$, single-electron detection efficiency $\epsilon$, a transmission of radiator and windows $T$, and a mirror reflectivity (if present) $R$, the proportionality factor $N_0$, called the *figure of merit*, is defined as

$$N_0 = \frac{\alpha}{\hbar c} \int \epsilon Q T R \, dE . \tag{13}$$

The energy limits in the integral are defined on the bottom edge by the photoionization threshold and on the top edge by the medium transparency. The larger the $N_0$, the better the detector.

If a detector is designed to detect photons in a spectral region far from where the radiating medium has absorption bands, $n$ can be taken to be independent of frequency, and the traditional equation used to describe Cherenkov counters results:

$$N = N_0 L \sin^2 \theta. \tag{14}$$

The quantity given by Eq. (14) is the mean or expected value of a Poisson distribution. For $\beta \to 1$ the Cherenkov angle tends to the asymptotic value $\theta_{\max}$ related to threshold $\gamma_t$ (Eq. 2) as

$$\sin^2 \theta_{\max} = \frac{1}{\gamma_t^2} \tag{15}$$

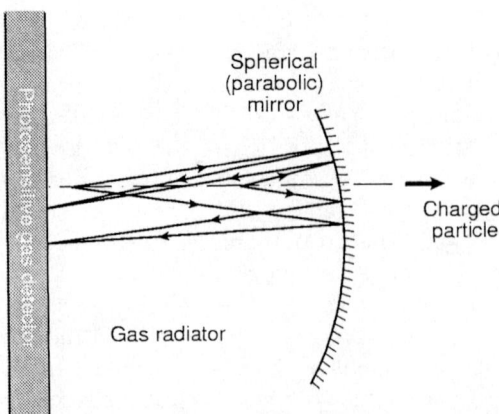

**FIGURE 6.** Focusing scheme: Cherenkov photons, collected by a mirror of focal length $f$, are focused onto the photon detector placed at the focal plane of the mirror. The resulting pattern is a circle of radius $r = f tan\theta_c$ regardless of the photon emission point along the particle track.

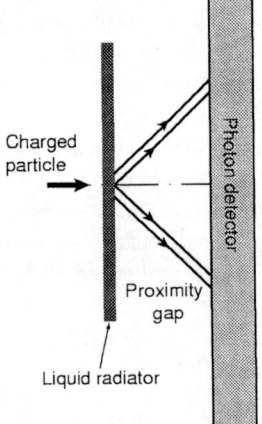

**FIGURE 7.** Proximity-focusing scheme: The detector volume, placed between the radiator and the photodetector, is known as the 'proximity gap' and is necessary to enlarge the Cherenkov cone to a more convenient size for the imaging.

with a maximum expected number of detected Cherenkov photons

$$N_{\max} = \frac{N_0 L}{\gamma_t^2}. \tag{16}$$

The fraction of Cherenkov photons at a given angle over the maximum yield is therefore expressed by

$$\frac{N}{N_{\max}} = \frac{\sin^2\theta}{\sin^2\theta_{\max}}. \tag{17}$$

For gas radiators, the following approximate expression is valid:

$$\frac{\sin^2\theta}{\sin^2\theta_{\max}} \simeq \frac{\theta}{\theta_{\max}} \simeq 1 - \frac{p_{\mathrm{th}}^2}{p^2} \tag{18}$$

implying that the Cherenkov angle and the number of detected photons depend in a universal way on the quantity $p/p_{\mathrm{th}}$. In Fig. 8 both the number of photons normalized to the asymptotic value $N_{\max}$ and the relative deviation of the Cherenkov angle from the asymptotic value are plotted against this scaling variable. The upper

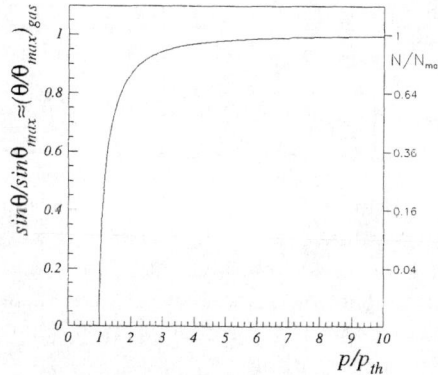

**FIGURE 8.** Variation of the Cherenkov angle sine and the number of photons normalized to their respective maximum values as a function of the scaling variable $p/p_{\mathrm{th}}$. In the case of a gas radiator the normalized ratio of the Cherenkov angle sines corresponds to the ratio between the Cherenkov angle and its maximum value. It is important to note that the maximum angle is reached much faster than the maximum number of photons as the particle momentum increases.

limit in the momentum range for particle discrimination is determined by the several sources of errors which limit the accuracy of the Cherenkov angle measurement. A complete discussion on this issue can be found in Ref. [2].

From Eq. (1), it derives:

$$\left(\frac{\sigma_\beta}{\beta}\right)^2 = (tan\theta\sigma_\theta)^2 + \left(\frac{\Delta n}{n}\right)^2. \tag{19}$$

The spread in particle direction due to the multiple scattering in the radiator, the finite spatial resolution of the photon detector and the aberration of the optics used allow a spread of angles $\Delta\theta_i$ to be detected. The chromatic aberration of the radiating medium, $\Delta n/n$, is usually the dominant contribution to the detector precision $\sigma_\beta/\beta$, especially if the RICH detector is designed to be operated in the ultraviolet region. As an example, a gas radiator has a chromatic dispersion in the UV band almost twice that in the visible region. These contributions are independent and add in quadrature: $\sigma_\theta^2 = \Sigma_i \Delta\theta_i^2$. Since each detected photoelectron gives a separate measurement, for $N$ photoelectrons, the Cherenkov angle resolution is improved:

$$\sigma_{\theta_c} = \frac{\sigma_\theta}{\sqrt{N}}. \tag{20}$$

A RICH detector with a figure of merit $N_0$ and a radiator characterized by the refractive index $n$ and total length $L$ measures the Cherenkov angles $\theta_1$ and $\theta_2$ of two particles of momentum $p$ and masses $m_1$ and $m_2$ respectively, with an accuracy described by the number of standard deviations $n_\sigma$ such that $\theta_2 - \theta_1 = n_\sigma \sigma_{\theta_c}$. From Eq. (9) it follows that the upper momentum limit $p_{m_1,m_2}$ for $n_\sigma$ standard deviation separation is

$$p_{m_1,m_2} = \left(\frac{\Delta m^2 (N_0 L)^{\frac{1}{2}}}{2n_\sigma\,\beta\,n\,\sigma_\theta}\right)^{\frac{1}{2}} = \left(\frac{\Delta m^2}{2n_\sigma\,\beta\,n\,\sigma_{\theta_c} sin\theta}\right)^{\frac{1}{2}} = \left(\frac{\Delta m^2 \sqrt{N}}{2n_\sigma\,\sigma_\theta tan\theta}\right)^{\frac{1}{2}}. \tag{21}$$

In Fig. 9, the upper momentum limit is plotted as a function of Cherenkov angle resolution for three different values of $n_\sigma$ in the case of a proximity-focusing detector employing a 1 cm thick layer of low-chromaticity $C_6F_{14}$ liquid as radiator (with $n = 1.28$ at $\lambda = 175$nm). The lower momentum limit is due to the decreasing number of detected photons towards threshold (Eq. 17) since it has immediate consequences on the pattern recognition.

A good RICH design allows the value of $p_{m_1,m_2}$ to be extended as far as possible once $n_\sigma$ has been fixed by physics requirements on the desired PID efficiency and the allowed contamination. By assuming that a Gaussian distribution is applicable to the measured Cherenkov angles, a particle contamination smaller than 1% requires a separation power larger than $n_\sigma = 4$, and of course if the ratio between the two particle populations is 1 to 10 then the actual contamination of the largest populated sample of particles in the other species is ten times larger, i.e. 10% for $n_\sigma = 4.2$.

Equation (21) entails that the largest momentum limit is achievable by increasing $N_0$ and decreasing $\sigma_\theta$. These two parameters are correlated with each other.

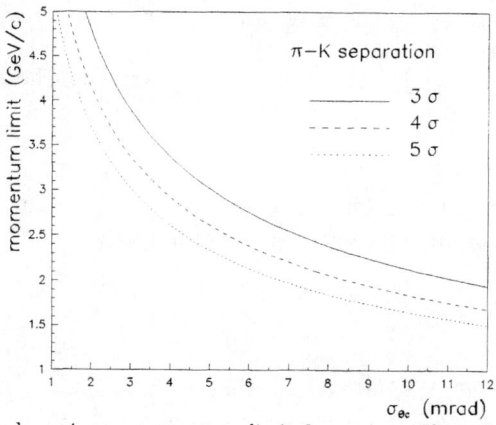

**FIGURE 9.** Estimated maximum momentum limit for a given Cherenkov angle resolution. A proximity-focusing RICH with $L = 1$ cm liquid perfluorohexane ($n = 1.28$ at $\lambda = 175$ nm) has been considered.

Indeed the new direction in the technique of Cherenkov light imaging is focused on achieving main advantages on both operational aspects and performance by designing RICH detectors that operate in the visible light region [16]. In fact, at longer wavelengths the detector figure of merit is larger due to the enlarged bandwidth for the relevant photoelectron yield (higher material transparency) and the angular accuracy for the single photon is smaller due to the reduced chromatic aberrations of materials in the visible region. In the following the detector components are analysed in the light of the most recent technological advances.

# DETECTOR COMPONENTS

## Photon detector

The basic task of the photon detector is to convert Cherenkov photons into a detectable electrical signal by means of a material with a high single-photon sensitivity defined by the quantum efficiency $QE$.

The low energy of the Cherenkov quanta implies that among the three interaction mechanisms of photons with matter, photoelectric absorption, Compton scattering and pair production, only the first one can be efficaciously exploited for practical purposes. Moreover it also implies that the photon detector must be able to detect with high efficiency the single electrons (photoelectrons) kicked off the atoms of the implemented photosensitive material that could be either a vapour (in this case the photoelectron production mechanism is called photoionization) or a thin solid layer (where more properly one refers to photoelectric production).

Besides the above-mentioned characteristics, the photon detector employed in a RICH device must also have:
- high localization accuracy able to match the errors in the Cherenkov angle;
- fast response;
- low noise;
- long-term stability;
- low cost in order to cover large surfaces.

Photon detectors can be divided into two classes: gaseous and vacuum-based detectors.

## Gaseous photon detector

This class gathers together multiwire proportional chambers (MWPC), multistep avalanche chambers (MSAC) and drift chambers with two-dimensional properties.

The pioneering work of T. Ypsilantis and J. Seguinot has shown that building such detectors is completely realistic by using vapours with a high $QE$ in the UV region [17]. The produced photoelectrons are detected by accelerating them in a uniform electric field towards a wire a few tens of $\mu m$ thick at high voltage. Close to the wire the electric field is very high and gives the drifting electrons enough energy to create an avalanche by knocking secondary electrons out of the gas atoms. The resulting ionization is large enough to be detected by a cathode electrode subdivided into pads and instrumented with sensitive electronics. Pad address gives an ambiguity-free two-dimensional image, allowing the reconstruction of overlapping rings from a multiparticle event.

At moderate amplification gain $(1 - 5 \cdot 10^5)$, single-electron pulse-height distribution has an exponential shape. In fact in the case of low electric field, the electron ionization is built up with several independent collisions with the gas atoms and therefore the probability $P(q)$ that an avalanche has a charge $q$ is obtained by the Furry distribution:

$$P(q) = \frac{e^{-q/\bar{q}}}{\bar{q}}, \tag{22}$$

where $\bar{q}$ is the mean charge of the avalanche.
The single-electron detection efficiency is therefore given by:

$$\epsilon = \int_{q_{th}}^{\infty} P(q)dq = e^{-q_{th}/\bar{q}}, \tag{23}$$

where $q_{th}$ is the threshold charge needed to remove the detector noise. The exponential form of $\epsilon$ is an unfavourable feature of gas detectors operated at low gains. In fact, a small decrease in the gas amplification implies a strong loss of efficiency. A more favourable pulse-height distribution (called the Polya distribution) occurs for higher gain values [18]:

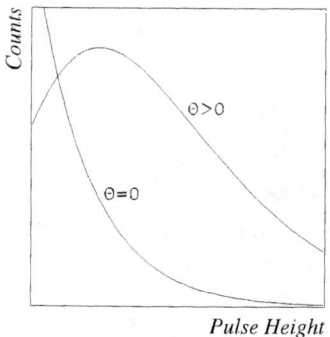

**FIGURE 10.** Qualitative behaviour of the single photoelectron pulse-height distribution for high ($\theta > 0$) and low ($\theta = 0$) gas amplification gains.

$$P(q) = \left(\frac{q(1+\theta)}{\bar{q}}\right)^\theta exp\left(\frac{-q(1+\theta)}{\bar{q}}\right), \qquad (24)$$

where the empirical parameter $\theta$ is related to the gas amplification mechanism. The Polya distribution takes the simple exponential form of the Furry distribution for $\theta = 0$ (Fig. 10), condition reached by decreasing the electric field. The peaked pulse-height distribution allows a more stable setting of the electronic threshold, but at higher gains the gas photon detectors experience a positive photon feedback caused by photons emitted by the de-excitation of gas molecules after the avalanche mechanism has occurred. Secondary photons initiate new avalanches after being converted by the photosensitive agent in the chamber. Blind electrodes are then implemented in order to shield the sensitive volume from the main avalanches (Fig. 11). Photon feedback is a clear limitation to the chamber gas gain. In fact, the light output from the avalanches grows exponentially as the chamber gain is increased beyond plateau.

Although benzene was used as a photosensor in the early prototypes, it was immediately replaced by photosensitive vapours with a lower photoionization threshold such as triethylamine (TEA) or tetrakis(dimethylamine)ethylene (TMAE) added to a regular gas mixture and flushed through the detector volume. As shown in Fig. 12, TMAE has a higher QE than TEA due to the strong electron donor properties of the attached dialkylamino groups (Fig. 13). The higher photoionization threshold of TEA (7.2 eV) severely restricts the choice of materials that can be used as UV windows, therefore fused silica that has transparency characteristics valid for TMAE must be replaced by more expensive $CaF_2$ windows. Moreover, extraordinary care must be taken to keep the contamination level of water and

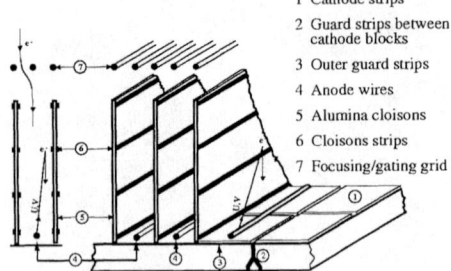

**FIGURE 11.** Schematic diagram of the blinds between anode wires used in the DELPHI RICH photon detector to control the photon feedback phenomenon.

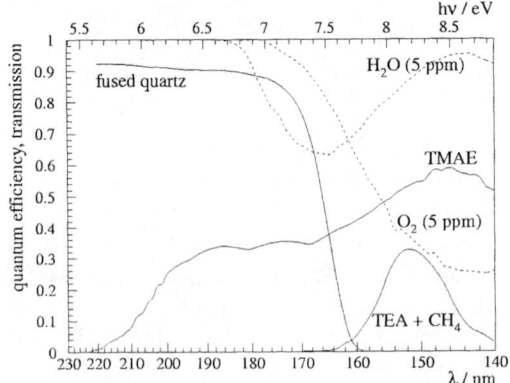

**FIGURE 12.** TEA and TMAE quantum efficiency as a function of the photon wavelength. Also shown are the transmission plot of a fused silica and the limit of transparency when 5 ppm of water and $O_2$ impurities are present in the gas volumes.

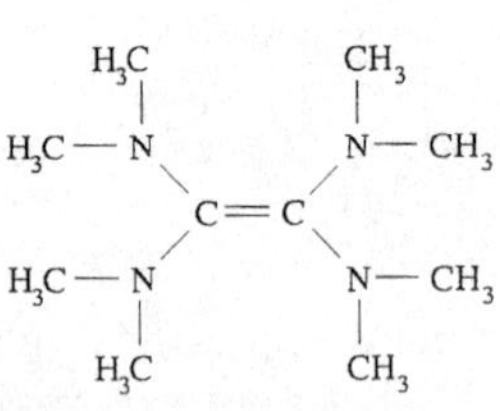

**FIGURE 13.** TMAE chemical structure.

oxygen, which absorb strongly in the UV, below 1 ppm in the gas volumes (radiator and drift gas), in order to avoid a reduction of detected photons (Fig. 12). In contrast to TEA, whose vapour pressure is 52 Torr at room temperature with an absorption length of 0.61 mm, TMAE vapour pressure of 0.30 Torr at room temperature is a disadvantage because it results in a long photon absorption length (3 cm) and therefore a large parallax error. Although the chemical reaction products of TMAE are highly electronegative and therefore absorb photoelectrons, the careful design of the detector and gas-handling system, and operation at higher temperatures adopted by the largest systems so far built (OMEGA [19], DELPHI [20] and SLD [21]) have enabled to reach large enough TMAE concentrations and stable operation.

An interesting feature of TMAE is its capability to emit light in the visible region as a result of the de-excitation of excited states formed by electron impact collisions with its molecules [22], during the avalanche amplification process in gaseous chambers. This excited state is caused by three basic mechanisms: photoabsorption, energy transfer from excited noble gas dimer and direct excitation by accelerated electrons [23]. The ratio of the amount of light produced to the amount of charge present is a function of the charge gain and light is more efficiently produced at low charge gain. The NA35 experiment exploited the light emission by TMAE in its optically read out RICH [24].

In recent years, a considerable effort has been made to prove that a thin film (100 nm $\div$ 1 $\mu m$) of CsI deposited onto the cathode plane of a gaseous detector is a valid alternative to the use of TMAE in large-area RICH detectors.

A specific R&D programme for the development of large-area CsI photocathodes was approved in 1992 by the DRDC at CERN under the name RD26 [25].

RD26 achieved a breakthrough in CsI deposition techniques by developing a successful standardized technology of evaporating photocathodes, as large as $50 \times 50$ cm$^2$, without the expensive and time-consuming implementation of masking techniques [26].

At CERN, the photocathodes are prepared in a large evaporation stand equipped with four DC heated tungsten crucibles operated simultaneously to achieve a uniform CsI layer. The best CsI quantum efficiency (Fig. 14) is obtained by depositing at least 250 nm of CsI onto a printed circuit board with a copper layer, accurately prepolished by mechanical and chemical treatments, and subsequentely covered with a thick (7 $\mu m$) chemically-deposited nickel layer, followed by a thinner (0.5 $\mu m$) layer of gold [29]. During the deposition, the pad substrate is held at 50°C. A 12-hour post-deposition heat treatment at 60°C, under vacuum, is necessary in order to achieve the final CsI QE [30] (Fig. 15).

In the near future, three experiments will implement a CsI RICH detector in their layout. Two of them (HADES [31] and COMPASS [32]) are building a system with a focusing scheme, whilst the third one (ALICE [33]), is envisaging a proximity focusing geometry.

The main advantages of a CsI RICH detector are an improved Cherenkov angle resolution since photoconversion is achieved in a single layer without parallax error,

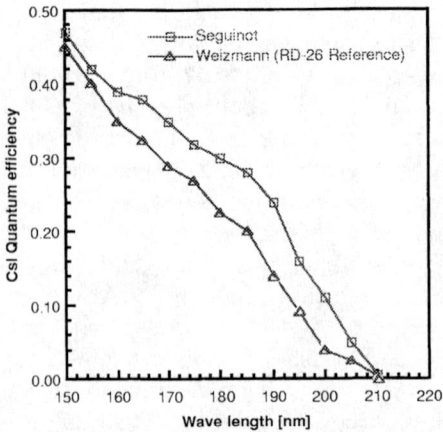

**FIGURE 14.** Measurements of CsI quantum efficiency in vacuum as a function of the photon wavelength performed by Seguinot et al. [27] and A. Breskin et al. [28].

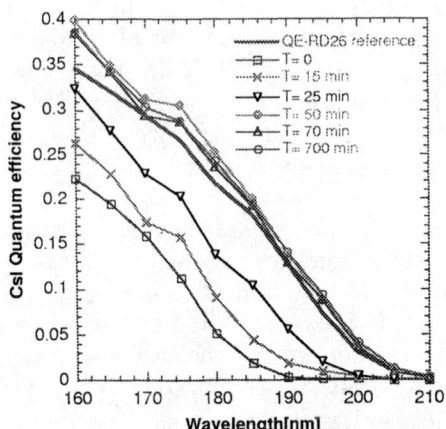

**FIGURE 15.** Heat treatment effect on QE of a CsI sample after evaporation at 60°C kept under vacuum.

and a simplified structure owing to the suppression of the photon detector window employed in the case of a TMAE RICH. This results in considerable cost-saving and a reduced total radiation length.

The possibility of using a thin anode-cathode gap (2 mm) (Fig. 16) simplifies the cumbersome Cherenkov pattern recognition in a high multiplicity environment because the chamber geometry approaches the ideal 2-D geometry and, furthermore, it reduces the background since ionizing particles traverse a small sensitive volume. These features are particularly mandatory in the case of ALICE RICH since an average multiplicity per event larger than 50 primary particles/$m^2$ is anticipated for LHC $Pb-Pb$ central collisions. A relevant feature of the photon detector shared

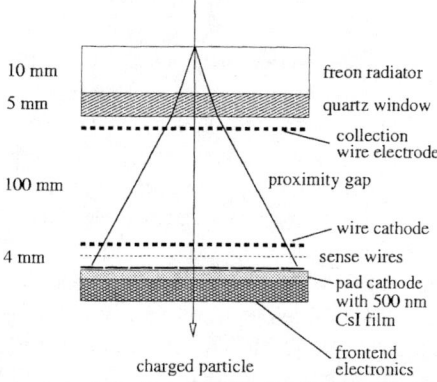

**FIGURE 16.** Schematic layout of the fast ALICE CsI RICH. The photodetector is a multiwire proportional chamber (MWPC), with anode wires of 20 $\mu$m diameter, 4 mm pitch and 2 mm anode-cathode gap. The MWPC is filled with pure methane at ambient temperature and pressure. Electrons released by ionizing particles in the proximity gap are prevented to enter the MWPC volume by a positive polarization of the electrode close to the radiator. A low noise and highly multiplexed VLSI analog electronics is fully integrated on the rear of the cathode plane, enabling the determination of the hit coordinates by centroid measurements.

by all the three CsI RICH designs is the "open geometry" i.e. the suppression of blinds that in TMAE photodetectors are specifically implemented to prevent spurious avalanches from feedback photons. In fact, the expected negligible background level from the CsI layer does not require any complex electrode structure [33].

## Vacuum-based photon detectors

Already in the past, RICH systems were succesfully designed and built to operate in the visible light region [34,35]. More recently, advances in technologies associated with the detection of visible light with devices of high granularity have stimulated fruitful new ideas [16]. With respect to the detector operational aspects, the main

benefits are as follows:
- no special handling for nasty photosensitive vapours such as TMAE;
- modest service and maintenance needs;
- savings in operating costs since gas circulation systems and expensive UV windows are no longer necessary;
- high segmentation flexibility and compactness.

The detector performance improves as follows:
- high rate capability and availability of the detector for triggering;
- a larger choice of materials as radiator, in particular the possibility of using aerogel;
- removal of background caused by incoming neutrons (neutrons create spurious hits in the hydrogenous gas mixtures used in RICH photon detectors as a consequence of proton recoils).

In 1991, T. Sugitate et al. [34] reported results from successful detection tests of Cherenkov rings focused onto an image intensifier coupled to a CCD camera. Although this technique seems very promising, CCDs are small and quite slow devices. This last constraint is severe, and in fact the acquisition rate of RICH detectors, barely relevant a few years ago, has now become a crucial issue.

The phototube (PMT) has the merits of robustness, low noise, high gain, and high rate capabilities, but it is sensitive to magnetic fields. "Quantacon-like" PMTs have a high single-photoelectron efficiency but a high cost per channel. A powerful application is represented by the experiment SELEX at Fermilab where almost 3000 PMTs have been employed to detect the Cherenkov light from a gaseous RICH device [35,36]. The experiments PHENIX [37] and BABAR [38] also detect the Cherenkov light with an array of PMTs.

In large-area applications, the multianode PMT (MaPMT) is more suitable since it offers the advantage of many channels with a single common power supply and a compact readout. MaPMTs, first used by S. Endo et al. [39], are now the baseline photodetectors for the Hera-B RICH [40]. The commercial tubes have a crosstalk of much less than 1% and a pad-to-pad variation in gain of less than 30% [41]. Use in a magnetic field of over 1.5 T is possible using fine mesh tubes, although at a high cost and with lower single-photon sensitivity.

Recently, hybrid photodevices have made considerable progress since their 're-discovery' to the stage where they are now being proposed in the LHCb experiment at LHC [42], as an alternative to the MaPMT and in the long baseline neutrino experiment AQUARICH [43]. They consist of an array of silicon pin diodes placed in a vacuum tube with a standard transmission photocathode kept at a negative voltage of several kV with respect to the silicon. Photoelectrons are accelerated by the electric field and penetrate the solid state diodes where thousands of electron-hole pairs are developed. Two electric field configurations are possible: proximity focusing and electrostatic focusing (Fig. 17). The latter has a small detector dead area but is very sensitive to magnetic fields. HPDs potentially offer outstanding features like high spatial resolution, stable gain, a wide dynamic range and an excellent single photoelectron response [44]. Nevertheless, for large area RICH device

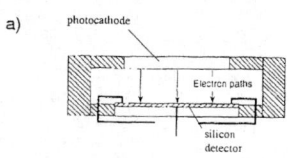

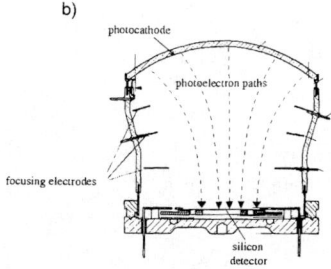

**FIGURE 17.** Different geometries of HPD detectors: a) proximity focusing, b) electrostatic focusing.

applications more R&D for implementing the FE electronics in vacuum is needed in order to avoid the large number of feedthrough lines. In addition, the commercially-available devices, in spite suitable performances, suffer from a large inactive area and high cost. The development of cheap hybrid devices with a large active area is presently underway at CERN [45].

Finally, it is worthwhile mentioning the Visible Light Photon Counters (VLPCs) because of their very high QE (85% for green light). VLPCs are based on doped SiAs crystals cooled at 7 K, biased at low voltage. Visible photons are guided through glass fibres into the intrinsic region of the detector where they create electron-hole pairs [46]. The resulting impact of one electron on a neutral crystal impurity starts an electron avalanche. Although they run with a speed of up to 30 MHz, the need of a cryogenic system and high costs have prevented them from being implemented in actual experiments so far.

## Radiator

The particle identification momentum range determines the choice of the radiator medium and a dual radiator geometry, i.e. a detector with both focusing and proximity focusing geometries, is in many cases mandatory to cover an extended momentum range for particle identification (Fig. 18). The choice of materials available able to feature as a Cherenkov radiator is quite limited. Since the intensity of the Cherenkov light emitted is much smaller than that given off in the scintillation process, an important requirement placed upon a radiator material is that it

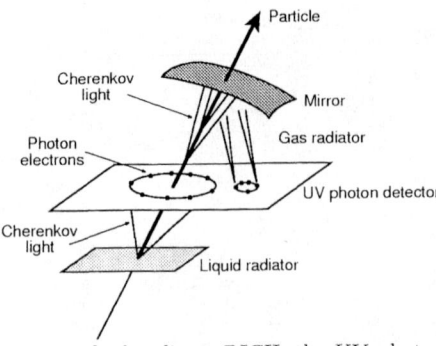

**FIGURE 18.** A schematic of a dual radiator RICH: the UV photon detector visualizes rings from both the liquid radiator and the gas radiator.

should not scintillate appreciably. Moreover, it must not have absorption bands in the wavelength range to be used.

The radiator volumes must be isolated from the photon detector, and at the same time, photons must be able to travel across the boundary with little loss. Since most of the Cherenkov light is emitted in the far ultraviolet region, (Eq.( 6), where the $\lambda^{-2}$ dependence is explicitly given), only a few substances are sufficiently transparent to permit good Cherenkov light transmission. Therefore, the choice of radiator is limited even further, particularly for liquid radiators.

UV transparency reduces the choice of window materials to: lithium fluoride, magnesium fluoride, calcium fluoride, natural quartz and fused silica. The latter is the only feasible choice in large RICH detectors, because it can be fashioned into large flat sheets, it has a high resistance to radiation and very good transparency, up to wavelengths of 160 nm (i.e. energy below 7.5 eV).

The radiator length required to produce at least 10 photoelectrons in a RICH detector with a given figure of merit is plotted in Fig. 19 as a function of the refractive index of the chosen medium.

Silica aerogel is the only existing material with optical properties suitable for filling the gap in the refractive index between liquids and heavy gases.

Aerogel is a manmade material that can have a density as low as three times that of air. It essentially consists of grains of amorphous $SiO_2$ with sizes ranging from 1 to 10 nm, linked together in a three-dimensional structure filled by trapped air. The huge number of such tiny primary particles determines an internal surface close to 1000 m$^2$/g which plays a fundamental role in the aerogel chemical and physical behaviour. There exists a simple relationship between the resultant index of refraction and the aerogel density $\rho$ in g/cm$^3$ [47]:

$$n = 1 + 0.21\rho . \tag{25}$$

Density values lying between 0.003 g/cm$^3$ and 0.55 g/cm$^3$ are in principle available,

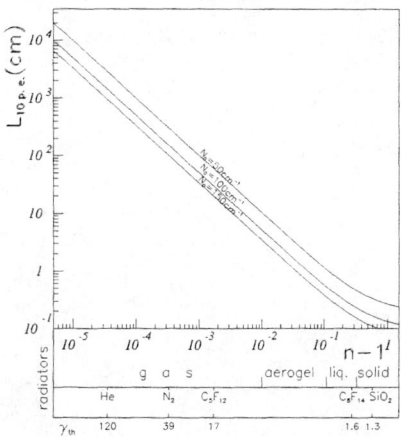

**FIGURE 19.** Radiator length corresponding to a yield of 10 photoelectrons as a function of the index of refraction of the medium for different RICH figures of merit.

corresponding to the refractive indices of n=1.0006 ($\gamma_t = 29$) and n=1.11 ($\gamma_t = 2.3$), respectively.

The granular structure of aerogel with a typical length scale of a few nm determines its optical properties. The behaviour of visible light in aerogel is dominated by Rayleigh scattering which increases as the fourth power of the frequency. When Rayleigh scattering occurs, the directionality of the Cherenkov radiation is completely lost. Therefore, the major concern associated with the design and construction of a RICH detector with an aerogel radiator is whether the Cherenkov photons that traverse the aerogel without any scattering are sufficient in number to allow the measurement of their emission angle with the expected accuracy.

Simple calculations show that the useful production of Cherenkov light is limited to the visible region.

Recently, hydrophobic, crack-free, very transparent aerogel samples became routinely available [48]. Loss of photons due to absorption and scattering processes in the bulk material has been minimized, as observed in test beam studies [49].

The "breakthrough" in the fabrication of aerogel has promoted more advances in the use of this material in Cherenkov detectors, as V.I. Vorobiov [50] and H. v. Hecke [51] pointed out in 1991 and 1993, respectively. Nonetheless, the major merit of the rapid progress of aerogel in real RICH devices must be ascribed to J. Seguinot and T. Ypsilantis who revised the van Hecke proposal in the light of currently available photodetector technology and envisaged an appealing application in the LHCb experiment [52]. Their design also inspired the upgrading of

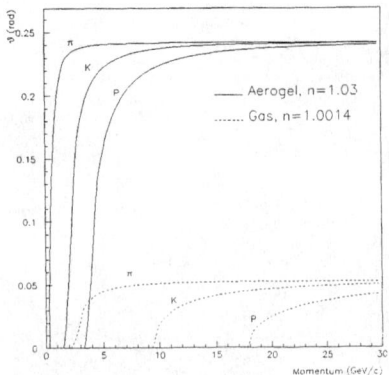

**FIGURE 20.** Momentum range covered by a RICH device with two radiators: gas $C_4F_{10}$ and aerogel with n=1.01

HERMES at DESY [53].
Both experiments use a dual radiator RICH with aerogel (n=1.03) and gas $C_4F_{10}$ (Fig. 20).

## CONCLUSIONS

A phenomenon discovered by the Russian scientist Cherenkov in 1934 has become in recent years the basic ingredient of the RICH devices able to identify particles in the vast momentum range 1 – 800 GeV/c. To be successfully exploited the Cherenkov radiation imaging technique requires a skilful team of physicists and engineers since many parameters must be controlled and mantained during the operation of the detector. Despite their complexity and high manpower costs, the outstanding physics results that have been achieved from the first generation of large RICH devices largely compensate the efforts made and justify the construction of new devices for future experiments.

## ACKNOWLEDGEMENTS

I wish to thank S. Majewsky, T. Ekelof and the Director of the VIII ICFA School on Instrumentation, Prof. M. Nizamettin Erduran, for having invited me to such an enjoyable and stimulating school.

# REFERENCES

1. A. Roberts, *Nucl. Instr. and Meth.* **9** (1960) 55;
   S.K. Poultney et al., *Rev. Sci. Instr.* **33** (1962) 574.
2. J. Seguinot, CERN-EP/89-92;
   T. Ypsilantis and J. Seguinot, *Nucl. Instr. and Meth.* **A343** (1994) 1-29 and 30-51.
3. The Proceedings of the 1st, 2nd and 3rd RICH Workshops have been published in *Nucl. Instr. and Meth.* **A343** (1994), **A371** (1996) and **A433** (1999), respectively.
4. O. Heaviside, Electrical Papers, II, 494 (Macmillan, London, 1892).
5. P.A. Cherenkov, *Acad. Sci. URSS*, **2**, 451 (1934); see also *Phys. Rev.* **52** (1937) 378 for English language publication of Cherenkov early work.
6. I.M. Frank and I.J. Tamm, *Dokl. Academie des Sciences de l'URSS*, **14** (1937) 107.
7. V.L. Ginzburg, *Zh. Eksp. Teor. Fiz. SSSR*, **10** (1940) 589.
8. P.A. Cherenkov, *Izv. Akad. Nauk. SSSR, Ser. Fiz. OMEN*, **1** (1937) 455.
9. J.V. Jelley, *Progress in Nuclear physics*, **3** (1953) 131.
10. O. Chamberlain et al., *Phys. Rev.* **100** (1955) 947.
11. J. Seguinot and T. Ypsilantis, *Nucl. Instr. and Meth.* **142** (1977) 377.
12. J. Seguinot and T. Ypsilantis, *Nucl. Instr. and Meth.* **A343** (1994) 1.
13. M. Adams et al., *Nucl. Instr. and Meth.* **217** (1983) 237.
14. C.W. Fabjan et al., *Nucl. Instr. and Meth.* **A367** (1995) 240.
15. J. Litt and R. Meunier, *Ann. Rev. Nucl. Sci.* **23** (1973) 158.
16. E. Nappi, *Nucl. Instr. and Meth.* **A409** (1998) 417;
    J. Seguinot and T. Ypsilantis, *Nucl. Instr. and Meth.* **A433** (1999) 1.
17. J. Seguinot et al., *Nucl. Instr. and Meth.* **173** (1980) 283;
    T. Ekelof et al., *Phys. Scripta* **23** (1981) 718.
18. J. Byrne, *Proc. R. Soc. Edinburgh* **66A** (1962) 33.
19. U. Muller et al., *Nucl. Instr. and Meth.* **A371** (1996) 27.
20. W. Adam et al., *Nucl. Instr. and Meth.* **A371** (1996) 12.
21. K. Abe et al., *Nucl. Instr. and Meth.* **A371** (1996) 8.
22. G. Charpack et al., *Nucl. Instr. and Meth.* **A258** (1987) 177.
23. M. Suzuki et al., *Nucl. Instr. and Meth.* **A254** (1987) 556.
24. J. Baechler et al., *Nucl. Instr. and Meth.* **A343** (1994) 213.
25. E. Nappi et al., RD26 proposal to DRDC, CERN/DRDC 92-3 and ADDENDUM to the DRDC PROPOSAL P35, CERN/DRDC/92-16.
26. E. Nappi et al., Status report of the CsI-RICH Collaboration, CERN/LHCC 96-20;
    F. Piuz, *Nucl. Instr. and Meth.* **A371** (1996) 96.
27. J. Seguinot et al., *Nucl. Instr. and Meth.* **A297** (1990) 133.
28. G. Malamud et al., *Nucl. Instr. and Meth.* **A343** (1994) 121.
29. F. Piuz, *Nucl. Instr. and Meth.* **A433** (1999) 178.
30. A. Buzulutskov at al., *Nucl. Instr. and Meth.* **A366** (1995) 410.
31. HADES, Proposal for a High Acceptance Di-Electrons Spectrometer, GSI, Darmstadt, 1993.
32. COMPASS proposal, CERN/SPSCL/96-14.
33. ALICE TDR 1, CERN/LHCC 98-19.
34. T. Sugitate et al., *Nucl. Instr. and Meth.* **A307** (1991) 265.

35. M. Maia et al., *Nucl. Instr. and Meth.* **A326** (1993) 256.
36. J. Engelfried et al.,The SELEX Phototube RICH detector, Fermilab-Pub-98/299-E, hep-ex/9811001.
37. PHENIX Conceptual Design Report, January 1993.
38. B.N. Ratcliff, *Nucl. Instr. and Meth.* **A371** (1996) 309.
39. S. Endo et al., A test of an improved RICH prototype, Institute for Nuclear Study (Tokyo) Annual Report, 1990.
40. P. Krizan et al., *Nucl. Instr. and Meth.* **A387** (1997) 146.
41. Y. Yoshizawa and J. Takenchi, *Nucl. Instr. and Meth.* **A387** (1997) 33.
42. LHCb Technical proposal, CERN/LHCC 98-4, LHCC/P4.
43. Letter of Intent for Long Baseline RICH, CERN-LAA 96-01, T. Ypsilantis et al., Nucl. Instr. and Meth. A 371 (1996) 330;
    P. Antonioli et al., Proceedings of the 36th Workshop on the INFN Eloisatron Project, Erice, Italy, 1-7 Nov. 1997 (World Scientific 1977).
44. R. DeSalvo, *Nucl. Instr. and Meth.* **A387** (1997) 92.
45. A. Go et al., *Nucl. Instr. and Meth.* **A433** (1999) 153.
46. M.D. Petroff and M. Atac, *IEEE Trans. Nucl. Sci.* **NS-36** (1989) 163.
47. G. Poelz and R. Reithmuller, *Nucl. Instr. and Meth.* **195** (1982) 491.
48. I. Adachi et al., *Nucl. Instr. and Meth.* **A355** (1995) 390.
49. R. De Leo et al., *Nucl. Instr. and Meth.* **A401** (1997) 187.
50. V.I. Vorobiov et al., Proc. of the Workshop on physics and detectors for DAPHNE, report INFN-Frascati, 1991.
51. H. van Hecke, *Nucl. Instr. and Meth.* **A343** (1994) 311.
52. J. Seguinot and T. Ypsilantis, *Nucl. Instr. and Meth.* **A368** (1995) 229.
53. R. De Leo et al., "Proposal to add a ring imaging Cherenkov detector to HERMES", INFN-ISS 96/9;
    E. Cisbani et al., Progress report on the Feasibility studies of a RICH detector for HERMES, INFN-ISS 96.

# Detectors for Particle Identification Time-of-Flight, dE/dx, and Transition Radiation

Marleigh Sheaff

*Physics Dept., University of Wisconsin, Madison, WI, USA 53705
and Dpto. de Fisica, CINVESTAV, Mexico, D.F., MEXICO*

**Abstract.** Three techniques that have been used successfully in a number of experiments to identify the species of the particles produced in interactions and decays are discussed; time-of-flight, mean ionization loss (dE/dx), and transition radiation. The physics principles on which each is based are given and also some examples of their use in experiments.

## INTRODUCTION

Identification of the particles produced in interactions and decays by species is fundamental to many measurements. By using the information that detectors specifically designed for this purpose provide to eliminate backgrounds it is often possible to increase the significance of a signal by a large amount.

The three techniques to be discussed in this article, time-of-flight, dE/dx (identification by specific ionization loss), and transition radiation, are not new, although they have evolved over time with technological advances in the field to provide better species discrimination. These techniques determine the mass and therefore the species of *charged* particles. They require that the momentum be independently measured in the same or a different subdetector. Also, they are non-destructive, which means the particles do not undergo significant energy loss in these detectors so that they usually survive to be observed in more downstream detectors (or detectors at larger radius in the case of central detector systems in colliders). In fact, an important design goal for these specialized detectors is to optimize performance relative to the amount of material the particles traverse in order to minimize the amount of multiple scattering as well as spurious interactions and conversions in the detector.

# TIME-OF-FLIGHT (TOF)

A relativistic particle traveling at velocity, $v$, traverses distance, $D$ in time, $t$, given by

$$t = \frac{D}{v} = \frac{D}{\beta c} = \frac{DE}{pc^2}, \qquad (1)$$

where $E$ and $pc$ are the particle energy and momentum. Then, substituting for the energy, we see that

$$t = t_o \left(1 + \frac{(m_o c^2)^2}{(pc)^2}\right)^{\frac{1}{2}}, \qquad (2)$$

where $t_o = D/c$ is the time it would take a particle traveling at the speed of light to traverse distance $D$.

Just how much time is this? For a flight path of 1.5 meters, a typical collider radius,

$$t_o = \frac{D}{c} = \frac{1.5}{3 \times 10^8 m/sec} = 5.0 \times 10^{-9} sec/m. \qquad (3)$$

This is *small*, only 5.0 *nanoseconds*, and, because of the dependence on the square of $m_o c^2/pc$, we approach this limit very quickly as the particle momentum increases. This is demonstrated graphically in Figure 1 for pions, kaons, and protons, where you can see that it is likely to be very difficult to use this technique with a flight path of only 1.5 m above a momentum of about 1 GeV/$c^2$.

From equation (2), for particles with two different masses, $m_1 c^2$ and $m_2 c^2$, and the same momentum, $pc$, the *difference* in the time-of-flight over a distance, $D$, is given by:

$$t_1 - t_2 = t_o \left[\sqrt{1 + \frac{(m_1 c^2)^2}{(pc)^2}} - \sqrt{1 + \frac{(m_2 c^2)^2}{(pc)^2}}\right]. \qquad (4)$$

Figure 2 shows this time difference for pions and kaons, kaons and protons, and protons and kaons as a function of momentum, again over a flight path of 1.5 m.

When $(pc)^2 \gg (m_1 c^2)^2, (m_2 c^2)^2$, we can substitute a series expansion for the two square roots,

$$t_1 - t_2 = t_o \left[\frac{(m_1 c^2)^2 - (m_2 c^2)^2}{2(pc)^2}\right]. \qquad (5)$$

The dependence of the time difference on the inverse of the momentum squared in this limit is clearly demonstrated in Figure 2.

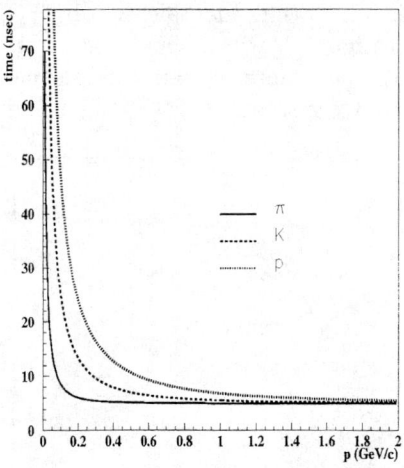

**FIGURE 1.** TOF versus momentum for $\pi$, K, and p over a 1.5 Meter Flight Path

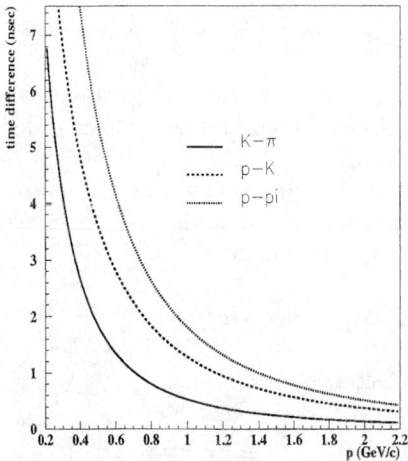

**FIGURE 2.** Differences in TOF versus momentum for $\pi K$, $Kp$, and $\pi p$ over a 1.5 Meter Flight Path

The resulting degradation in species identification by TOF with increasing momentum is apparent in Figure 3, which shows data from the Mark II collaboration [1]. The scintillation counter system used in the TOF detector for this experiment had an average resolution of 300 ps for hadrons.

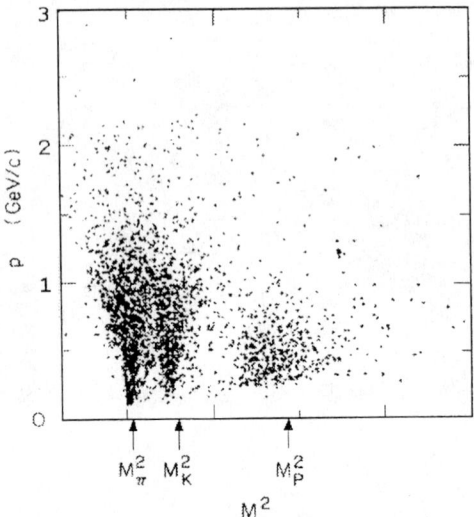

**FIGURE 3.** Scatterplot of $p$ versus $m^2$ for $\pi$'s, K's, and p's using the Mark II TOF data.

The identification power of a TOF detector system can be quantified in terms of the *significance* of the $\Delta t$ measurement, which is defined as $\Delta t/\sigma(\Delta t)$, where $\sigma(\Delta t)$ is the error, i.e, the *standard deviation*, in the measured time difference. If $\sigma_{t_1}$ and $\sigma_{t_2}$ are the time resolutions of the two detectors used to measure the time at the beginning and at the end of the particle trajectories, then the error in the time measurement is

$$\sigma(\Delta t) = \sqrt{\sigma_{t_1}^2 + \sigma_{t_2}^2}. \qquad (6)$$

In the case where the detectors are the same, e.g., two identical scintillation counters, $\sigma_{t_1} = \sigma_{t_2} = \sigma_t$, and this becomes

$$\sigma(\Delta t) = \sqrt{2\sigma_t^2} = \sqrt{2}\,\sigma_t. \qquad (7)$$

In many experiments the initial time can be measured using the beam timing devices, which have much better time resolution than the scintillation counters or other detectors used to measure $t_2$, so that $\sigma_{t_1}$ can be ignored relative to $\sigma_{t_2}$, and

$$\sigma(\Delta t) = \sigma_{t_2} = \sigma_t \qquad (8)$$

The most common detector used to measure the time in a TOF detector system is the plastic scintillation counter (which is made from organic scintillator material embedded in a solid plastic solvent) with attached photomultiplier tube (PMT) and base (which contains the high-voltage divider chain) as shown schematically in Figure 4 [2].

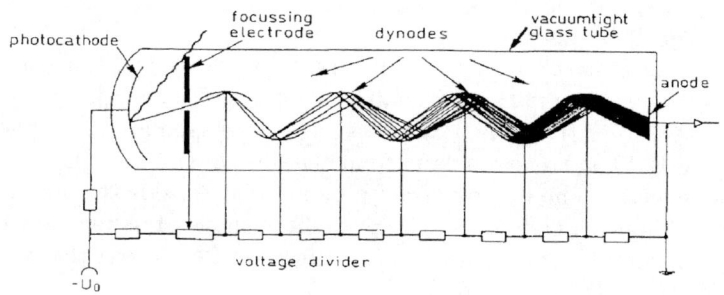

**FIGURE 4.** Schematic of a typical PMT and attached base.

What are the sources of uncertainty in the times we measure with this device? One source is the finite rise and decay times of the flourescence in the scintillator material. Since this is a *stochastic* process the *exact* times at which the photons are emitted can't be predicted. We can use "fast" organic scintillator to keep the uncertainty as small as possible. The signal output can be mathematically described by

$$N(t) = N_o f(\sigma, t) e^{-\frac{t}{\tau}}, \tag{9}$$

where $f(\sigma, t)$ is a Gaussian with standard deviation $\sigma$ and $\tau$ is the decay constant [3]. For some common "fast" plastic scintillators, these fit parameters are:

| Scintillator | $\sigma$(ns) | $\tau$(ns) |
|---|---|---|
| NE102A | 0.7 | 2.4 |
| NE111 | 0.2 | 1.7 |
| Naton 136 | 0.5 | 1.87 |

There is a very large difference in pulse signal size due to statistical fluctuations. Two MeV of energy is deposited per centimeter on average by a minimum ionizing particle (MIP), which has unit charge. Since the average amount of deposited energy needed to produce one photon is 100 eV, the average number of photons produced per centimeter in the scintillator is 20,000. A large fraction of these escape out the surfaces of the counter material because they are emitted at angles smaller than the angle of total internal reflection and thus are lost before reaching the PMT. Also, the quantum efficiency of the PMT photocathode is relatively low, some 10-30% for a good bi-alkali tube, and depends on the frequency of the light. To reduce the large fluctuations in the number of photons produced, TOF systems are built with relatively *thick* scintillation counters. (See the discussion of the ionization loss, or Landau, distribution in the section on dE/dx that follows.) Five centimeters is typical. But of course we can't overlook the impact this has on the material budget. Because they do represent a considerable amount of material, TOF detectors are usually placed outside of the tracking detectors and just in front of the calorimeters in collider detector systems.

Another source of uncertainty is the difference in the arrival times at the PMT for photons emitted in the scintillator. Light is emitted in all directions. How much travels directly to the PMT depends on the solid angle the photocathode of the PMT subtends looking from the photon emission point. The pulse shape depends on how many "fast" photons (those inside the dotted region shown in Figure 5) and "slow" photons (those outside this region, which undergo one or more bounces due to total internal reflection on the way to the PMT) are included in the signal.

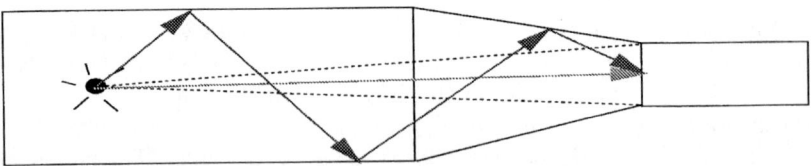

**FIGURE 5.** Paths of Scintillator Photons Impinging on the PMT Photocathode.

Better time resolution can be achieved with a faster signal risetime. We can increase the solid angle and thus the number of "fast" photons by eliminating the light guide and gluing the PMT directly to the scintillation counter. This also reduces the number of bounces the "slow" photons undergo. We will also get more "fast" (as well as "slow") photons if we glue a PMT on *both* ends of the scintillation counter as shown in Figure 6 [1]. The time information from the time-to-digital converters (TDCs) at both ends of the counter provides a means by which the dependence of the time resolution on the hit position in the counter can be corrected. The pulse height information from the analog-to-digital converters (ADCs) provides a further means by which the time slew due to pulse height differences can be corrected.

If $t_1$, $t_2$ are the arrival times of the signals at the two ends of the counter, then the position, $x$, at which the particle entered the counter, measured in a coordinate system with the $x$ axis aligned along the length of the counter and with $x = 0$ at the center of the counter, is given by

$$x = \left(\frac{t_1 - t_2}{2}\right) v_{eff}, \qquad (10)$$

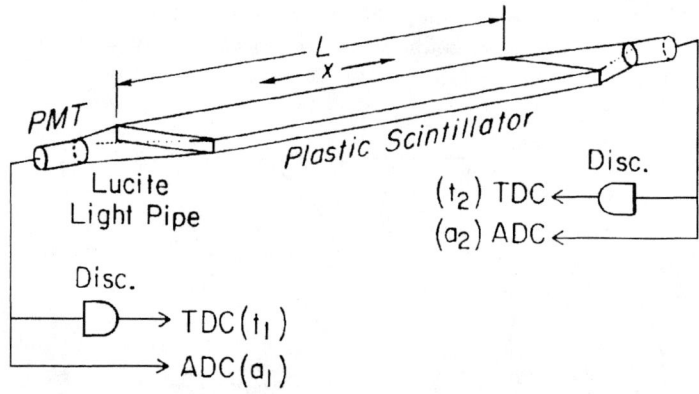

**FIGURE 6.** TOF counter with PMTs attached at both ends with times read out via TDCs and pulse heights read out via ADCs.

where $v_{eff}$ is the effective propagation velocity of light in the scintillator. $v_{eff}$ depends on both the counter geometry and the index of refraction of the material from which it is made, and can be measured by illuminating the counter with hits and measuring $t_1$ and $t_2$ for many values of $x$ while precisely measuring $x$ at each position using another detector. The effective velocity is then the slope of the straight line fit to $x$ versus $(t_1 - t_2)/2$. The time, $t$, at which the particle entered the counter is given by the average of $t_1$ and $t_2$ minus the time it takes signals to travel from the center to the ends of the counter:

$$t = \left(\frac{t_1 + t_2}{2}\right) - \frac{L}{2v_{eff}}, \qquad (11)$$

where $L$ is the length of the counter. This time is independent of the position at which the particle hit the counter.

There is also some time "jitter" introduced by the PMT. The difference in the kinetic energies of the photoelectrons emitted in the photocathode due to its finite thickness is small relative to the difference in their time of traversal from the photocathode to the first dynode [2]. This time difference can be reduced significantly by increasing the voltage on the focusing electrode shown in Figure 4. Most PMT bases provide a means by which this can be accomplished.

Further time uncertainty is produced by the electronics circuits. This is illustrated in Figure 7 [4]. Any rapid fluctuations in pulse signal size from the sources listed above as well as from electronics noise produced in the circuits themselves contributes to time "jitter" as shown in Fig. 7a). As indicated on the figure, the effect is smaller for signals with faster rise times. There is also "time walk" for fixed threshold discriminators, the type most commonly implemented, due to differences in pulse signal size as shown in Fig. 7b). By reading out at both ends of the counter and taking the time average, we reduce this uncertainty.

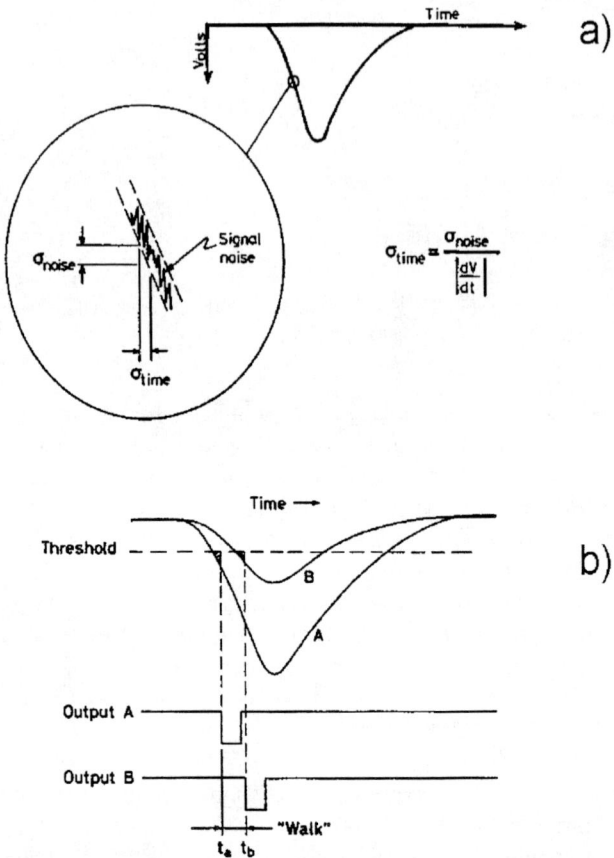

**FIGURE 7.** Time uncertainty introduced by the electronics; a) shows "jitter" while b) demonstrates "time walk" in a fixed threshold discriminator.

Two methods have been used to correct offline for the dependence of the recorded time on the pulse signal size. One is to use the integrated charge in the pulse, $a$, measured using the ADC [1,5]. Then,

$$t_{corr} = t_{meas} - W\left(\frac{1}{\sqrt{a_o}} - \frac{1}{\sqrt{a}}\right) - \frac{x}{v_{eff}}, \tag{12}$$

where $a_o$ is a reference pulse height and W is a parameter of the fit. The term $x/v_{eff}$ corrects for the position where the track hit the counter and requires that $x$ be measured using another subdetector, usually by extrapolation of the associated track reconstructed using the tracking detectors. The improvement in time resolution that can be achieved is significant as is demonstrated in Figure 8 [6]. $\sigma(\Delta t)$ goes from 0.28 ns before correction to 0.15 ns after correction, which reduces the ambiguity between $\pi$'s and K's seen in the plot at TOFs near 1.5 ns by a large factor. The TOF is measured in this case by two scintillation counters placed 5.5 m apart. This relatively long flight path allows good separation of the pions and kaons at a beam momentum of 1.4 GeV/c as indicated on the figure.

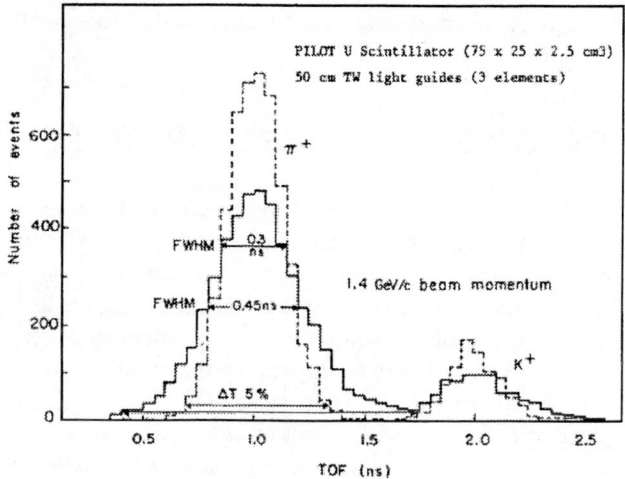

**FIGURE 8.** TOF measurements for $\pi^+$ and $K^+$ before (solid curve) and after (dashed curve) correction for the dependence of the time measurement on the pulse height using equation (12). $\sigma(\Delta_t)$ is reduced from 0.28 ns to 0.15 ns by the correction.

The second method uses two discriminators, one having a low threshold and the other a high threshold, on each PMT. Each is input to a TDC and the time between the registered hits is used to measure the pulse rise time. The pulse is then extrapolated back to its initial start time, i.e., the time at which it deviated from zero. The CLEO experiment used a combination of the two methods to correct the data provided by their barrel TOF counters, which were located at radius $\sim 1$ m relative to the $e^+e^-$ beams. The machine rf provided the timing signal to which the times measured in the scintillators were referred. The time resolution achieved for pions was 154 picoseconds rms, which provided at least $2\sigma$ pion-kaon separation out to 1.07 GeV/c [7].

Better time resolution could be achieved if parallel plate spark chambers, which have better time resolution than scintillation counters, were to be used to measure the times instead. E. g., the ALICE experiment planned for the Large Hadron Collider expects to use resistive plate chambers (RPCs) in their TOF system. The plates will be made from the same (resistive) glass that is used for welders' face masks and will be placed 1 mm apart. The gaps will be filled with Helium-Ethane at atmospheric pressure to limit the charge in the discharge and thus to protect the glass surface from damage. In tests of 8 mm x 8 mm prototypes, time resolutions of better than 70 ps have aready been achieved [8].

Another way to achieve better time resolution is to take more time measurements along the particle trajectory. The significance of the TOF measurement is improved by a factor $1/\sqrt{N}$, where N is the number of measurements, but this comes at the expense of more material through which the particles must pass.

## MEAN IONIZATION LOSS, OR DE/DX

Tracking detectors designed to make position measurements along the trajectory of a charged particle can simultaneously be used to identify the particle type by collecting the charge deposited in each detector cell along the track and using the charges collected to measure the mean ionization loss per unit path length, dE/dx. This applies to any detector in which particles lose energy to ionization, e.g., gas-filled wire chambers or silicon strip detectors. Such detectors need to be highly segmented for precise tracking in a high-rate environment. Thus, each wire or strip will collect only a small fraction of the total charge deposited. Because the ionization energy loss is a statistical process with large fluctuations, many measurements are needed along each track to get a precise mean. Since the pulse height on each wire or strip must be read out, this technique requires the use of analog readout electronics. For gas-filled detectors, the statistics are improved by operating at higher gas pressure, since this results in more primary ionizations along the particle path. The precision of the mean is improved by using the method of "truncated mean" whereby a certain fraction of the signals of largest size are removed in taking the average. This eliminates samples which come from points way out on the tail of the Landau distribution, named for the physicist who first calculated the shape of the typical energy loss distribution. As is evident in the schematic representation of a Landau distribution, shown in Figure 9 [9], the curve is significantly skewed toward higher energy losses because of the production of delta rays, ionizations where the electron is kicked out with high enough energy to itself form ionizations along its path. This "tail" will be longer when the mean energy loss for the cell or strip is small relative to the maximum possible energy loss in a single collision in the material. This is the case for the highly segmented cells in a typical tracking detector, which is why it is necessary to use the "truncated mean" calculation. When the mean energy loss is much larger than the maximum energy transfer possible, the curve becomes more "Gaussian-like" and the mean and most probable energy loss shown on the curve will also be closer together. By removing the 30-50% of the signals of largest size, we shift the mean that we estimate closer to the most probable energy loss, which gives us a mean that we can estimate more reliably on an event-by-event basis.

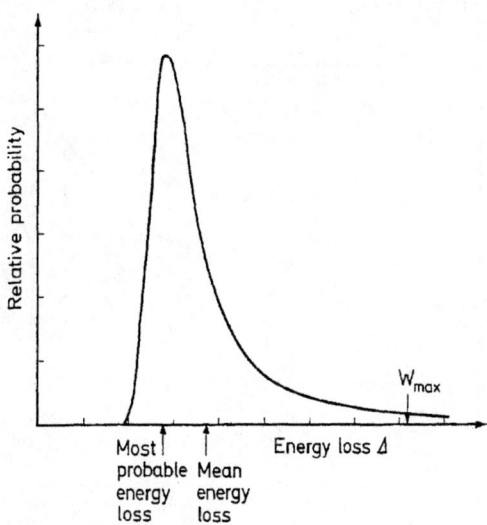

**FIGURE 9.** Schematic showing a typical Landau distribution. The mean energy loss, most probable energy loss, and maximum energy loss are indicated.

The *mean* energy loss, dE/dx, of heavy particles ($m \gg m_e$) of unit charge in matter is described by the Bethe-Bloch equation [10],

$$dE/dx = D_e \left(\frac{1}{\beta}\right)^2 n_e \left[ ln\frac{2mc^2\beta^2\gamma^2}{I} - \beta^2 - \frac{\delta(\gamma)}{2}\right], \qquad (13)$$

where

$$D_e = 2\pi r_e^2 m_e c^2 = 5.0989 \times 10^{-25} MeV - cm^2. \qquad (14)$$

$n_e$ is the number of atomic electrons per unit volume, $r_e$ is the classical radius of the electron, and $m_e c^2$ is the electron rest mass. $I$ is the mean ionization potential of the material. This is a universal function of $\beta\gamma$ for all particle masses. The energy loss decreases with increasing momentum as $1/\beta^2$ at low momenta and reaches a minimum at $\beta\gamma \sim 4$. Above that point, the energy loss rises logarithmically due to the $ln\gamma^2$ dependence. This is called the "relativistic rise" region. This rise does not continue indefinitely, but saturates when it reaches the so-called "Fermi plateau" above which it remains constant. This is because the so-called "density effect", represented by the term $\delta(\gamma)/2$ in the equation, also increases with $\beta\gamma$ and offsets the rise. This effect depends on the density of the medium as the name indicates, which limits the rise to a few percent in liquids and solids. The relativistic rise can be as high as 50-70% in high-Z noble gases, however [11]. Figure 10 [10] compares the density effect, $\delta$, for different materials. Since the density of a gas increases with increasing

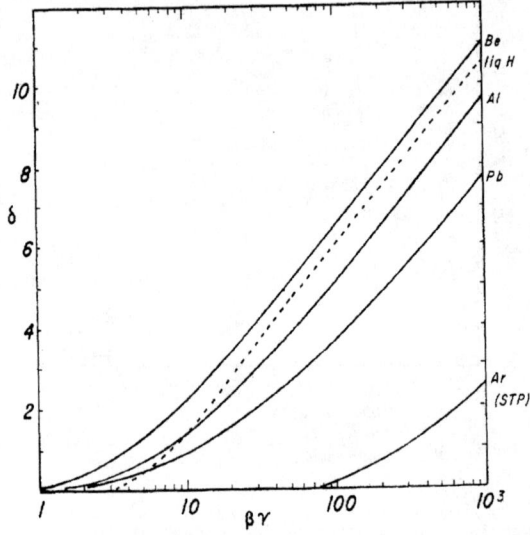

**FIGURE 10.** Density effect versus $\beta\gamma$ for some typical materials.

pressure, the rise will be suppressed by an increase in pressure as shown in Figure 11 [10]. The characteristic shape of Bethe-Bloch versus $\beta\gamma$ is shown in Figure 12 [12] along with curves indicating the different contributions.

When plotted versus momentum for the different particle masses, as shown in Figure 13 [13], the curves all have the same shape but are offset from each other because $\beta\gamma = p/m$ reaches the minimum of $\sim 4$ at a different value of momentum for each different mass. Figure 14, also from Reference [13], shows the relative difference in ionization loss for $\pi$'s and K's as a function of momentum. The large dip in this distribution is the so-called "crossover region", momenta at which the ionization loss difference is very small because the relativistic rise portion of the $\pi$ dE/dx curve crosses the K curve as it falls toward its minimum. Experiments which use the dE/dx technique for

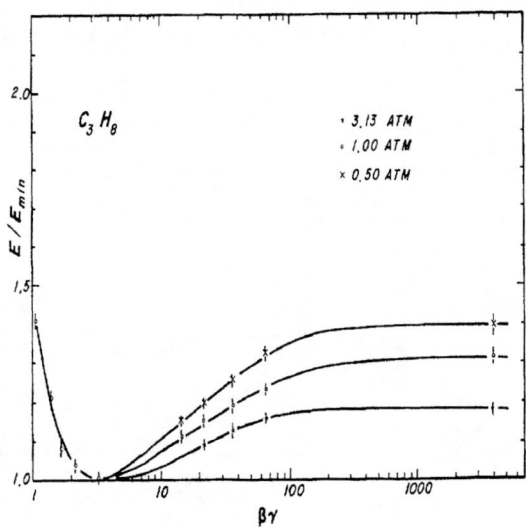

**FIGURE 11.** Dependence of the relativistic rise on gas pressure. The rise is seen to be suppressed more strongly as the gas pressure is increased because the gas density increases with increasing pressure.

particle identification must use a second technique such as TOF to identify the particles in this region.

dE/dx was used successfully for particle identification in the OPAL jet chamber at LEP [12]. The chamber has 24 sectors, each with 159 long anode wires aligned along the beam axis as shown schematically in Figure 15 [12]. Over 73% of the solid angle, the tracks cross all 159 wires. Although 30% of the samples are removed in calculating the truncated mean, the sampling statistics are thus still large enough to allow a precise calculation of the mean ionization loss. The results are further improved by careful analysis, including quality cuts on the tracks and corrections for systematic shifts in the relevant parameters. The detector was run at a pressure of 4 bars, which was chosen to give optimum separation power. Lower pressure would give fewer primary ionizations per unit path length, but higher pressure would cause the curves to reach the Fermi plateau at lower momenta, and thus would degrade the particle separation that could be achieved in the relativistic rise region.

Figures 16 and 17, both from Reference [12], document the performance of the OPAL jet chamber. Figure 16 a) shows the dE/dx distributions as a function of momentum for tracks seen in the chamber. The curves exhibit the features expected from the Bethe-Bloch formula and show clear separation of the particle species except in the crossover region. The intense distribution

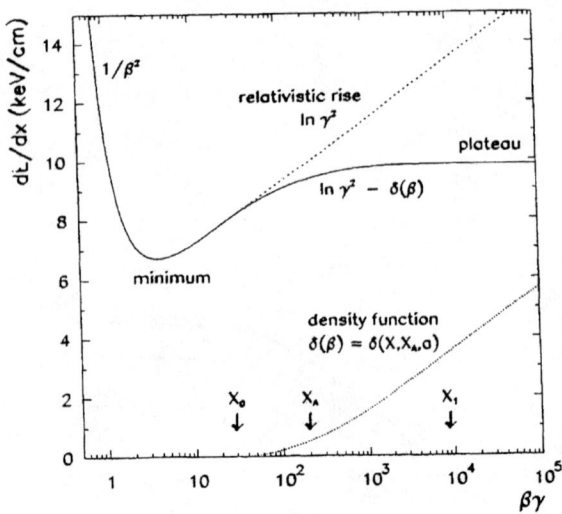

**FIGURE 12.** Characteristic shape of the mean ionization energy loss (Bethe-Bloch) versus $\beta\gamma$. The contributions from the relativistic rise and density terms are indicated. The mean ionization potential, $I$, has been adjusted to match the OPAL experimental data. See Reference [12] for details.

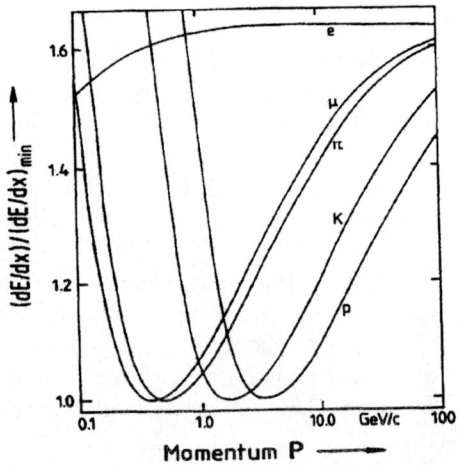

**FIGURE 13.** Specific ionization loss $(dE/dx)/(dE/dx)_{min}$ versus momentum by particle type.

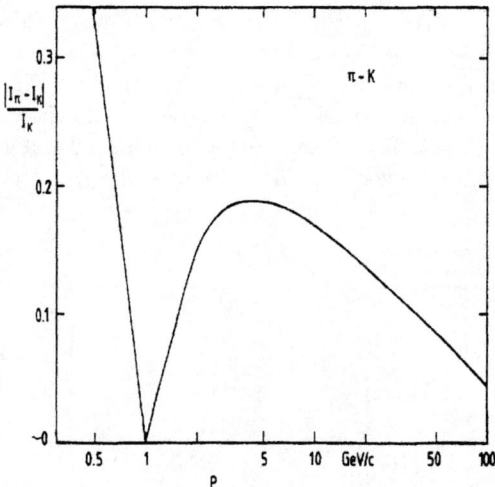

**FIGURE 14.** Relative difference in dE/dx for $\pi$s and Ks versus momentum showing the "crossover region" near 1 GeV/c where this technique can't be used to separate the two species.

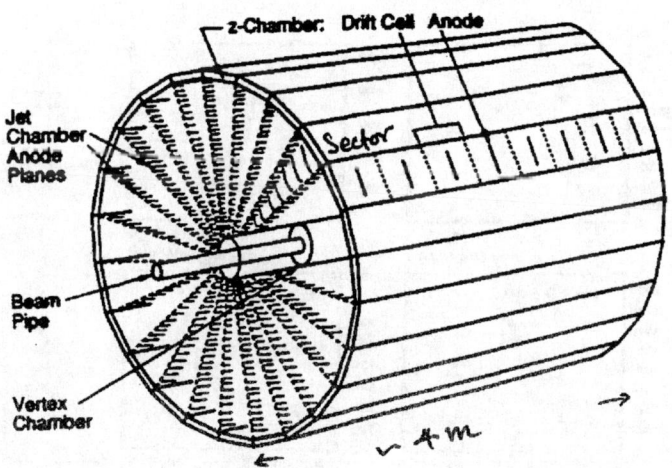

**FIGURE 15.** Schematic showing the main features of the OPAL jet chamber.

of points at high momentum in the figure are calibration data from dimuon events collected simultaneously with the event data. Figure 16 b) shows the same data in slices of momentum. Figure 17 is a plot of the significance of the separation for pairs of particle species in the detector. As shown in this figure, at least 2 standard deviation $\pi$-K separation is achieved out to momenta as large as 20 GeV/c, which is well out in the relativistic rise region.

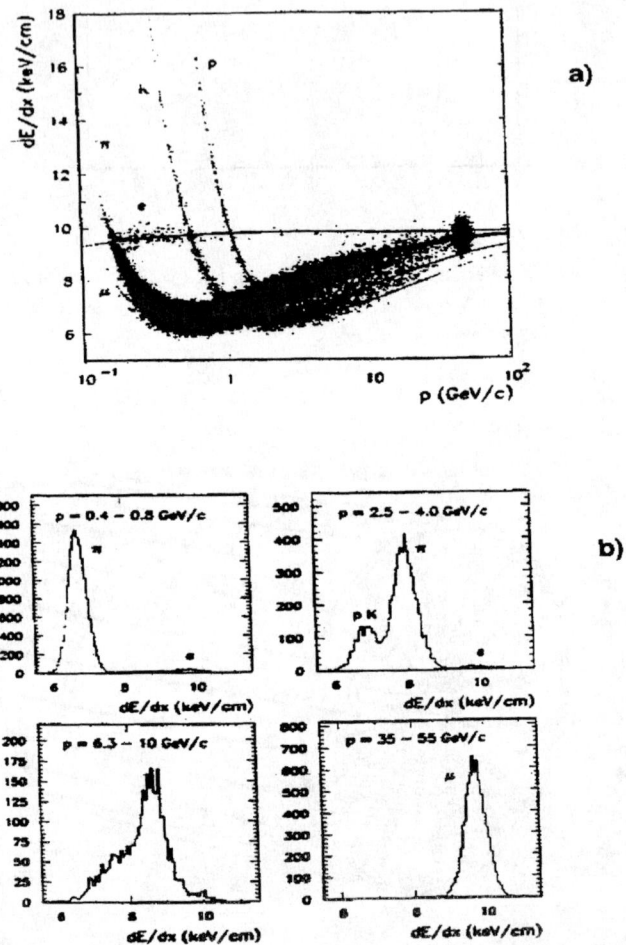

**FIGURE 16.** dE/dx data from the OPAL experiment. The dE/dx distributions are shown in a) superposed on curves showing the dE/dx curves predicted by calculations for Ar/CH$_4$ at 4 bars pressure. b) shows the same data in slices of momentum.

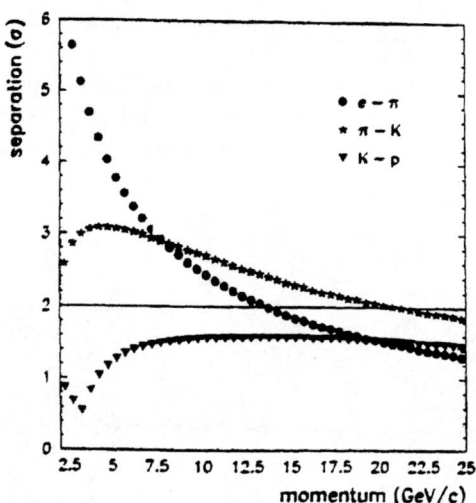

**FIGURE 17.** Significance of the separation (dE/dx)/($\sigma_{fracdEdx}$) for e-$\pi$, $\pi$-K, amd K-p in the OPAL jet chamber.

## TRANSITION RADIATION DETECTORS

Like Cerenkov Radiation, Transition Radiation (TR) is a relativistic effect that depends on the fact that the velocity of a highly relativistic particle in a particular medium can exceed the velocity of light in the medium, which is given by $c/n$, where $c$ is the velocity of light in vacuum and $n$ is the index of refraction of the medium. The phenomenon was first predicted by two Russian physicists, Ginsberg and Frank, in the late 1940's, although more than 20 years went by before a Transition Radiation Detector (TRD) was used successfully in an accelerator-based high energy physics experiment [14].

TR is the radiation that must be emitted to satisfy the boundary conditions in the solution to Maxwell's equations as a relativistic particle crosses the boundary between two media with different plasma frequencies (or, equivalently, different dielectric constants) [15]. An approximate expression for the energy radiated per unit solid angle per unit frequency interval at a single such interface between medium 1 and medium 2 is:

$$\frac{d^2W}{d\omega d\Omega} = \frac{\alpha}{\pi^2}\left|\frac{\Theta}{\gamma^{-2}+\Theta^2+\frac{\omega_{p1}^2}{\omega^2}} - \frac{\Theta}{\gamma^{-2}+\Theta^2+\frac{\omega_{p2}^2}{\omega^2}}\right|^2. \qquad (15)$$

This approximation is valid for $\gamma \gg 1$, $\Theta \ll 1$, and $\omega^2 \gg \omega_{p1}^2, \omega_{p2}^2$, where $\gamma$ is the Lorentz factor of the particle, $\Theta$ and $\omega$ are the angle and the frequency of the emitted radiation, and $\omega_{p1}, \omega_{p2}$ are the plasma frequencies of the two media.

As shown schematically in Figure 18, the effect is strongly peaked in the forward direction and it is symmetric, i. e., the radiation emitted is the same whether the particle enters medium 2 from medium 1 or enters medium 1 from medium 2.

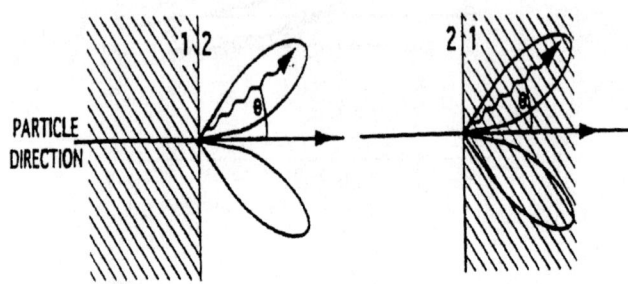

**FIGURE 18.** TR emission at a single surface between two media with different plasma frequencies.

Notice also that $\omega_{p1}$ should be very different from $\omega_{p2}$ for a large effect. This is true of the materials used in constructing the TRD's that have been used successfully in experiments to date. E.g., a typical foil material is polypropylene, $CH_2$, for which $\omega_{p1}=21.8$ ev, and a typical gap material is air, for which $\omega_{p2}=0.7$ ev.

TR is a small effect. This is because the total energy radiated at each interface depends linearly on the fine structure constant, $\alpha=1/137$. But, by stacking many thin foils ($\sim 200$, typically) of material 1, separated by gaps of material 2, and choosing the foil and gap thicknesses to get positive interference between the radiation emitted at the incident and exit surfaces, it is possible to enhance the effect enough to build a practical detector.

For a single foil in a gas, as pictured in Figure 19, the amplitude of the radiation emerging from the second boundary is

$$\vec{E}(\omega, \vec{\Theta}) = \vec{e}_2(\omega, \vec{\Theta}) + \vec{e}_1(\omega, \vec{\Theta})\, e^{-\sigma_{foil} + i\phi_{foil}}, \qquad (16)$$

where $\vec{e}_j(\omega, \vec{\Theta})$ is the amplitude of the radiation emitted at surface j, $\sigma_{foil}$ is the attenuation of the radiation in the foil (Let's neglect that for now - in any case, it is small in a single foil!), and $\phi_{foil}$ is the phase lag due to the difference in the wave and particle velocities in the medium.

Now,

$$\vec{e}_1(\omega, \vec{\Theta}) \simeq \frac{\Theta}{\gamma^{-2} + \Theta^2 + \frac{\omega_{p1}^2}{\omega^2}} - \frac{\Theta}{\gamma^{-2} + \Theta^2 + \frac{\omega_{p2}^2}{\omega^2}} = -\vec{e}_2(\omega, \vec{\Theta}). \quad (17)$$

Substituting this expression into equation (16) and squaring to get the intensity of the radiation emerging from the foil, we see that

$$\left(\frac{d^2W}{d\omega d\Omega}\right)_{single\ foil} = \left(\frac{d^2W}{d\omega d\Omega}\right)_{single\ interface} \times 4\sin^2\left(\frac{\phi_{foil}}{2}\right). \quad (18)$$

We therefore get positive interference for $\phi_{foil} = \pi, 3\pi$, etc.

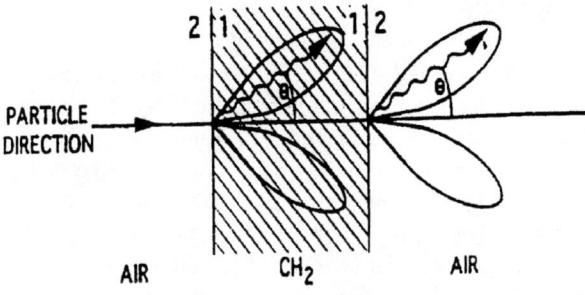

**FIGURE 19.** TR emission from a single foil in a gas.

It can be shown (Reference [15] - Equation 2.10) that the phase lag in the foil,

$$\phi_{foil} \simeq \frac{(\gamma^{-2} + \Theta^2 + \frac{\omega_{p,CH_2}^2}{\omega^2})\omega l_{foil}}{2}, \quad (19)$$

where $l_{foil}$ is the foil thickness. If we define

$$z_{foil} = \frac{2}{(\gamma^{-2} + \Theta^2 + \frac{\omega_{p,CH_2}^2}{\omega^2})\omega}, \quad (20)$$

then

$$\phi_{foil} \simeq \frac{l_{foil}}{z_{foil}}. \quad (21)$$

$z_{foil}$ is the well-known "formation zone". If $l_{foil}$ is much smaller than this, we get no contribution from the foil because the interference term is near zero.

Thus, TR, like Cerenkov radiation, is a bulk property of matter. There is a minimum thickness of material that the particle must traverse below which radiation will not occur at the interfaces.

The procedure described above for calculation of the radiation emitted from a single foil can be generalized to an n-foil set by grouping the foil amplitudes two-by-two as above and adding the amplitudes from the n foils coherently. It is possible to make a series approximation to the expression that results to facilitate computer calculation. The computer program [16] used to simulate the performance of the E769 TRD [17] to aid in the selection of its design parameters used such an approximation. Integration over the angle of the radiation is performed because the angles are all very small. In fact, the radiation is detected in the same cell as the particle track itself in most cases. While the particle track could be bent away from the radiation by use of a magnetic field, this requires some distance for a stiff track, and the region must be evacuated to avoid absorption of the TR x-rays in material upstream of the detector.

Figure 20 shows the expected yield of TR per keV as a function of the energy of the radiation in keV for a foil set made from 200 $CH_2$ foils [18]. The foils are each 82 $\mu$m thick, and they are separated by 1.4 mm helium gaps. The particle gamma assumed in the calculations is $\gamma = 10^4$, which corresponds to a 5 GeV electron. Two curves (solid lines) are shown for the foil set, one including the effect of foil self-absorption and the other neglecting it to demonstrate the severe attenuation of the radiation in the foil stack itself at lower x-ray energies. These curves provide graphic illustration of the modulation due to the interference term. They have been divided by 400 so that they can be compared directly with the single interface curves (dashed lines) which are also shown on the plot. Thus, the total amount of radiation per unit energy emitted by the foil set is 400 times the number shown on the abscissa.

The total intensity, $W$, radiated by a foil is given by

$$W = \frac{\alpha}{3}\frac{\hbar(\omega_1 - \omega_2)^2 \gamma}{\omega_1 + \omega_2} \simeq \frac{2\alpha}{3}\hbar\omega_1\gamma, \tag{22}$$

since $\omega_1 \gg \omega_2$. It therefore increases linearly with the plasma frequency of the foil material and with $\gamma$ of the particle. The increase with $\gamma$ does not go on indefinitely, however, but saturates at approximately

$$\gamma_{sat} \approx \gamma_{sat}(\omega_{max}) \approx 0.6\omega_1(l_1 l_2)^{\frac{1}{2}}/c. \tag{23}$$

Figure 21, from Reference [18], shows the growth with $\gamma$ of the x-ray yield per interface in the absense of absorption for a single $CH_2/He$ interface and for three many-foil radiator sets with different foil and gap thicknesses. The x-ray yield is larger for foil sets with thicker foils and wider gaps. This is because the yield increases linearly with $\gamma$ up to $\gamma_{sat}$ as shown in the plot, and $\gamma_{sat}$ depends on the square root of the product of $l_1$ and $l_2$. The higher yield can

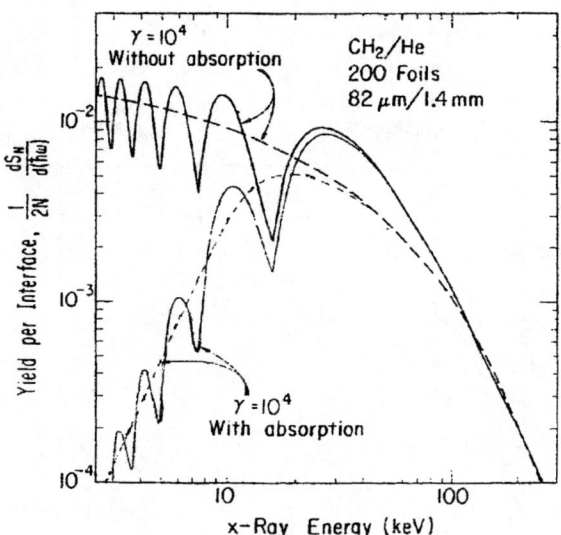

**FIGURE 20.** TR yield per interface for a foil set made from 200 $CH_2$ foils each 82 $\mu$m thick separated by 1.4 mm He-filled gaps showing the modulation that comes about from interference (solid curves). TR yield for a single such interface (dashed curves). The upper curves do not include attenuation in the foil material while the lower ones do.

be understood as the increased bandwidth of the spectrum emitted because of the increase in the maximum frequency, $\omega_{max}$, of the radiation emitted with increasing thickness of the foils. $\omega_{max}$ also depends on the square of the plasma frequency of the radiator material. For $l_2 \gg l_1$,

$$\omega_{max} = \frac{l_1 \omega_1^2}{2\pi c}. \qquad (24)$$

The E769 TRD was designed to identify pions in an incident 250 GeV/c positive hadron beam, for which $\gamma \sim 1800$. This was the first time that a TRD was used for beam tagging in the hadron beam for a running particle physics experiment. The E769 detector is a good illustration of a typical, practical TRD. It was made from 24 identical modules, one of which is shown schematically in Figure 22 [17]. Each contains a radiator made from 200 12.7 $\mu$m polypropylene ($CH_2$) foils stacked alternately with nylon net spacers, which are 180 $\mu$m thick. The nylon net was cut away in the region of the beam since it was found to attenuate the TR x-rays by a factor of approximately 2. The radiator volume was flushed with helium during the E769 run but was run with air during tests carried out with the TRD during Fermilab experiment E791, since, although the plasma frequency of helium is smaller than that of air, the difference in the TR output of the radiator stack is not significant.

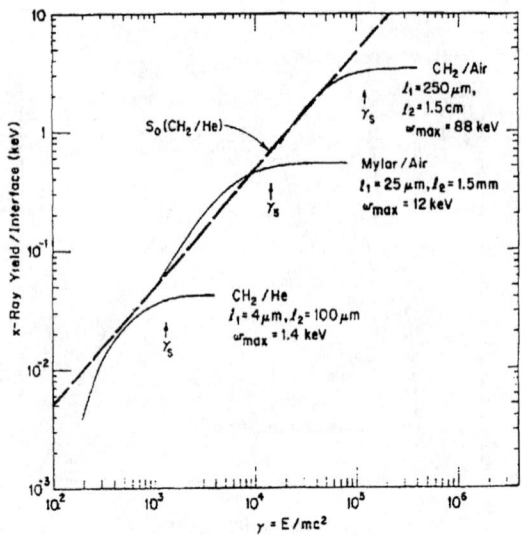

**FIGURE 21.** TR yield per interface versus $\gamma$ for three different foil sets configured as noted on the figure and for a single interface. $\gamma_{sat}$ is indicated for each foil set.

The radiator is followed by a two-plane proportional chamber with single cell depth .635 cm, and active area 76 mm wide by 65 mm high. The 64 sense wires (anodes) are spaced at 1 mm and all are oriented horizontally since the chambers were not used to measure position. The wires are 10.2 $\mu$m gold-plated tungsten and the cathodes are 12.7 $\mu$m mylar with 140 of aluminum sputtered onto both sides. The chamber gas used was xenon bubbled through methylal at $0°C$, which results in a mixture that is approximately 90% xenon. There is a .3175 cm buffer volume filled with nitrogen in front and in back of the two-plane chamber. The gas volumes were maintained at equal pressure to keep the chamber gains uniform across the planes.

Figure 23 [17] shows the results of Monte Carlo simulations of the TR expected from the 200-foil radiator set. The upper curve shows the average number of TR photons expected to be produced in the absence of self-absorption and without requiring detection in the two-plane xenon-filled chamber. The lower curve shows the severe attenuation that self-absorption produces for low energy x-rays on the leading edge of the distribution and the sizeable loss of high energy x-rays out the back of the two-plane chambers on the falling edge of the distribution. The mean free path for x-ray absorption in xenon goes through a minimum at 4.8 keV, where it is about 2 mm, then rises abruptly at energies above that. Because it is comprised of many layers, each with a relatively small number of foils in the radiator stack followed by two

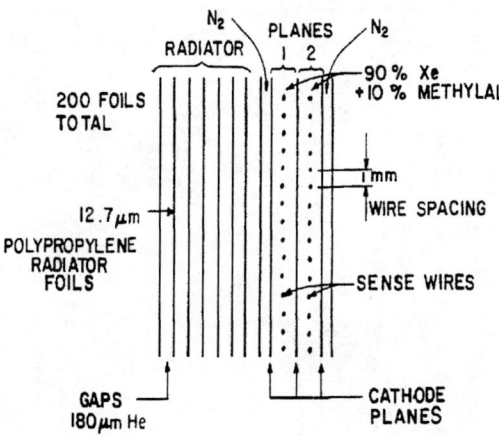

**FIGURE 22.** Schematic of one module of 24 total in the E769 TRD as viewed from the side.

chamber planes that are shallow in depth, this detector is an example of a "fine-sampling TRD [19,20]. This means that at most one x-ray is likely to be captured per plane per event. This is the reason that a simple digital read-out by means of a latch on each detector plane, indicating one or more hits above the TR threshold, although not optimum, sufficed. Also, because of the short integration time of the electronics circuits used, which shaped the pulses from the very localized ionization of an $Fe_{55}$ source to 26 ns full width at half maximum, this TRD discriminates using the technique of "cluster counting" [20,21]. This has been shown to give better separation between species than the method of total charge collection.

The length of the detector as built was 2.79 m. The total amount of material in the detector was 8.7% of an interaction length and 16.9% of a radiation length including two .3175 cm scintillation counters used for gating. It would be difficult to reach the 90% efficiency for pions coupled with a factor of 30 in background rejection (in this case protons, since the kaons were separately tagged by means of a Differential Isochronous Self-Focusing Cerenkov counter [DISC] [22]) that was achieved with this detector with much less material than this. The method by which the pion sample was selected is illustrated in Figure 24 [17], which shows the distribution of TRD planes hit per event for all events in which the beam particle was not tagged by the DISC as a kaon from a typical E769 data run. As shown by the curves in the figure, the proton and pion peaks were each fit with a double binomial on a run-by-run basis. A plane count cut was chosen such that 90% of the integrated pion distribution lay above it. Then, the background above this cut was calculated using the

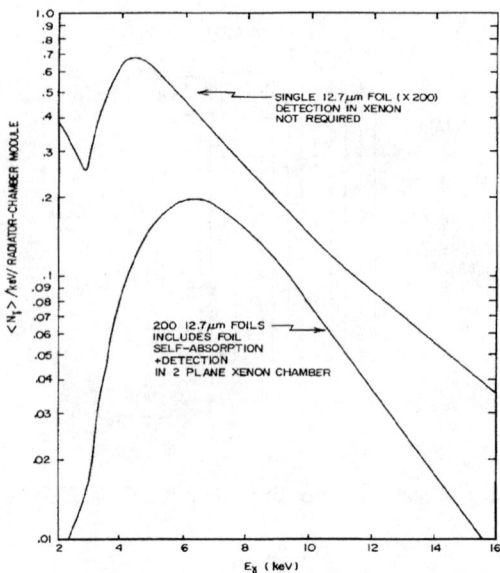

**FIGURE 23.** TR x-rays per keV expected from the 200-foil E769 radiator set according to Monte Carlo simulations. The upper curve shows the radiation without the self-attenuation expected in the foils and without requiring the x-rays to be detected in the 2-plane Xe-filled chamber. The lower curve shows the radiation expected when these are included.

proton curve. The technique was verified using plane count distributions made for the protons and pions separately during special runs in which the DISC pressure was set to tag them. Further details about the E769 detector are contained in [17].

Figure 25 shows the expected average number of TR photons radiated and detected per module of the E769 TRD for electrons, pions, kaons, and protons incident as a function of particle energy. The numbers shown have been calculated using the simulation package developed for modeling this detector [16], which was found to reliably predict the actual detector performance. The measured efficiency for the x-ray capture signal to be above the 4 keV threshold set on the electronic readout circuit, which was 83%, has been included in the numbers shown. Comparisons to References [15] and [18] indicate that saturation is not modeled correctly in the simulations, so it has been put in by hand at the gamma value corresponding to that for pions at an energy of 500 GeV. This seems prudent, since no experimental data are available from the E769 detector at higher pion energies than this. Tests run using the TRD during Fermilab E791 indicated that saturation does not occur below this value, although this is somewhat above the saturation energy for pions of 420 GeV predicted using Equation (23).

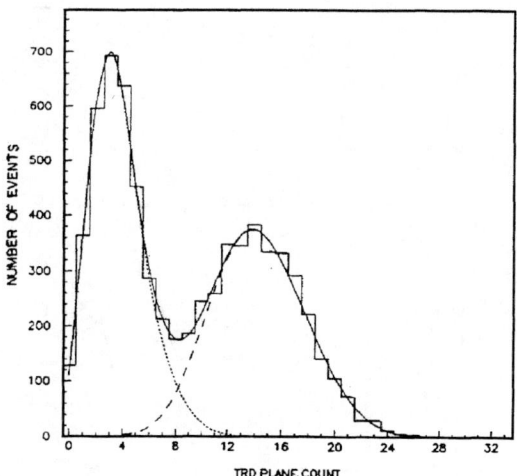

**FIGURE 24.** Distribution of TRD planes hit for events where the beam particle was not identified in the DISC as a kaon. The peak to the left are the protons, the one to the right the pions. The binomial fits used to separate the two species on an event-by-event basis are shown on the figure.

Since, as shown in Figure 25, electrons radiate TR and hadrons do not over a two order of magnitude range in momentum, the momentum window over which it is possible to discriminate electrons from hadrons is large. A TRD can be designed to saturate in the vicinity a few GeV for electrons, so that above that value, the efficiency for electrons is constant. In the same detector, since the effect goes as $\gamma$, pions do not radiate appreciable TR until they are at energies near that same value of $\gamma$, in the range of hundreds of GeV. When used in combination with an electromagnetic calorimeter, the two can provide a background rejection of $\sim 10^{-4}$ with good selection efficiency for electrons. This is demonstrated in Figure 26, which shows the electrons from candidate events for the beta decay of $\Sigma^-$ hyperons, $\Sigma^- \to n e^- \nu$, from Fermilab experiment E715 before and after requiring identification of the electron in the E715 TRD [23,24]. The branching fraction for the decay mode of interest is three orders of magnitude smaller than that of the non-leptonic decay mode, $\Sigma^- \to n\pi^-$, which has a branching fraction close to 100%. Thus, in order for this experiment to obtain a beta decay sample with only a few percent hadron contamination it was necessary to achieve a better rejection of pions than could be accomplished by the use of a calorimeter alone.

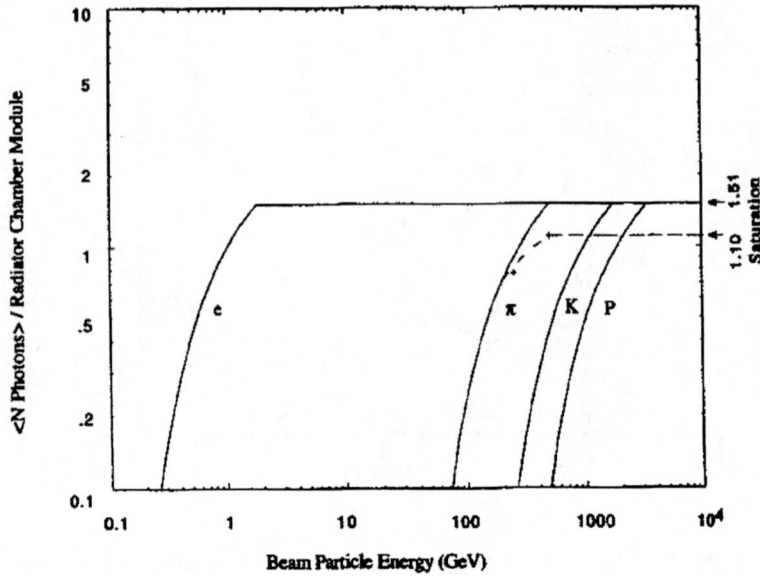

**FIGURE 25.** Average number of TR photons expected to be radiated and detected in one module of the E769 TRD as predicted by Monte Carlo simulations for various particle types versus particle energy (solid curves). Saturation has been put in by hand (See text for details.) The dashed curve shows the effect of using a latch for readout so that at most only one hit per plane could be recorded. The two small crosses shown on the plot demonstrate that the experimental results for pions at both 250 and 500 GeV are in good agreement with the predictions.

# REFERENCES

1. Atwood, W. B., *Proceedings of the Summer Institute on Particle Physics*, Stanford Linear Accelerator Center, July 28 - August 8, 1980, pp. 287-308, available from the National Technical Information Service, U. S. Dept. of Commerce, 5285 Port Royal Road, Springfield, VA 22161.
2. Grupen, C., *Particle Detectors*, Cambridge: Cambridge University Press, c1996, pp. 165-167.
3. Leo, W. R., *Techniques for Nuclear and Particle Physics Experiments*, 2nd Edition, New York: Springer-Verlag, c1994, ch. 7, p. 164.
4. Leo, W. R., *Techniques for Nuclear and Particle Physics Experiments*, 2nd Edition, New York: Springer-Verlag, c1994, ch. 17, pp. 325-326.
5. Fernow, R., *Introduction to experimental particle physics*, New York: Cambridge University Press, c1986, ch. 7, pp. 172-176.

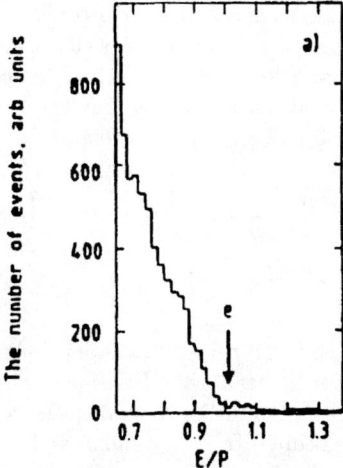
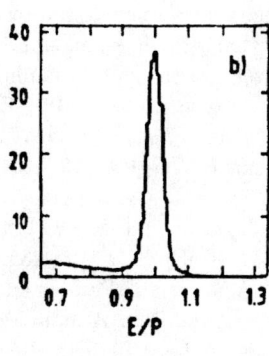

**FIGURE 26.** Distribution of Fermilab E715 $\Sigma \to pe\nu$ candidates versus E/p before, a), and after, b), imposing the electron selection criteria using the TRD data.

6. Agostini, G. D', et al., *NIM* **185**, 49 (1981).
7. Kubota, Y., et al., *NIM* **A320**, 66 (1992).
8. Nappi, E., private communication.
9. Leo, W. R., *Techniques for Nuclear and Particle Physics Experiments*, 2nd Edition, New York: Springer-Verlag, c1994, ch. 2, p. 50.
10. Fernow, R., *Introduction to experimental particle physics*, New York: Cambridge University Press, c1986, ch. 2, pp. 38-41.
11. Allison, W.W.M., and Cobb, J.H., *Ann. Rev. Nucl. Part. Sci.* **30**, 253 (1980).
12. Hauschild, M., et al., *NIM* **A314**, 74 (1992).
    Lou, X., "dE/dx Measurement with the OPAL Jet Chamber", talk given at the session on Hadron ID at the *Workshop on B Physics at Hadron Accelerators*, Snowmass, Colorado, June 21 - July 2, 1993.
13. Kleinknecht, K., *Detectors for Particle Radiation*, Cambridge University Press, c1986, p. 143.
14. Cobb, J., et al., *NIM* **140**, 413 (1977).
15. Artru, X., et al., *Phys. Rev.* **D12**, 1289 (1975).
16. The simulation package has two parts. The first is a calculation of the average number of TR photons as a function of energy produced by the radiators including detection in the xenon chamber planes. See L. C. Myrianthopoulos et al., "Status Report on the University of Maryland Transition Radiation System for Beam Particle Identification", University of Maryland, MdDP-TR-80-106, May 1980, unpublished, for details. Since the Monte Carlo part of that program

was appropriate for the total charge method as opposed to "cluster counting", the method used in the E769/791 TRD, the part of the package that generates the detector hits was written by our group. The simulations take into account the fact that signals were simply latched and not scaled, so that two or more clusters seen in the same chamber plane would be recorded as one.

17. Errede, D., et al., *NIM* **A309**, 386 (1991).
18. Cherry, M. L., *Phys. Rev.* **D17**, 2245 (1978).
19. Dolgoshein, B., *NIM* **A252**, 137 (1986).
20. Ludlam, T., et al., *NIM* **180**, 413 (1981).
21. Fabjan, C., et al., *NIM* **185**, 119 (1981).
22. Benot, M., et al., *NIM* **105**, 431 (1972).
23. Kulikov, A.V., et al.,"Performance of the E715 Transition Radiation Detector", talk given at the First Annual Meeting (New Series) of the Division of Particles and Fields of the American Physical Society, Santa Fe, New Mexico, Oct 31 - Nov 3, 1984. Published in the proceedings edited by T. Goldman and Michael Martin Nieto, World Scientific Publishing Co. Pte. Ltd.
24. Hsueh, S.Y., et al., "Measurement of the Electron Asymmetry in the Beta Decay of Polarized $\Sigma^-$ Hyperons", Phys. Rev. Lett. 54 (1985) 2399.

# Future Particle Detector Systems

## Allan G. Clark

*Département de Physique Nucléaire et Corpusculaire, Université de Genève, Switzerland.*

**Abstract.**
Starting with a short summary of the major new experimental physics programs, we attempt to motivate the reasons why existing general-purpose detectors at Hadron Colliders are what they are, why they are being upgraded, and why new facilities are being constructed. The CDF and ATLAS detectors are used to illustrate these motivations.

Selected physics results from the CDF experiment provide evidence for limitations on the detector performance, and new physics opportunities motivate both machine and detector upgrades. This is discussed with emphasis on the improved physics reach of the CDF experiment at the Fermilab Tevatron ($\sqrt{s} = 2$ TeV).

From 2005, the Large Hadron Collider (LHC) at CERN will become operational at a collision energy of $\sqrt{s} = 14$ TeV, seven times larger than at the Tevatron Collider. To exploit the physics capability of the LHC, several large detectors are being constructed. The detectors are significantly more complex than those at the Tevatron Collider because of physics and operational constraints. The detector design and technology of the aspects of the large general-purpose detector ATLAS is described.

## INTRODUCTION

Several major experimental particle physics programs are being completed. The confirmation with very high precision of the Standard Electroweak and QCD models by these experiments has prompted new physics questions which can only be answered by the construction of intense new specialised accelerators, and associated performant detectors.

In particular, the CDF and D0 experiments at the Fermilab $p\bar{p}$ Tevatron collider, the four experiments at LEP (Aleph, Delphi, L3 and Opal), and the SLD detector at the SLAC Linear Collider, have provided extensive data on the $W$ and $Z$ properties that confirm the Standard Electroweak Model [1]. At the same time, the Super-Kamiokande [2] experiment has reported evidence for neutrino oscillations that can be interpreted as a $\nu_\mu$-$\nu_\tau$ mass difference $\Delta m^2 \approx 10^{-3}$ eV.

An incomplete list of major experiments now starting includes the following:

a) At SLAC and KEK, $b$-quark factories have been constructed and have been equipped with the large general purpose detectors BaBar (SLAC) [3] and Belle (KEK) [4]. These high-intensity $e^+e^-$ accelerators are optimised for the production of $\bar{b}b$ pairs at the $\Upsilon$ resonance. The subsequent $b$-quark decays will be used for detailed studies of the CP-violating CKM matrix elements, and for rare $b$-quark decays that might signal new physics processes. The upgraded CLEO detector [5] at Cornell will remain competitive with these detectors. Although operating in a more difficult hadronic environment, the HERA-B experiment at DESY [7] will also collect data from $\bar{b}b$ pair production, but at higher collision energy.

b) At Frascati, the DAPHNE $e^+e^-$ accelerator is optimised towards high-intensity $\phi$-meson production, allowing accurate CP-violation measurements in the K-meson sector [6].

c) The Tevatron $p\bar{p}$ Collider is being increased in both collision intensity and collision energy ($\sqrt{s} = 2$ TeV) and the CDF and D0 experiments are being upgraded to profit from the machine improvements. These experiments will independently study $b$-quark production and decay (including CP-violating terms), and will search at the present high-energy limit for new physics processes (including Higgs production).

d) The Large Hadron Collider (LHC) is now being constructed at CERN. This proton-proton collider will have a collision energy of $\sqrt{s} = 14$ TeV, and a peak luminosity of $\mathcal{L} = 10^{34}$ cm$^{-2}$s$^{-1}$. The general purpose ATLAS and CMS experiments [8,9] will search for evidence for the production of the Higgs boson, which results from symmetry breaking in the Standard Model, and possibly new phenomena that are extensions of the Standard Model.

e) At the same time, there have been extensive development programs at CERN and DESY in Europe, and in the U.S. and Japan, towards high energy Linear $e^+e^-$ Colliders, which should be able to map in a more detailed way the physics beyond the Standard Model.

f) Long baseline $\nu_\mu$ beams are being constructed at Fermilab [10] and CERN [11,12] for $\Delta m^2$ measurements and data are already being collected by Super-Kamiokande using a lower energy $\nu_\mu$ beam from KEK [13]. Even more interesting is the possibility of using muon storage rings to produce intense $\nu_\mu$ beams and $\nu_e$ beams for use in long baseline experiments.

Since the construction of the LEP and Tevatron detectors, there have been major detector developments, resulting in part from developments in micro-electronics, and in part from the extreme requirements of LHC detectors. All the above experiments benefit from such developments. These lectures concentrate on the development of detectors at hadron colliders. Several authoritative accounts exist and are recommended [14,15]. These lectures instead attempt to describe the motivation of detector or accelerator improvements. For reasons of brevity, this written version of the lectures mainly emphasises the CDF experiment at the Tevatron Collider, and the ATLAS experiment at the CERN LHC.

In the following sections, we briefly describe the existing CDF detector, and some key physics results obtained by the detector (together with the D0 detector). We note deficiencies or limitations of the CDF detector, in the light of future exploitation at the Tevatron or elsewhere. We then describe the Fermilab Tevatron and CDF detector upgrades, and the ATLAS detector specification.

# MOTIVATION FOR NEW HADRON COLLIDER DETECTORS

## The CDF Run 1 detector

Figure 1 shows the layout of one quadrant of the CDF detector at the Fermilab Tevatron $p\bar{p}$ Collider. The detector aims to allow a reliable and efficient identification (efficiency $\epsilon$) of leptons and quarks over the largest possible acceptance and with the best possible rejection against fake particle signatures:

- The identification of hadronic jets that are characteristic of quark or gluon (parton) fragmentation, and a measurement of the parton energy;

- The identification and accurate momentum measurement of charged leptons $(e^{\pm}, \mu^{\pm}, \tau^{\pm})$;

- The identification of photons and the efficient separation of photon and $\pi^0$ signatures;

- The tagging with good efficiency of long lived quark ($b$- and $c$-quarks) and $\tau$ lepton decays from a measurement of their offset vertices;

- The measurement of missing transverse energy resulting from the non-detection of the neutrinos of weak quark of lepton decays.

A 1.9 m long berrylium beam pipe surrounds the $p\bar{p}$ interaction region. Surrounding the beam pipe, a silicon micro-strip detector (SVX) covers approximately 60% of the geometrical acceptance. The detector, of length 51 cm and outer radius 7.87 cm, consists of four radial layers that are divided into 12 azimuthal wedges. The inner layers have an axial readout pitch of $(\delta r$–$\phi) = 60$ $\mu$m, with the outer layer $(\delta r$–$\phi) = 55$ $\mu$m. With a total of 46080 readout channels, the single hit resolution is $\sigma = 13$ $\mu$m, and the transverse impact resolution for high-$p_T$ tracks is $\sigma = 17$ $\mu$m (before multiple scattering contributions). The state-of-the-art custom electronics used for this detector is now obsolete.

Surrounding the SVX is a vertex drift chamber (VTX) covering the pseudo-rapidity range $|\eta| < 3.25$ in the radial range $8 < r < 22$ cm. Tracks that are reconstructed by the VTX are used to constrain the $p\bar{p}$ interaction vertex (or vertices in the case that more than one primary interaction occured during a beam crossing) with a resolution in $z$ of 1 mm.

The central tracking chamber (CTC) of CDF is 3.2 m long with an outer radius of 1.32 m. A total of 84 concentric layers of drift chamber wires are used to reconstruct charged tracks. Five axial super-layers each contain 12 axial layers of sense wires, and $r-z$ information is obtained using 4 stereo super-layers ($\pm 3^0$) of 6 wires each.

A super-conducting solenoid of length 4.8 m, radius 1.5 m, and magnetic field 1.4 Tesla allows the momentum evaluation of reconstructed charged tracks. Using the CTC and SVX information, the momentum of reconstructed tracks can be determined with a precision $\delta p_T/p_T \sim 0.001 \cdot p_T$. In conclusion, the Run 1 inner detector allows excellent charged particle identification, with the possibility of long-lived quark or lepton tagging over a restricted acceptance.

As shown in Figure 1, the solenoid is surrounded by electromagnetic and hadronic calorimeters covering the full azimuthal range and a polar region $|\eta| < 4.2$. The calorimeters are segmented into towers projected towards the interaction point. The electromagnetic (hadronic) calorimeters in the central region (towers of $15^0$ deg and 0.1 in $\eta$) consist of alternating layers of lead (steel) absorber, and scintillator with the signal read by phototubes. A pre-radiator using proportional chambers preceding the calorimeter and at a depth of $\sim 6X_0$ in the electromagnetic calorimeter accurately measures the point of shower initiation. In the forward regions, lead (iron) plates are interleaved with gaseous proportional chambers having a cathode pad readout. The performance of the calorimeters is shown in Table 1. The difference in performance of the central and forward detectors is marked (as will be shown). Not shown in this description is the fact that insensitive regions (for example between each tower), and gaps in the acceptance (for example to allow services to pass) deteriorate the performance of the calorimeter and especially the evaluation of missing transverse energy.

Outside the calorimeters is the central muon system (CMU, CMP and CMX). The central calorimeter acts as a hadron absorber, allowing muons of $p_T > 1.5 \, \text{GeV}/c$ to be identified. The CMU consists of 4 gaseous drift chamber layers, followed by 0.6 m of absorbing steel, and a further 4 drift chamber layers. It

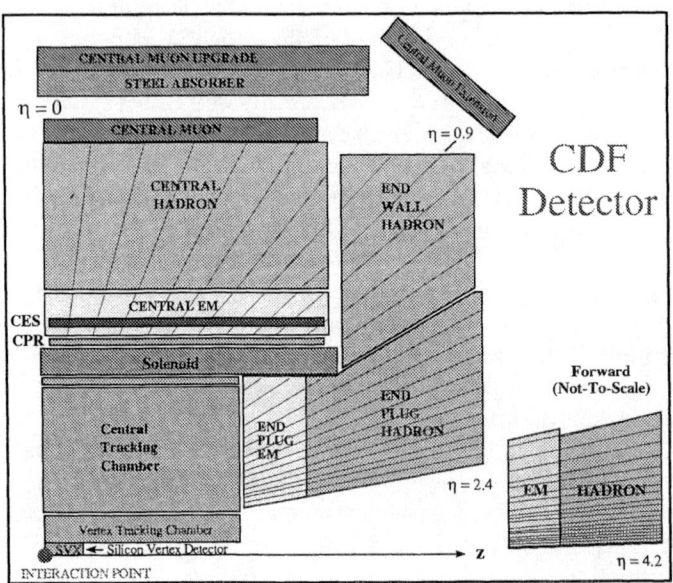

**FIGURE 1.** A side-view cross-section of the CDF detector layout during Run 1.

| System | Range | Energy Resolution | Thickness |
|---|---|---|---|
| CEM | $|\eta| < 1.1$ | $13.7\%/\sqrt{E_T} \oplus 2\%$ | $18\ X_0$ |
| PEM | $1.1 < |\eta| < 2.4$ | $22\%/\sqrt{E} \oplus 2\%$ | $18 - 21\ X_0$ |
| FEM | $2.2 < |\eta| < 4.2$ | $26\%/\sqrt{E} \oplus 2\%$ | $25\ X_0$ |
| CHA | $|\eta| < 0.9$ | $50\%/\sqrt{E_T} \oplus 3\%$ | $4.5\ \lambda_0$ |
| WHA | $0.7 < |\eta| < 1.3$ | $75\%/\sqrt{E} \oplus 4\%$ | $4.5\ \lambda_0$ |
| PHA | $1.3 < |\eta| < 2.4$ | $106\%/\sqrt{E} \oplus 6\%$ | $5.7\ \lambda_0$ |
| FHA | $2.4 < |\eta| < 4.2$ | $137\%/\sqrt{E} \oplus 3\%$ | $7.7\ \lambda_0$ |

**TABLE 1.** Properties of the calorimetry components used by the Run 1 CDF detector.

covers 84% of the region $|\eta| < 0.6$, and 71% of the region $0.6 < |\eta| < 1.0$. The CDF muon system has known advantages and disadvantages. With a relatively small absorption length $\lambda_0$, the threshold for $\mu^\pm$ detection is low, but information from the CTC is required to evaluate the track momentum, and to reduce the background of decays and hadronic punch-through. Furthermore, the CDF acceptance for muon identification is limited.

The CDF experiment uses three levels of trigger. The choice of trigger is motivated by physics interest, and by the need to minimise dead time for interesting physics signatures while respecting limits of data storage. Because of technological advances and stringent machine constraints, this is an area of marked development for future systems. A $t\bar{t}$ event as detected by the CDF detector is shown in Figure 2. These figures indicate the complexity of final states in typical $p\bar{p}$ interactions. Given that the event of Figure 2 has a production cross section of $\sim 6$ pb as opposed to a total inelastic cross section of $\sim 40$ mb, it is obviously necessary to select event signatures of interest in real time; the event of Figure 2 was selected by requiring an isolated electron of $> 8$ GeV in the central electromagnetic calorimeter at level-1, followed by matching with a high-$p_T$ charged track at level-2, and a detailed event reconstruction in the level-3 processor farm. More generally, the CDF level-1 trigger requirements relevant to these lectures are for an isolated $e^\pm$ or $\mu^\pm$ of $p_T > 8(6)$ GeV, or calorimeter (localised or otherwise) energy depositions beyond selected energy thresholds. With a bunch crossing rate of 280 kHz (3.5 $\mu$s), and a level-1 accept rate of $< 10$ kHz, there is no level-1 dead time. At level-2, decisions must be made within $\approx 20$ $\mu$s to reduce the rate to a few tens of Hz. The level-3 trigger is a buffered commercial processor system, with a selected rate determined by the off-line storage capacity.

As suggested above, a typical selected high $p_T$ interaction from CDF includes leptons, jets and $E_T^{\rm miss}$. The jets, and certain meson or baryon decay signatures, may be characteristic of the production of a specific quark flavor. Nevertheless, given event signatures may be experimentally faked by other physics processes. As evident from Figure 2, the physics of interest at Hadron Colliders is that of parton spectroscopy. The final states of quark interactions may be complex and in the case of Hadron Colliders may be more complex because of overlapping and underlying event structures.

During the period 1987-1996, each of the CDF and D0 detectors collected an integrated luminosity of $\int \mathcal{L} dt \sim 110$ pb$^{-1}$. This data sample enabled the observation of the final states of physics processes having a production cross-section in the region of 1-10 pb. A number of important physics results have been obtained that motivate a critical review of the machine and detector performance, and an improvement programme for each.

## Selected CDF physics results from Run 1

In the previous sub-section, the CDF detector was briefly described. Despite its undoubted limitations, it has provided data leading to several outstanding physics results. The D0 detector, which has not been described, has been equally successful, despite a different category of limitations. In this sub-section, we summarize selected results demonstrating the success of the CDF detector, the detector limitations, and possible detector improvements.

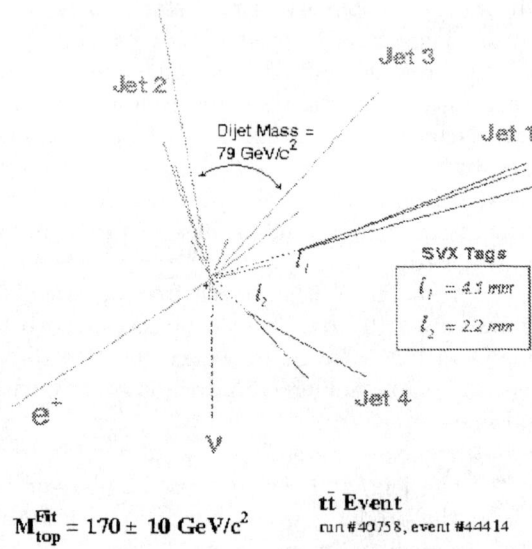

**FIGURE 2.** Details of a top quark event.

## Top Quark Physics.

Figure 3 shows the dominant production mechanism for $t\bar{t}$ production at the Tevatron Collider. In Run 1, following the identification of the top quark by CDF and D0 [16,17], the mass $m_t$ and cross-section $\sigma_{t\bar{t}}$ were measured for the different allowed final state signatures (see Figure 4 and Figure 5) [18].

Given the small $\sigma_{t\bar{t}}$, only a few hundred $t\bar{t}$ events have so far been produced. After accounting for the detector acceptance and identification efficiency, that number is further reduced. The CDF experiment has focused on several experimental signatures for $t\bar{t}$ studies:

- In the di-lepton channel, the produced $t$ and $\bar{t}$ quarks each decay into a $W$ and a $b$-quark, with a subsequent leptonic decay of the $W$. The final state therefore consists of 2 charged leptons, $E_T^{\mathrm{miss}}$ from the decay $\nu$'s, and 2 $b$-jets. Because of the small leptonic branching ratio $BR(W \to l\nu)$ the acceptance is low ($< 5\%$).

- In the lepton-plus-jets channel, one $W$ decays leptonically and the other decays to two quarks. Hence, the nominal signature is a charged lepton, $E_T^{\mathrm{miss}}$ (the $\nu$ from $W$-decay), and 4 jets from the 2 $b$-jets and 2 $W$-decay quarks. Before geometrical acceptance, approximate 30% of the data sample is in this category. To remain efficient, only 3 central ($|\eta| < 2$) jets are experimentally required. The background to this channel is large, mainly from the $W$-plus-

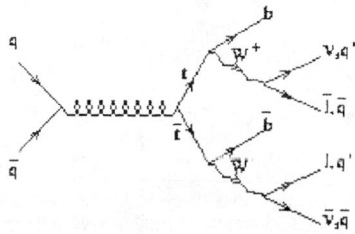

| Decay mode | Branching ratio |
|---|---|
| $t\bar{t} \to (q\bar{q}'b)(q\bar{q}'b)$ | 36/81 |
| $t\bar{t} \to (q\bar{q}'b)(e\nu b)$ | 12/81 |
| $t\bar{t} \to (q\bar{q}'b)(\mu\nu b)$ | 12/81 |
| $t\bar{t} \to (q\bar{q}'b)(\tau\nu b)$ | 12/81 |
| $t\bar{t} \to (e\nu b)(\mu\nu b)$ | 2/81 |
| $t\bar{t} \to (e\nu b)(\tau\nu b)$ | 2/81 |
| $t\bar{t} \to (\mu\nu b)(\tau\nu b)$ | 2/81 |
| $t\bar{t} \to (e\nu b)(e\nu b)$ | 1/81 |
| $t\bar{t} \to (\mu\nu b)(\mu\nu b)$ | 1/81 |
| $t\bar{t} \to (\tau\nu b)(\tau\nu b)$ | 1/81 |

**FIGURE 3.** Details of the different top pair production graphs with their branching ratio.

multijet QCD processes. The background is reduced by tagging the $b$-jets,

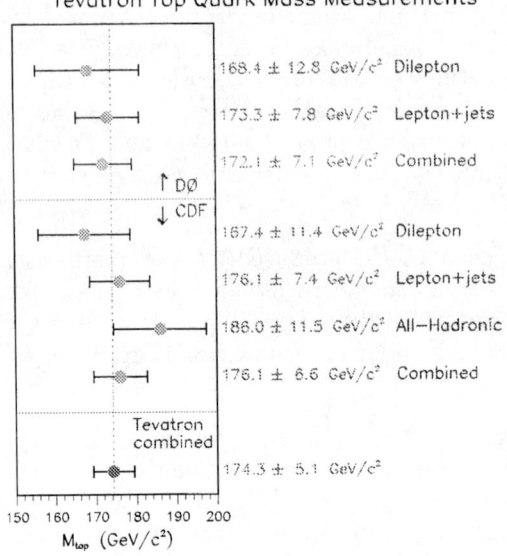

**FIGURE 4.** Details of the top mass measurements at CDF and D0 for Run 1.

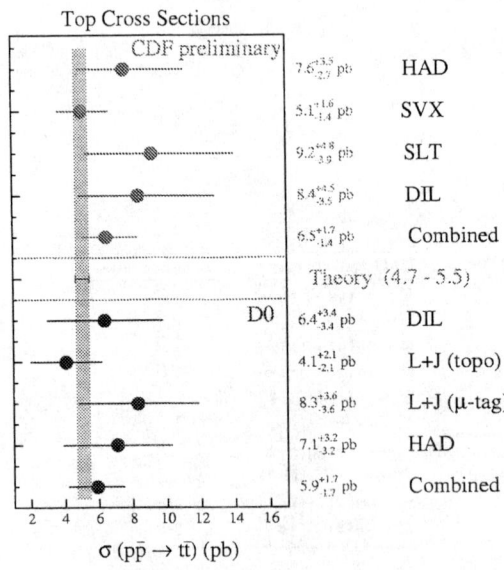

**FIGURE 5.** Details of the $t\bar{t}$ production cross-section measurements at CDF and D0 for Run 1.

using (in the case of CDF) either an offset vertex reconstructed in the SVX ($\epsilon_{tag} < 40\%$), or the detection of the low-$p_T$ lepton from a semileptonic $b$-quark decay ($\epsilon_{tag} < 20\%$).

- In the all-jets channel, potential $t\bar{t}$ candidates are identified with a large irreducible QCD background after kinematic selections followed by $b$-tagging via the SVX.

An accurate evaluation of $m_t$ and $m_W$ provides fundamental information on the symmetry breaking sector of the Standard Model (or deviations thereof) and a more accurate measurement of $m_t$ alone justifies detector improvements. Figure 6 shows the existing data on $m_t$ and $m_W$, and their sensitivity to the Higgs mass in the context of the Standard Model. At present, the statistical and systematic uncertainties on $m_t$ are similar and an improved measurement must address both uncertainties. Figure 5 compared the expected precision of QCD next-to-leading order calculations for $\sigma_{t\bar{t}}$ with existing data. An improved measurement should aim to provide a constraint on the QCD expectations and any deviation might signal new physics.

Future top quark measurements for an upgraded detector include the following (assuming an initial integrated luminosity $\int \mathcal{L} dt \sim 2 fb^{-1}$ at $\sqrt{s} = 2$ TeV during 2001-2 and $\int \mathcal{L} dt > 10 fb^{-1}$) before LHC turn-on [19,20]:

- A measurement of $m_t$ with a precision of $\delta m_t \sim 2$ GeV/$c^2$;

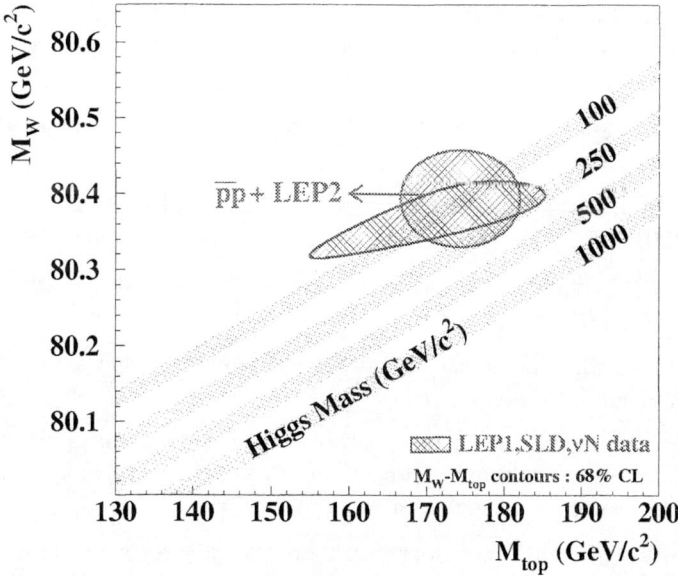

**FIGURE 6.** World average $W$-mass and Tevatron top-mass measurements, with predictions for various Higgs masses.

- A measurement of $\delta\sigma_{t\bar{t}}/\sigma_{t\bar{t}} < 0.06$.

Other t-quark physics studies are under way, but with poor statistical accuracy. Such measurements include:

- A measurement of $BR(t \to Wb)$ where a measurement uncertainty of $< 2\%$ is expected;
- A measurement of the ratio of the di-lepton to single lepton decay rates; the upgraded detector will aim for a measurement uncertainty of $< 5\%$;
- A search for $t\bar{t}$ resonances at masses of up to 1 TeV/$c^2$; because of its large mass, the t-quark is an excellent probe for physics beyond the Standard Model, and new processes can be expected to modify the shape of the $t\bar{t}$ spectrum, the transverse momentum $p_T(top)$, or the centre-of-mass production angle;
- The isolation of single top production via the processes $qg \to t\bar{b}q'$ or $q'\bar{q} \to t\bar{b}$; in Run 2, a measurement of the cross-sections to $\sim 10\%$ to evaluate $V_{tb}$ should be feasible;
- Rare decay studies, excluded at present but possible with adequate statistics (for example $t \to c\gamma$ or $t \to Zc$).

The Tevatron machine performance must be upgraded to deliver the integrated luminosity necessary for these measurements (next section). Direct consequences of this for the detectors include:

- A machine bunch spacing of 132 ns instead of 3.5 $\mu$s implies the need to use fast electronics shaping in regions of high occupancy, and the pipelining of data at both the front-end electronics and data acquisition stages;
- The drift chambers of CDF (the VTX and CTC) must be replaced because of their large drift regions;
- Higher radiation levels impose the need to use recent technology developments on radiation hard electronics for regions close to the beam pipe;
- Because of radiation damage, it is necessary to operate the silicon detectors at low temperature.

In addition to these necessary detector developments, other significant detector improvements can be made:

- The silicon detector was shown to be of crucial importance in Run 1. Any detector improvement should address the limited pattern recognition capability and acceptance for $b$-tagging of this device;
- In analysis of Run 1 data, primary lepton signatures were generally required by CDF in the central region, because of the limited quality of electromagnetic calorimetry (electrons) in the forward direction, and the limited muon coverage.

## Symmetry Breaking in the Standard Model

A primary goal of any improved Tevatron Collider detector or LHC detector must be the identification of the spontaneous symmetry breaking mechanism for the electroweak interaction. The simplest model for this mechanism is the standard Higgs model, which requires a neutral Higgs boson. The search for this particle, indirectly or directly, is a benchmark. The lower bound changes 'daily', according to the beam energy of the CERN LEP2 accelerator. With beam energies of 102 GeV, the LEP experiments exclude $M_{H^0} < 108$ GeV/$c^2$ [21].

Cross-section limits from CDF [22] are shown in Figure 7 and indicate the need for increased data samples to be competitive with LEP prior to the LHC turn-on. While the data of Figure 7 correspond to the reaction ($p\bar{p} \rightarrow VH^0$; $V = W$ or $Z$; $H^0 \rightarrow b\bar{b}$), the higher rate process $gg \rightarrow H^0$ may also be accessible if it can be identified at the trigger level. For this and other reasons, any upgrade should include the features described above and the ability to trigger on $b$-quarks. The rate of different production processes is shown in Figure 8, at $\sqrt{s} = 2$ TeV and at LHC energies ($\sqrt{s} = 14$ TeV). By combining the CDF and D0 data, a window of opportunity for identifying either the Standard Model or neutral MSSM Higgs exists before the LHC turn-on [23].

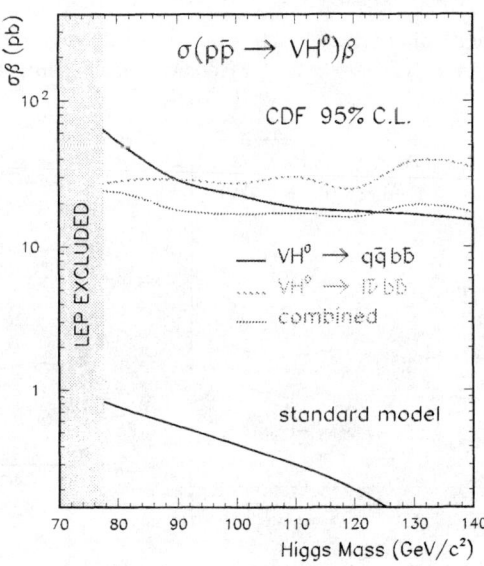

**FIGURE 7.** 95% CL upper-limit for Higgs production at CDF for Run 1.

## CP violation in the b-quark sector.

Although dedicated $B$-factories are now coming into operation, studies of the $B$-physics sector at the Tevatron by both CDF and D0 have already been very rich and will remain competitive. The introduction of the SVX at CDF has enabled competitive measurements on time dependent $B_d^0$ mixing, rare decay limits, and in particular the CP-violation parameter $sin(2\beta)$ [24,25].

In $p\bar{p}$ interactions at $\sqrt{s} = 1.8$ TeV, $b\bar{b}$ pairs are produced with a cross section of $\sim 100$ $\mu$b; the $b$ quarks then hadronise into all species, including b-baryons and the mesons $B_s$, $B_d$ and $B_c$. The high cross-section implies the need for a high rate trigger and data acquisition. The SVX is important for tagging the flavor of the B-mesons (that is whether it originates from a $b$-quark or a $\bar{b}$-quark), and excellent tracking and momentum resolution is required for reconstructing the mass of B-meson or baryon states. The B-hadron spectrum is relatively soft; for example, for the decay $B^0 \to J/\psi K_S^0$ with a $p_T > 2$ GeV/c requirement on the muons of $J/\psi \to \mu^+\mu^-$, the mean $p_T$ of the B is about 10 GeV/c. Good pattern recognition and momentum resolution of low $p_T$ tracks with only a limited deterioration from multiple scattering places strong constraints on the full tracking system, and in the case of electrons on the calorimeter performance.

During Run 1, a measurement of the relative rates of the decays $B^0 \to J/\psi K_S^0$ and $\bar{B}^0 \to J/\psi K_S^0$ following $b\bar{b}$ production enabled a measurement by CDF of $\sin(2\beta) = 0.79^{+0.41}_{-0.44}(stat+sys)$. To accurately map the elements of the CKM matrix, both the B-factories and the Tevatron would like to measure $\delta \sin(2\beta)/\sin(2\beta) < 0.08$.

Another measurement of interest is the quantity $\sin(2\alpha)$, which is related to CP violation in the decays $B_d^0/\bar{B}_d^0 \to \pi^+\pi^-$. Because of the small branching ratio of

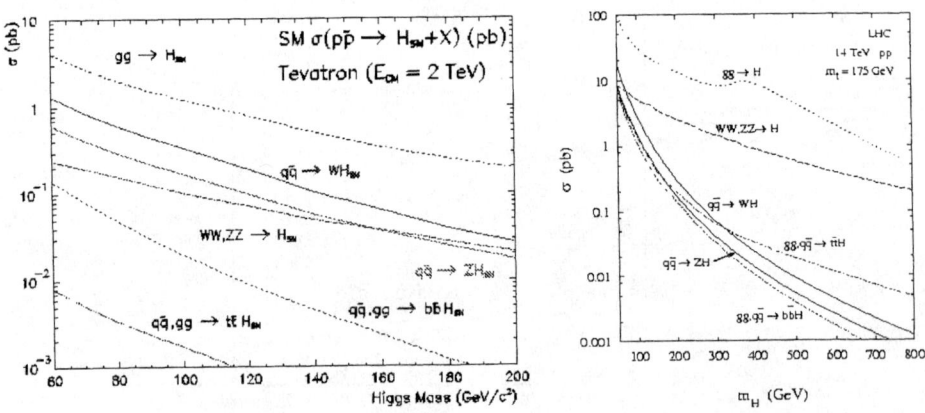

**FIGURE 8.** Standard Model Higgs production cross-section predictions for the Run 2 Tevatron ($\sqrt{s} = 2$ TeV, left) and for LHC ($\sqrt{s} = 14$ TeV, right).

this decay, and the large background, a real time trigger on offset vertices in the SVX is required. A real time trigger will be implemented in the upgraded CDF and D0 detectors. Because of potential backgrounds from (for example) $B_s^0 \to K^+K^-$, the best possible $K--\pi$ separation is desirable, a time-of-flight system will be implemented to complement ionisation measurements that are already made in the central gaseous tracking systems.

# TEVATRON COLLIDER AND CDF DETECTOR UPGRADES.

## Tevatron Collider Upgrades

Table 2 shows the technical performance of the Tevatron during Run 1. By the end of Run 1, the original machine design specifications had been significantly exceeded. The most important improvement during Run 2 will be an increase of the delivered luminosity.

The Fermilab accelerator complex during Run 1 is shown in Figure 9. Protons from a Linac are accelerated to 8 GeV in the Booster before injection in 12 bunches into the Main Ring which is above the superconducting Tevatron ring. The 12 bunches are re-combined into a single bunch, and accelerated to 150 GeV before injection into the Tevatron. To create $\bar{p}$'s, protons of 120 GeV from the Main Ring are focused onto a Be target. The $\bar{p}$'s are extracted into a 'debuncher' ring where stochastic cooling is used before storage in an accumulator ring. When enough $\bar{p}$'s have been stored, a fraction will be re-transferred to the Main Ring and the Tevatron Ring before acceleration of the $p$'s and $\bar{p}$'s to $\sqrt{s} = 1.8$ TeV. The Run 1 accelerator complex had several limitations:

- The Main Ring passed through the D0 detector and near the CDF detector, and backgrounds prevented $\bar{p}$ accumulation during detector operation;

- The emittances of the different machine components were not well matched.

The Main Ring was replaced in 1999 by a new high-intensity Main Ring of 100 GeV which is better adapted in emittance to Tevatron requirements. This allows the storage of $\bar{p}$'s with increased efficiency, and during Tevatron operation. Together with an increased $p$-bunch intensity, it allows an increase of the Tevatron intensity for Run 2 by a factor 10. Further improvements can be envisaged by improved targeting and $\bar{p}$'s accumulation. In order to prevent bunch-bunch effects, the bunches are spread out in the machine, with 396 bunches instead of 6. Therefore, to profit from the increased luminosity, the detectors must be sensitive to interactions occurring each 128 ns; detectors having long collection times experience pile-up, and trigger or readout latencies impose the need for pipelines in the front-end electronics and the data acquisition.

Table 2 also compares the $\bar{p}p$ luminosity with that achievable at higher luminosity by the LHC $pp$ Collider.

|  | Tevatron | | | LHC |
| --- | --- | --- | --- | --- |
|  | Run 1B (1993-6) | Run 2 (2001-) | Run 3 (TEV33) | LHC (2005-) |
| $p$/bunch | $2.32 \times 10^{11}$ | $2.7 \times 10^{11}$ | $2.4 \times 10^{11}$ | $1.05 \times 10^{11}$ |
| $\bar{p}$/bunch | $5.5 \times 10^{10}$ | $5.5 \times 10^{10}$ | $1.0 \times 10^{11}$ | – |
| Total $\bar{p}$ | $3.3 \times 10^{11}$ | $1.98 \times 10^{12}$ | – | – |
| Beam Energy (TeV) | 0.9 | 1.0 | 1.0 | 7.0 |
| Bunches | 6 | 36 – 108 | 108 | 2835 |
| Bunch length (rms) in cm | 60 | 18 | 26 | 6 |
| $\mathcal{L}$ (typ) (cm$^{-2}$s$^{-1}$) | $1.58 \times 10^{31}$ | $2.03 \times 10^{32}$ | $1.04 \times 10^{33}$ | $< 10^{34}$ |
| $\mathcal{L}$ (max) (cm$^{-2}$s$^{-1}$) | $2.5 \times 10^{31}$ |  |  | $10^{34}$ |
| $\mathcal{L}$ (int/week) (pb$^{-1}$) | 3.2 | $\approx 41$ | $\approx 210$ | $10^5$/yr |
| Bunch spacing (ns) | 3500 | 396 – 132 | 132 | 25 |
| <Interactions>/crossing | 2.5 | 5.3 – 1.8 | 9.1 | 23 |

**TABLE 2.** Performance parameters of the Run 1 Tevatron, the Run 2 Tevatron and the LHC.

## Detector Improvements

Figure 10 schematically shows the evolution of the CDF detector between Run 1 and Run 2. An elevation view of one half of the CDFII detector is shown in Figure 11 and the tracking system is shown in Figure 12. The performance improvements motivated above, and developed in [19,20], are as follows, bearing in mind the requirement of detector live-time at 128 ns bunch crossing intervals:

- The reconstruction of charged tracks with efficiency $\epsilon > 95\%$ over the range $|\eta| < 2$, with $\delta p_T/p_T^2 < 0.001$ for $|\eta| < 1$ and $\delta p_T/p_T^2 < 0.004$ for $|\eta| < 2$;

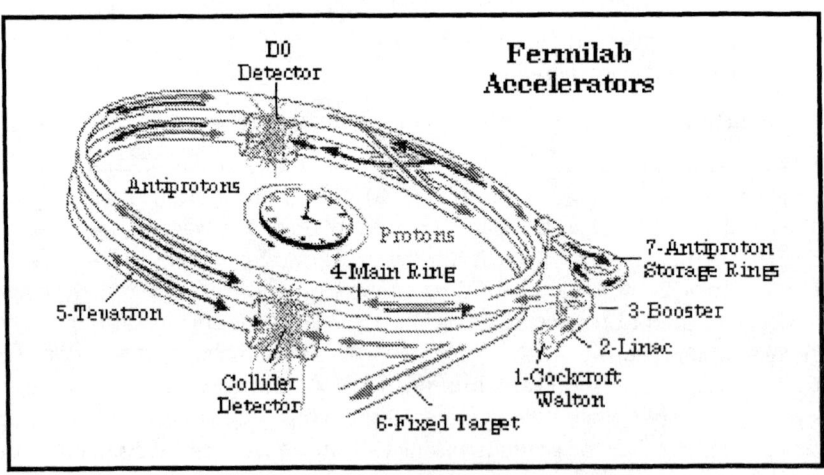

**FIGURE 9.** The Fermilab Run 1 machine configuration

- The efficient reconstruction of tracks from offset vertices, even in dense track environments (jets), over the full $|\eta|$ range;
- Trigger on, identify, reconstruct and measure the kinematics and charge of electrons and muons with high efficiency over the full range $|\eta| < 2$;
- Trigger on, identify, reconstruct and measure the kinematics of photons with high efficiency over the full range $|\eta| < 2$;
- Trigger on and reconstruct jets over the range $|\eta| < 3$, and measure the jet energies with good resolution and a scale precision of better than 2.5%;
- Measure particle types using both dE/dx and time-of-flight techniques in the central region, $|\eta| < 1$.

These aims are largely achieved in the CDFII detector, by the following upgrades:

- The replacement of the plug and forward calorimeters by a scintillator-based calorimeter having the same performance as the central region, a very fast time response, and only a limited dead space;
- The replacement of the central gaseous tracking chamber (CTC) by a similar design (COT) able to operate at high luminosity;
- The implementation of a scintillator layer in the radial space between the COT and the solenoid cryostat, providing a time of flight measurement resolution for charged tracks of 100 ps;

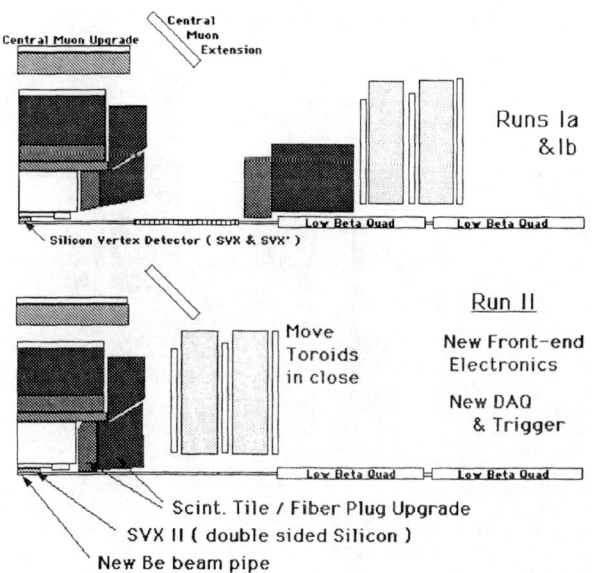

**FIGURE 10.** Evolution of CDF detector between Run 1 and Run 2

- The replacement of the SVX and VTX by a large silicon system with space-

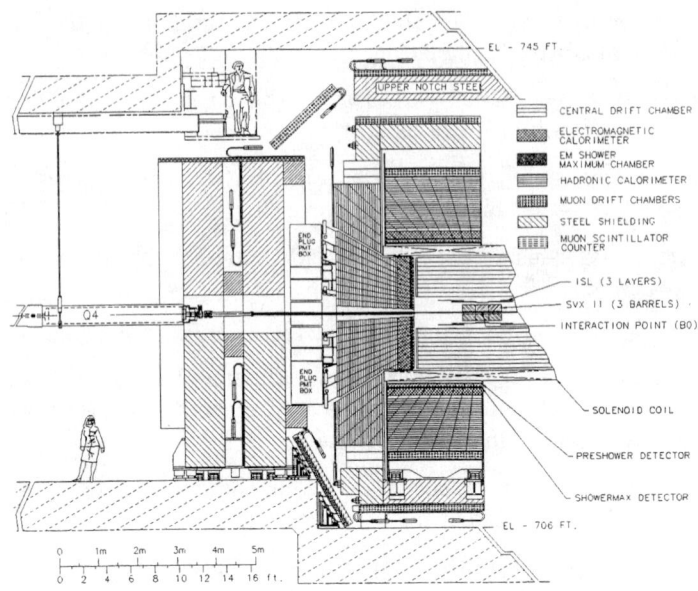

**FIGURE 11.** Elevation view of Run 2 CDFII detector

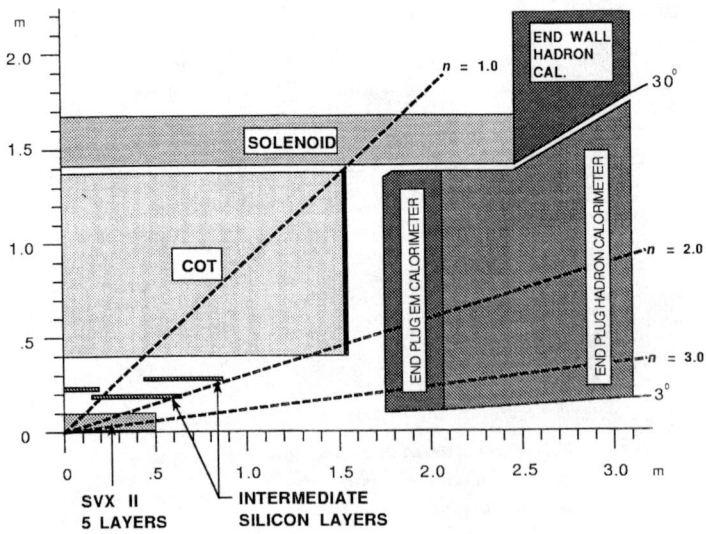

**FIGURE 12.** The CDFII tracking detector

point readout by custom radiation resistant 128-channel front-end chips;

- The implementation of high-$p_T$ tracking trigger in level-1 and a sophisticated level-2 trigger for the detection of offset vertices over the full range $|\eta| < 2$;

- An increase in the muon acceptance;

- The upgrade of the full readout electronics chain and data acquisition to operate with short bunch spacing.

In the following subsections, the silicon tracking and trigger upgrade is described in more detail. The upgrade of other detector elements is described in References [19,20].

### Silicon Tracking - a challenge

Among the most challenging system developments at CDFII is the silicon tracker, SVXII, which is an order of magnitude larger than that of the SVX. A similar silicon tracker is being developed for the D0 experiment. A micro-vertex layer ('layer-00') against the beam pipe establishes the track impact with high precision (even for low $p_T$ tracks). This is followed by 5 layers of silicon detectors in the radial range $2.4 < r < 10.7$ cm (SVX II), and an additional two layers covering the forward region (ISL). Full stand alone tracking and $b$-tagging is possible in the interval $|\eta| < 2$. Figure 13 shows the silicon tracker layout, and Table 3 indicates the specifications of the tracker.

|  | Layer-00 | SVXII | ISL |
|---|---|---|---|
| Layers | 1 | 5 | 2 |
| Length (cm) | 94.2 | 96 | 174 |
| $r_{min}$ (cm) | 1.35 | 2.5 | 20 |
| $r_{max}$ (cm) | 1.62 | 10.5 | 28 |
| Material ($X_0$) |  | 0.035 | 0.02 |
| Readout length (cm) | 15.7 | 14.5 | 21.5 |
| Readout pitch ($\mu m$) | 25 | 60-65 r-$\phi$ | 110 r-$\phi$ |
|  | - | 60-150 stereo | 146 stereo |
| Resolution ($\mu m$) | 6 | 12 axial | 16 axial |
| Modules | 36 | 360 | 300 |
| Channels | 13824 | 405504 | 307200 |

**TABLE 3.** Silicon tracker specifications.

The 'layer-00' of the SVX system is a layer of axial strip silicon detectors located at a radius 1.35<r<1.62 cm from the beam line. Each sensor, is 7.84 cm long (and read out in pairs of length 15.7 cm) with a read out pitch of 25$\mu$m. The strip resolution will be only 6$\mu$m. This layer significantly improves the impact resolution and pattern recognition for low- $p_T$ charged tracks. In particular, for tracks not passing through hybrid material, ($\sigma_{impact} = 9 \oplus 34/p_T$ μm, $p_T$ in GeV/c) is reduced to ($\sigma_{impact} = 6 \oplus 22/p_T$ μm). The improvement is even more marked if additional material is traversed by the track. The silicon sensors and readout electronics will suffer radiation damage at the level of LHC detectors described later. The sensors are therefore designed to operate at high bias voltages, as developed for LHC operation.

The main SVX detector is built in three cylindrical barrels, of total length 96 cm. There are 5 layers of silicon sensors: three layers combine an r–$\phi$ measurement with a $90^0$ stereo measurement, and 2 layers combine an r–$\phi$ measurement with a $1.2^0$ stereo measurement. The silicon sensors are supported on ladders that are mounted between precision beryllium bulkhead. Each strip is connected to a 128-channel radiation hardened readout chip having a preamplifier, a 46-cell dual port pipeline, and an ADC. A highly parallel fibre system reads the full detector in $\sim 10\mu s$. The radiation level at the inner layer is $\sim 0.5$ MRad/fb$^{-1}$. The silicon sensors will therefore suffer radiation damage that necessitates low-temperature operation (see below for the case of the ATLAS silicon tracker).

The ISL design is similar to that of the SVX, and the same readout system is used.

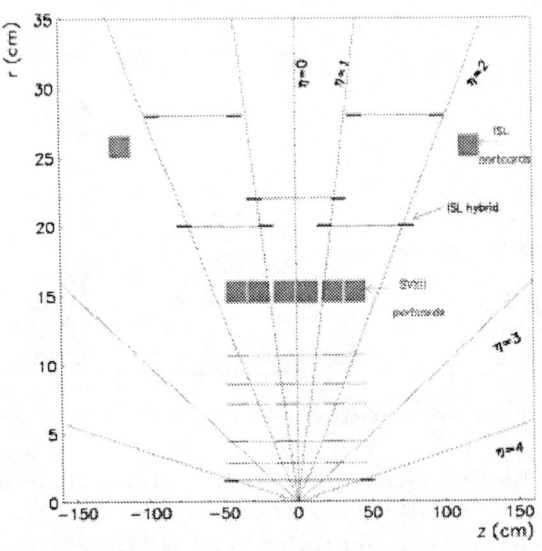

**FIGURE 13.** CDFII silicon tracking detector in the $r - z$ plane.

## The CDFII silicon trigger, SVT

Because of the reduced beam spacing, the CDF front-end electronics readout has been completely revised for Run 2. Given a 132 ns bunch spacing and a 4 $\mu$s level-1 decision time, all the front-end readout is buffered, as described above for the SVX. Data from the COT tracker, the muon chambers and the calorimeters are used for a level-1 decision which is itself pipelined. From a 7.6 MHz beam crossing rate, the maximum level-1 accept rate of 40 kHz must be respected. Given a level-2 accept, the data of each front-end system is read into level-2 buffers and processed asynchronously with an average decision time of $\sim 20\mu$s for a maximum accept rate of 300 Hz.

The level-1 trigger of CDFII includes a track trigger, enabling the matching of high-$p_T$ tracks to a muon stub or calorimeter cluster, or to be used alone for triggers such as $B^0 \to \pi^+\pi^-$. It can also be used with the SVT as an offset vertex trigger. This hardware device reconstructs 2-D tracks using measurements from the SVXII and the COT to allow for the first time in a Hadron Collider environment a B hadron trigger. Because of the silicon readout time of about 3 $\mu$s, its use at level-1 is impossible; however 2D tracks of $p_T > 1.5$ GeV/c in the COT are selected at level-1. These tracks, together with the silicon information, reconstruct tracks with precise track parameters: $p_T$, $\phi$ and $d$. SVT tracks are then used to select events compatible with a secondary vertex.

To achieve high speed (20 $\mu$s total processing time, including 10 $\mu$s for level-2 decision logic) and provide flexibility, the SVT has adopted the basic architecture of separating the pattern recognition and track fitting into two pipelined stages. The pattern recognition is made using Associative Memory (AM) [27], a highly parallel template matching system. A coarse resolution can be used at this stage to reduce the memory size and to improve tracking efficiency. The roads found by the AM system are passed via a hit buffer to a track fitting stage, where the full resolution of the silicon detector is exploited.

The principle of AM template matching pattern recognition is to store legitimate hit patterns of all tracks of interest in a memory. Hits from an event are compared to these pre-stored patterns. If most hits of a pattern are present, a track candidate (road) is found. Hence, pattern recognition is complete as soon as the last hit of an event is read. The roads are sequentially sent with limited accuracy to the track fitting stage. A full custom VLSI chip with 0.7 $\mu$m technology has been fabricated and tested to 40 MHz; it foresees 5 silicon layers, and the COT information.

The Track Fitter, however, uses the full hit information to accurately evaluate the track parameters $\vec{p} = (p_T, \phi, d)$ by a fast linear approximation technique [26]. The Track Fitter has been built and tested. For a typical event of 5 roads and with 1.5 combinations per road the processing time is $\sim 2.8$ $\mu$s.

The pattern recognition and track fitting algorithms of SVT have been tested on actual data from CDF Run I using a bit-level SVT simulation program. The fitted tracks have almost off-line precision. The resolution of impact parameter is about 35 $\mu$m for tracks of 2 GeV/c. The resolution of $\phi$ is 1 mrad, and that of $p_T$ is

$0.003 p_T^2$ where $p_T$ is in GeV/c. To avoid a significant deterioration of the trigger rate and impact parameter resolution, however, the detectors must be aligned in $z$ to better than ≈100 μrad. Detail simulations show that a Level-2 trigger requiring two SVT tracks with impact parameter (>100μm) can reduce the Level-1 2-track trigger rate by a factor of $10^4$.

In summary, the silicon tracker developments that are being implemented in the CDFII detector will open a new ability to reliably detect rare physics processes at the Tevatron. Similar silicon tracker developments at D0, and other upgrades including the implementation of a central solenoidal magnetic field, will make the two detectors equally performant.

# EXPERIMENTS AT THE LARGE HADRON COLLIDER

## The LHC experimental environment

As shown already in Table 2, at the Large Hadron Collider a total 2835 bunches of protons will circulate in each of two rings that will be constructed in the CERN LEP tunnel [29]. The bunches will collide at four intersection regions with a very high luminosity, $\mathcal{L} = 10^{34}$ cm$^{-2}$s$^{-1}$. Several features of the LHC machine have enormous consequences for the detector designs:

- The bunch spacing of 25 ns means that interactions of one bunch crossing occur before all particles from interactions of a previous bunch crossing have traversed the detector. In order to prevent the pile-up of interactions over several bunch crossings, a fast detector signal response, small detector dead time, and extensive signal pipelining prior to initial trigger decisions is required. The level-1 trigger latency of the ATLAS detector (the time for trigger information to be read, a trigger decision to be made, and read out authorisation to be received at the front-end pipeline buffers) is 2.5 μs.

- To achieve the design luminosity, the colliding proton bunches are short (∼6 cm) and intense (∼1.05×10$^{11}$ protons per bunch). The total inelastic, non-diffractive cross-section at LHC energies is expected to be ∼80 mb, corresponding to an interaction rate of ∼ 10$^9$ Hz. Figure 14 shows the production cross-section for several Standard Model processes as a function of $\sqrt{s}$. Typically, in the case of the Higgs particle with mass $m_H = 500$ GeV, about 17 K events are expected per operating year (10$^7$s) at design LHC luminosity, compared to a total of $1.7 \times 10^{16}$ events from inelastic interactions. The LHC experiments must identify rare processes at this level.

- At the design luminosity, ∼23 interactions occur in each bunch crossing. This results in ∼ 10$^4$ tracks in the detector each 100 ns, the typical duration of a pulse in the detectors. The individual detector elements must therefore be highly granular in order to minimise the contribution of pile-up in a given detector cell.

- The high flux of particles from proton-proton interactions (and to a lesser extent from beam losses) places the detectors and associated electronics in a high-radiation environment. Only radiation resistant detectors and read-out electronics can be used. This is discussed below for the case of the ATLAS silicon tracker.

## What justifies the ATLAS or CMS experiments and their design?

Among the four experiments (ATLAS [8], CMS [9], LHC-B [30] and ALICE [31]) being constructed for operation at the LHC, the design philosphy has been determined by the physics goals of the experiments and the relevant available technology able to meet the experimental requirements. In the case of the LHC, the LHC-B and ALICE detectors are optimised to their specific roles - respectively a study of the $b$-quark physics sector, and the study of heavy ion collisions. The ATLAS and CMS detectors are large general purpose detectors, and differences in their design reflect choices of the collaborations within technologies able to meet the design requirements - sometimes for non-scientific reasons (for example a participating laboratory may have developed a given detector technique) and sometimes on the basis of scientific judgement.

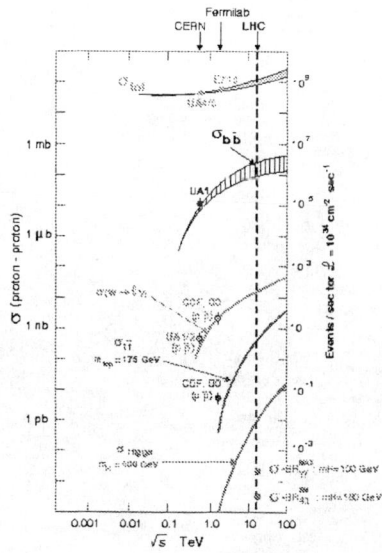

**FIGURE 14.** The expected total inelastic cross-section, and the production cross-section for typical Standard Model processes, as a function of $\sqrt{s}$.

Figure 15 shows a 3-dimensional view of the ATLAS detector, and Figure 16 shows the CMS detector. Superficially, the experiments resemble the CDF experiment. Both the CMS and ATLAS experiments have:

- Hermetic and highly granular electromagnetic and hadronic calorimeters for $e, \gamma$ and jet identification, and for the measurement of $E_\mathrm{T}^\mathrm{miss}$;

- Large, efficient and accurate tracking detectors able to operate at the highest luminosities inside a solenoidal magnetic field, for lepton measurement, $b$-quark tagging, enhanced $e/\gamma$ identification, $\tau$ lepton and heavy flavour vertexing and reconstruction capability;

- Large acceptance muon detectors outside the hadronic calorimeter;

- A low $p_T$ trigger capability with the possibility to read out large data volumes following a trigger.

What justifies the machine and detectors, given the important discoveries at CDF? The answer is clear, even on the basis of the selected physics results described above for CDF. A far more complete justification is given in Reference [32].

- As already described, the top quark has been identified and a few hundred $t\bar{t}$ events have been produced. The mass has been measured with an accuracy of $\sim \pm 5$ GeV/$c^2$, and the top pair production cross-section has been measured to be $\sim$6 pb (as expected from the Standard Model within experimental and

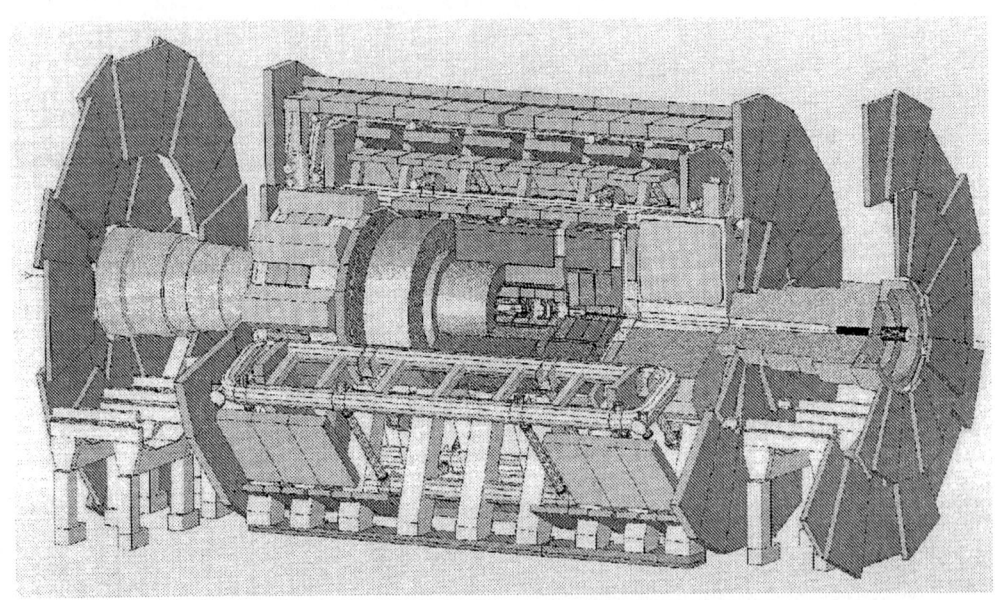

**FIGURE 15.** A 3-dimensional schematic of the ATLAS detector.

theoretical uncertainties). At the LHC, the pair production cross-section will be largely from the process $(gg \to t\bar{t})$ rather than $(q\bar{q} \to t\bar{t})$ and will be $\sigma_{t\bar{t}} \approx 800$ pb. Assuming an integrated luminosity of only $10^{40}\text{cm}^{-2}\text{yr}^{-1}$, a total of $\sim 8\times 10^9$ $t\bar{t}$ pairs will be produced. This will allow a measurement of the top mass with a precision of $\pm 1-2$ GeV/$c^2$, a determination of $|V_{tb}|$ to an accuracy of $\sim \pm 1\%$, limits on new physics such as resonant $t\bar{t}$ production, and limits or even the discovery of rare decays. What fun!

- The reach of LEP2 and the Tevatron experiments for the identification of the Standard Model Higgs particle has been discussed. Defining the discovery potential as the number of signal events divided by the statistical uncertainty of the evaluated background, Figure 17 shows that quantity for the ATLAS detector as a function of the Higgs mass in several decay channels. If the Higgs particle has a mass of $m_H < 1$ TeV/$c^2$, it will be identified at the LHC.

- Even at low luminosity, extensive $B$-physics studies are possible at both LHC-B and the general purpose detectors. The previously discussed CP-violation parameters $\sin(2\alpha)$ and $\sin(2\beta)$ can be measured to a precision of respectively $\pm 0.05$ and $\pm 0.01$, and the limits on forbidden or rare $B$-decays can be improved by an order of magnitude compared with CDF.

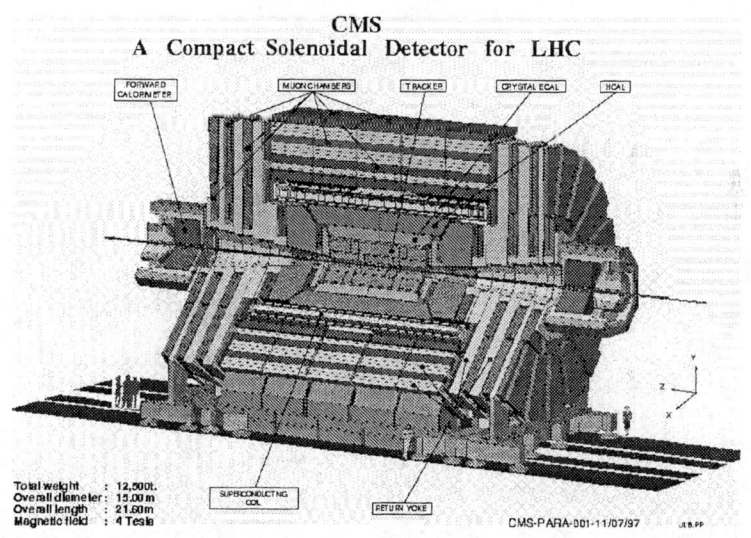

**FIGURE 16.** A 3-dimensional schematic of the CMS detector.

# A general description of the ATLAS detector

In order to meet the requirements imposed by the LHC as described above, the ATLAS detector has been designed with the following sub-system components. Details are available in the ATLAS Technical Proposal [8] and a number of Technical Design Reports referenced below.

## *The Inner Tracker*

A 6.8 m length Inner Tracker (ID) [33,35] surrounds the ATLAS interaction region of the CERN LHC. Its external active radius of 1.07 m is defined by the 5.3 m long superconducting coil of the 2 Tesla solenoidal magnet [34], whose inner cryostat radius is r = 1.15 m. A three dimensional view of the Inner Tracker is shown in Figure 18. Looking radially outwards from the interaction region, the ID consists in the central (forward) regions of:

- 3 silicon pixel layers (4 disks) in the radial range $4 < r < 20$ cm providing space points of typical granularity $\sim 50 \times 400$ $\mu m^2$;

- 4 silicon micro-strip layers (9 disks) in the range $26 < r < 56$ cm, with each layer or disk providing both an $r - \phi$ and a 40 mrad stereo measurement of granularity $\sim 80$ $\mu m \times 12.8$ cm;

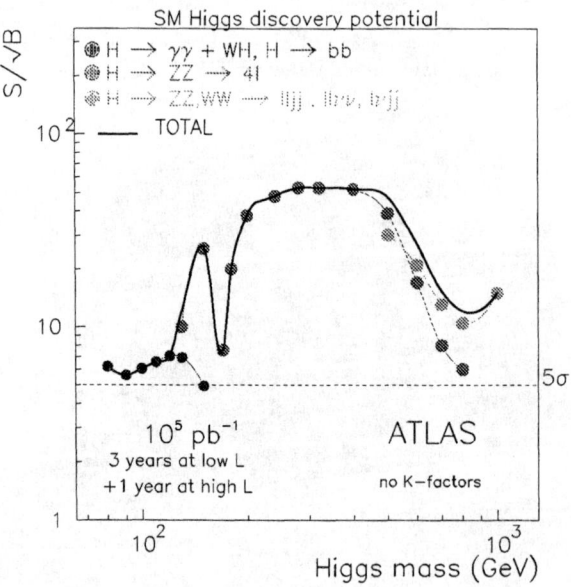

**FIGURE 17.** The Higgs particle discovery potential for ATLAS.

- A straw tube tracker (the TRT) of lower granularity which provides typically 36 points with $r - \phi$ precision $\sim 160$ $\mu$m, and transition radiation information for electron identification.

The main parameters of the Inner Tracker are shown in Table 4, and a simulated event for the process $(p\bar{p} \to H^0 + X; H^0 \to b\bar{b})$ as seen by the Inner Tracker at both high and low luminosity, is shown in Figure 19

The **Pixel Detector**, possibly the most innovative detector development of the ATLAS and the CMS experiments, must operate very close to the beam line, and will receive a radiation dose of up to $3.5 \times 10^{14}$ equivalent 1 MeV neutrons per cm$^{-2}$ per year. Throughout a 10 year operational period, it must maintain good pattern recognition capability, and good $b$-tag or $b$-trigger capabilities.

As described below, both the sensors and the electronics must remain operational in high radiation conditions. The sensor characteristics most sensitive to radiation are the effective doping concentration, the leakage current, and the charge collection efficiency. In particular, the increase in acceptor-like defects leads to an inversion of the conduction band from $n$ to $p$ and to an increase of the depletion voltage. For this reason, the baseline substrate uses n$^+$ readout implants with an n–type bulk, to enable partially depleted operation after radiation. Because of an increase in leakage current, the whole detector must be maintained at a low temperature ($<$-7°C). An enormous effort is continuing to ensure the detailed design of pixel sensors that maintain stable operation at high voltage.

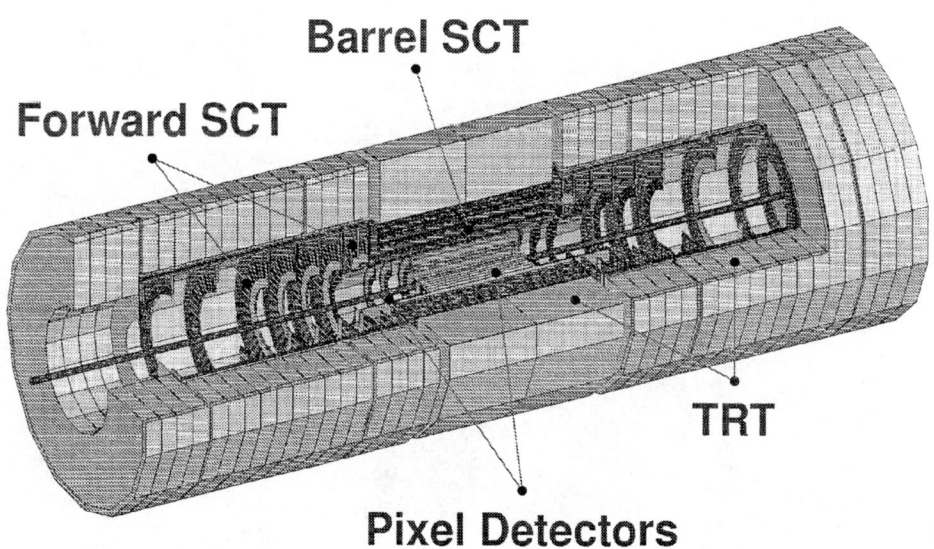

**FIGURE 18.** Three dimensional cut-away view of the ATLAS Inner Tracker.

In the case of ATLAS, the full detector will be composed of ∼1500 pixel barrel silicon modules and ∼1000 pixel forward silicon modules. Each module sensor is divided into individual readout elements of approximate granularity 50 $\mu$m×400$\mu$m;

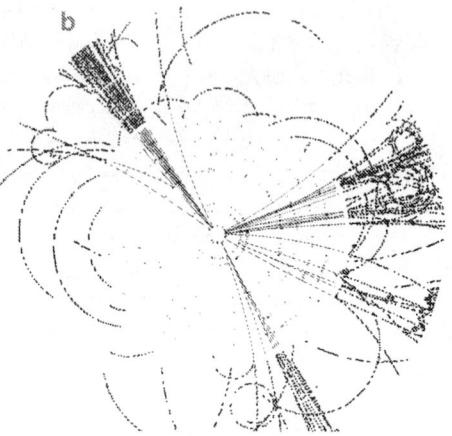

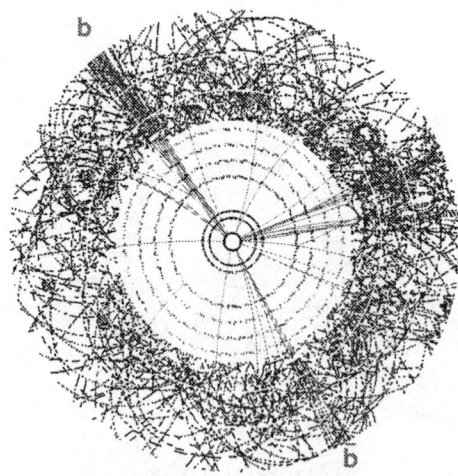

**FIGURE 19.** Display of a simulated $H^0 \to b\bar{b}$ with $m_H = 400$ GeV in the ATLAS barrel inner detector, at both low luminosity (left) and the full design luminosity (right).

| Position | Area (m$^2$) | Resolution $\sigma(\mu m)$ | Channels ($10^6$) | coverage ($\eta$) |
|---|---|---|---|---|
| Pixel Detector | | | | |
| 1 repl. barrel layer | 0.2 | $R\phi = 12, z = 66$ | 16 | ±2.5 |
| 2 barrel layers | 1.4 | $R\phi = 12, z = 66$ | 81 | ±1.7 |
| 4 ec disks/side | 0.7 | $R\phi = 12, R = 77$ | 43 | 1.7-2.5 |
| SCT Detector | | | | |
| 4 barrel layers | 34.4 | $R\phi = 16, z = 580$ | 3.2 | ±1.4 |
| 9 ec wheels/side | 26.7 | $R\phi = 16, R = 580$ | 3.0 | 1.4-2.5 |
| TRT Detector | | | | |
| axial barrel straws | | 170/straw | 0.1 | 0.7 |
| radial end-cap straws | | 170/straw | 0.32 | 0.7-2.5 |

**TABLE 4.** The main parameters of the Inner Tracker. The quoted resolutions are typical values (the actual resolution in each detector depends on $|\eta|$).

the full detector will have $\sim 1.5 \times 10^8$ individual elements. Given the very high track multiplicity in the region of the pixel detectors, the granularity of pixels represents an enormous advantage because of the small track occupancy. Because of the resulting small capacitive load on each channel of the front-end electronics, the noise remains low even after radiation damage; the pixel specification requires an rms noise of only 200 $e^-$, with a similar value for the threshold dispersion.

A key element of the development is the readout electronics. Following a level-1 trigger, all hit channels must be read out. To do this, a highly parallel column based readout with data compression is used. Integrated front-end electronics circuits are bump-bonded onto the silicon substrate of each sensor. A special 'end of colum' (EOC) circuit collects 24 columns of 160 rows from attached front-end circuits, and a Module Control chip (MCC) collects the data from the EOC chips and transfers the data to the end of the detector. The MCC circuits also control the clock and trigger signal distribution. The full electronics chain must be designed in radiation hard technology, and must balance the conflicting requirements of short signal shaping, low noise and low power. Prototype modules for the detector have been built and tested in a beam, using prototype electronics fabricated in a standard CMOS process, with encouraging performance. Intensive development is continuing on the design and fabrication of radiation–hard electronics using the DMILL technology [36]. A prototype module is shown in Figure 20.

An extremely challenging aspect is the mechanical support for the pixel detector. Despite the low operating temperature and high power consumption, the detector thickness must be minimised to ensure good impact resolution, and precision must be maintained following temperature cycling. A carbon structure using integrated cooling tubes, and having a coefficient of thermal expansion close to 0 ppm, is foreseen.

The **Silicon micro-strip Tracker (SCT)** system is an order of magnitude larger in surface area than the SVXII tracker of CDF and in addition it must face radiation levels changing the silicon substrate bulk from $n$ to $p$ type as for the pixels. Although the operational environment encountered by the pixel detector exists at a less extreme level, the scale of the project is overwhelming. In particular, costly 'hi-tech' solutions that are being developed for the pixel detector may not be appropriate to the SCT for economic reasons. The design and development status of the SCT, as well as the expected performance of the pixel+SCT system will be briefly described in a following section.

The **Transition Radiation Tracker** is based on the use of straw tube detectors, which can operate at very high rate because of their small diameter and the isolation of the sense wires within individual gas envelopes. Electron identification capability is added by employing Xe gas to detect transition-radiation photons created in a radiator between the straws. Each straw is 4 mm in diameter, giving a fast response and good mechanical properties for a maximum straw length of 150 cm. The barrel contains $\sim$50,000 straws, divided in two at the centre to reduce the occupancy and read out at each end. The end-caps contain 320,000 radial straws with the read-out at the outer radius. The total number of electronic channels is $\sim$420,000. Each channel provides a drift-time measurement, giving a spatial resolution of 170 $\mu$m per straw, and two independent thresholds. These allow the detector to discriminate between tracking hits, which pass the lower threshold, and transition radiation hits, which pass the higher threshold.

**FIGURE 20.** A prototype module for the pixel detector.

## Calorimetry

The calorimeters [37–39] are crucial to measure the energy and direction of photons, electrons, isolated hadrons and jets, as well as the missing transverse energy $E_T^{miss}$; they are the leading detectors for the triggering and measurement of many important physics channels. As already noted, a fast detector response (<50 ns) and a fine granularity are required to minimise the impact of the pile-up on the physics performance. High radiation resistance is also needed, given the high particle fluxes expected.

The ATLAS calorimetry covers the range $|\eta| < 5$ using techniques most suited to the requirements and radiation environment. A view of the ATLAS calorimetry is shown in Figure 21. The rapidity coverage and calorimeter granularity is summarised in Table 5. The EM calorimeter system is contained in a cylinder of outer radius 2.25m and a length of 6.65 m along the beam axis. The barrel hadronic calorimeter system has an outer radius of 4.23 m and a total length of ~12 m. The EndCap EM, Hadronic and Forward calorimeters are housed in the same cryostat.

| System | $\eta$ coverage | Granularity ($\Delta\eta \times \Delta\phi$) |
|---|---|---|
| EM barrel | $|\eta| < 1.475$ | $0.003 \times 0.1$ (s1) |
| | | $0.025 \times 0.025$ (s2) |
| | | $0.05 \times 0.025$ (s3) |
| Presampler | $|\eta| < 1.8$ | $0.025 \times 0.1$ |
| Hadronic barrel | $|\eta| < 1.8$ | $0.1 \times 0.1$ |
| Hadronic EC | $1.5 < |\eta| < 2.5$ | $0.1 \times 0.1$ |
| | $2.5 < |\eta| < 3.2$ | $0.2 \times 0.2$ |
| FWD Calo | $3.2 < |\eta| < 4.9$ | $0.2 \times 0.2$ |

**TABLE 5.** The ATLAS calorimeter system.

The **electromagnetic calorimeter** is a Lead Liquid-Argon (LAr) detector with accordion geometry. The principle of the calorimeter is shown in Figure 22. For $|\eta| < 1.8$ it is preceded by a presampler detector installed immediately behind the cryostat cold wall, and used to correct for the energy lost in the material (inner detector, cryostats, coil) upstream the calorimeter. The total thickness of $\sim 25\ X_0$ in the barrel and 26 $X_0$ in the end-caps allows full containment of the electromagnetic showers. There are a total of $\sim 200000$ readout channels.

Full size detector modules have been constructed and tested in a beam. The electron energy resolution has a sampling term $(9.90\text{-}10.4)\%/\sqrt{E}$, a local constant term $(0.27\text{-}0.35)\%$ and a noise term of $(280\text{-}520)$ MeV over the full $\eta$ range. The design constant term for the full detector (which dominates at high energies) is 0.7% which puts stringent requirements on the detector construction, the dead material in front and the calibration precision.

The high EM granularity allows powerful $e/\gamma$ identification and the rejection of jet backgrounds. Thin strips in the first sampling allow $\pi^0$ rejection of $> 3\sigma$ at

50 GeV $E_T$, and a total jet rejection for $E_T >20$ GeV of $\sim 5000$. The narrow

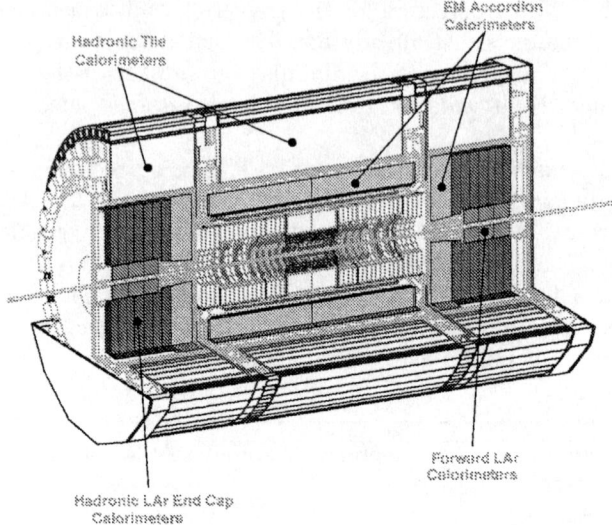

**FIGURE 21.** Three dimensional cut-away view of the ATLAS calorimetry.

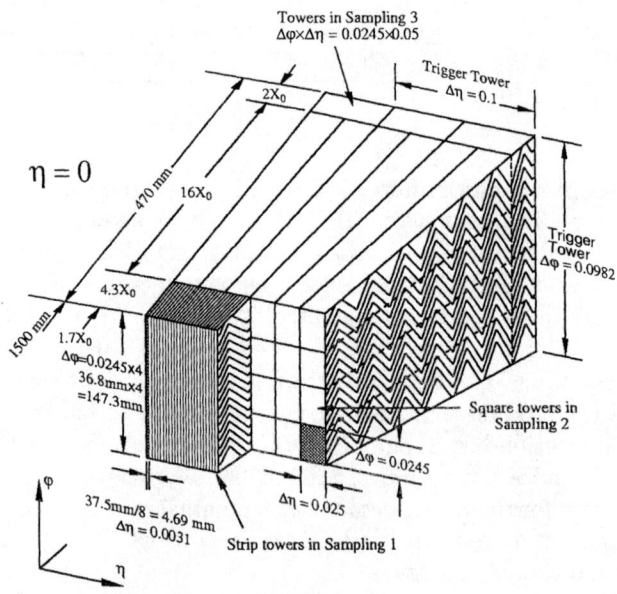

**FIGURE 22.** A drawing of the accordian structure of the electromagnetic calorimeter.

strips also contribute to the photon angular measurement in the $\eta$ direction, with an accuracy of about 50 mrad/$\sqrt{E}$.

The **hadronic tile calorimeter** covers the range $|\eta| < 1.6$. The cylindrical hadronic barrel calorimeter has an inner radius of 2.28 m and an outer radius of 4.23 m. It is divided into a central and two extended barrels. It is based on a sampling technique with plastic scintillator plates (tiles) embedded in an iron absorber matrix and read out by wave length shifting fibres. The tiles are placed in a plane perpendicular to the beam axis and staggered in depth, simplifying the mechanical construction and the fibre routing. The calorimeter is segmented in three layers, respectively 1.4, 4.0 and 1.8 $\lambda$ thick at $\eta = 0$. Azimuthally, the barrel and extended barrels are divided into 64 modules, and in $\eta$, the read-out cells are formed by grouping fibres to a photomultiplier pseudo-projectively. The calorimeter is placed behind the EM calorimeter ($\sim 1.2\,\lambda$) and the solenoid coil, making a total active calorimeter thickness (EM + Tile) of 9.2 $\lambda$ at $\eta = 0$ and a total amount of material in front of the muon system of 11 $\lambda$ at $\eta = 0$.

The required hadronic energy resolution is driven by the measurement of jet energies. From test beam runs with both EM and hadronic prototypes, the pion resolution is measured to be
$\frac{\Delta E}{E} = \frac{(38.3 \pm 4.6)\%}{\sqrt{E}} + (1.62 \pm 0.29)\% \oplus \frac{(3.06 \pm 0.18)}{E}$,
well within the specifications for ATLAS of
$\frac{\Delta E}{E} = \frac{50\%}{\sqrt{E}} + 3\%$.

In the range $1.5 < |\eta| < 4.9$ a **Liquid Argon Hadronic calorimeter** takes over because of the higher radiation: the end-cap hadronic calorimeter extends to $|\eta| = 3.2$ and the range $3.2 < |\eta| < 4.9$ is covered by the high-density forward calorimeter. Both calorimeters are integrated in a single cryostat housing with the EM end-caps. Each hadronic end-cap calorimeter consists of two independent wheels, of the same diameter. The wheels have 25 mm (50 mm) copper plates, separated by a gap between plates of 8.5 mm; the gaps are equipped with 3 electrodes, making 4 drift spaces of $\sim$1.8 mm. The first wheel is divided in two longitudinal read-out segments (of 8 and 16 layers respectively in depth), while the second wheel has one segment of 16 layers. The read-out cells are fully projective in $\phi$ and pseudo-projective in $\eta$. The thickness of the active part of the end-cap calorimeter is $\sim$12 $\lambda$.

The **extreme forward calorimeter** is integrated in the end-cap cryostat, with the front face at about 5 meters from the interaction point. It is important for a measurement of forward jets and of $E_T^{\text{miss}}$, but is particularly challenging due to the high level of radiation. Because of space constraints, the detector is extremely dense. It has three longitudinal sections, the first made of copper, and the other two of tungsten. Each consists of a metal matrix with regularly spaced longitudinal channels filled with rods. Liquid Argon fills the gap between the rod and matrix; The gaps are 250, 375 and 500 $\mu$m in succesive sections.

## The Muon Spectrometer

Final-state muons [40–42] provide a robust signature for many key physics issues. The design has therefore been optimised using benchmark processes such as the Higgs, and (in particular for the trigger) also at lower $p_T$ because of the interest for $b$-quark tagging, and CP violation studies. ATLAS has designed a high-resolution muon spectrometer with stand-alone triggering and momentum measurement capability over a wide range of transverse momentum, pseudorapidity, and azimuthal angle. A view of the muon spectrometer is shown in Figure 23.

The muon spectrometer exploits the magnetic deflection of muon tracks in a system of three large superconducting air-core toroid magnets (one barrel and two end-caps) instrumented with separate-function trigger and high-precision tracking chambers. In the pseudorapidity range $|\eta| < 1$, a large barrel magnet consists of eight coils surrounding the hadron calorimeter. For $1.4 < |\eta| < 2.7$, muon tracks are bent in smaller end-cap magnets inserted each end of the barrel. In the interval $1.0 < |\eta| < 1.4$, referred to as transition region, deflection is provided by a combination of barrel and end-cap magnetic fields. The field is largely orthogonal to the muon trajectories, minimizing the degradation of resolution from multiple scattering.

Different contributions to the barrel momentum resolution are shown in Figure 24. The resolution (typical for an open geometry) is limited by energy loss fluctuations at low momenta and by detector resolution at high momenta; the multiple scattering effect is approximately momentum-independent. The momentum resolution is 2-3% over most of the kinematic range excepting very high momenta where it increases to $\sim$10% at $p_T = 1$ TeV. The dependence on $|\eta|$ is shown in Fig-

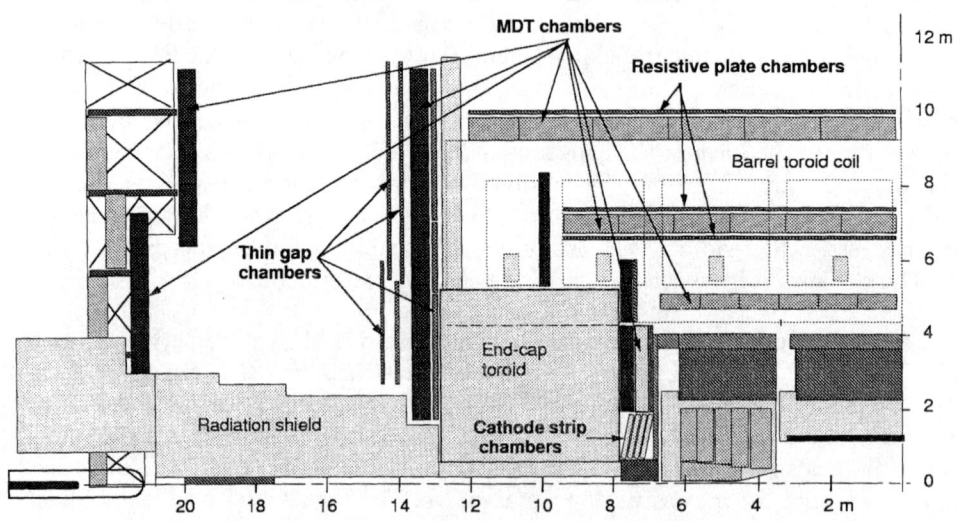

**FIGURE 23.** Layout of the ATLAS muon spectrometer.

ure 22 for $p_T = 100$ GeV/c muons; the resolution is uniform except at obstructions from the barrel magnet elements and an enhancement around $|\eta| = 1.5$, due to a degraded bending power.

Over most of the pseudorapidity range, a precision measurement of the track coordinates in the principal bending direction of the magnetic field is provided by **Monitored Drift Tubes (MDT)**. They are aluminium tubes of 30 mm diameter and 400 $\mu$m wall thickness, with a 50 $\mu$m diameter central W-Re wire. The tubes are operated with a non-flammable $ArCH_4N_2$ mixture at 3 bar pressure. The envisaged working point provides for a highly linear space-time relation with a maximum drift time of $\sim 500$ ns, a small Lorentz angle, and good ageing properties due to small gas amplification. The single-wire resolution is typically 80$\mu$m. To improve the resolution of a chamber beyond the single-wire limit and to achieve adequate redundancy for pattern recognition, the MDT chambers are constructed from $2 \times 4$ monolayers of drift tubes for the inner and $2 \times 3$ monolayers for the middle and outer stations. Full size prototype modules have been built and tested in the beam, showing good performance within the specifications for ATLAS.

**Cathode strip chambers (CSC)** are used in the first station of the end-cap region and for pseudorapidities $|\eta| > 2$, to provide the finer granularity that is required to cope with the demanding rate and background conditions. They are multiwire proportional chambers with cathode strip read-out. The precision coordinate is obtained by measuring the charge induced on the segmented cathode by the avalanche formed on the anode wire. With an anode wire pitch of 2.54 mm and a cathode read-out pitch of 5.08 mm, r.m.s. resolutions of better than 60 $\mu$m have been measured in prototypes. These detectors also provide good time resolution

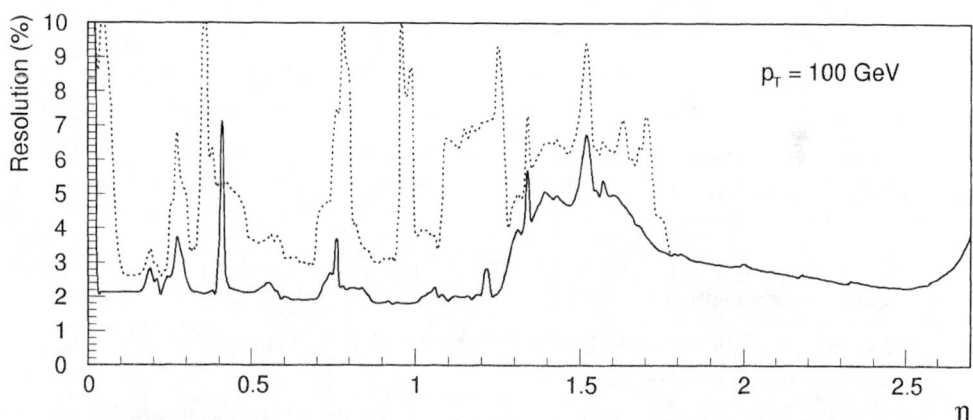

**FIGURE 24.** Momentum resolution for $p_T = 100$ GeV/c as a function of $\eta$ averaged over all azimuth angles. The solid curve applies to a standard sector; the dotted curve corresponds to one of the bottom sectors, where barrel toroid coils are captured inside the support structure for the inner parts of the detector.

(7 ns), good two-track resolution, and low neutron sensitivity.

The **trigger chambers** for the ATLAS muon spectrometer serve a threefold purpose: bunch crossing identification requiring a time resolution better than the LHC bunch spacing of 25 ns, a trigger of well defined $p_T$ implying a granularity of $\sim 1$ cm and a measurement of the second coordinate in a direction orthogonal to that measured in the precision chambers with a typical resolution of 5-10 mm. The system employs two different types of detectors, Resistive Plate Chambers (RPC) in the barrel and Thin Gap Chambers (TGC) in the end-cap region. They cover an area of $\sim 3650$ m$^2$ in the barrel and 2900 m$^2$ in the end-cap region, each chamber containing at least two detector layers. The number of channels is $\sim 350,000$ in the barrel and $\sim 440,000$ in the end-caps.

## Design and development of the SCT - a case study.

The layout of the silicon tracker (pixels and SCT) [33] is shown in Figure 25. The layout has been designed to operate according to the following performance criteria during a 10 year period for the pattern recognition and track precision of charged hadrons and leptons:

- The track reconstruction efficiency of isolated $\pi^{\pm}$ tracks with transverse momentum $p_T > 5$ GeV/c should be $\epsilon > 95\%$ over the full acceptance $|\eta| < 2.5$, even in the absence of some layer, and with a fake rate <1%;

- Tracks with $p_T > 1$ GeV/c should be reconstructed with $\epsilon > 90\%$ and with a fake rate of <10% within a $(\eta - \phi < 0.25)$ cone around a high $p_T$ track;

- The track reconstruction efficiency during operation at low luminosity should be $\epsilon > 95\%$ for all tracks with $p_T > 0.5$ GeV/c;

- A 2-track resolution of <200 $\mu$m at a radius of 30 cm should be maintained to limit track losses in $b$-jets to <5%.

Primary electron track reconstruction will be affected primarily by the detector material, which is not directly specified as an upper limit. The aim of the detector is to:

- Reconstruct high $p_T$ electron tracks ($> 7$ GeV/c), and low $p_T$ ($> 0.5$ GeV/c) tracks in the vicinity of the high $p_T$ track with an efficiency of $\epsilon > 90\%$;

- Reconstruct all electrons of $p_T > 1$ GeV/c with an efficiency of $\epsilon > 70\%$ during low luminosity operation.

Of particular importance is the measurement precision of found tracks, and of primary vertices or secondary vertices from $b$-decays. Listed specifications require:

- Including the TRT, a momentum resolution at $p_T = 500$ GeV/c of $\sigma(1/p_T) < 0.3/p_T$ for $|\eta| < 2.0$ and <0.5 for $|\eta| < 2.5$ is required (there is no listed requirement at lower $p_T$ where multiple scattering is important);

- There should be an adequate impact parameter precision, and a resolution of primary vertices with at least 4 tracks of $\sigma(z) < 1$ mm;

- The reconstruction of secondary vertices, with an efficiency for tagging $b$-jets of $\epsilon > 0.5$, for a non-$b$ hadronic rejection of at least 50 is required.

To fulfill these criteria, the current SCT design has 4 barrel layers of double sided modules in the radial range $30 < r < 52$ cm, with 9 disks in each of the forward and backward directions over the radial range $26 < r < 56$ cm. In the barrel, each module consists of 4 single-sided silicon detectors with an active area $61.6 \times 62.0$ mm$^2$. On each side of the double sided module, 768 strips of pitch 80 $\mu$m and length 12.32 cm and aligned axially or at 40 mrad to the axial direction. The module construction is similar in the forward direction. The silicon strips will be read out by binary front-end electronics. The expanded view of an ATLAS SCT barrel module is shown in Figure 26.

Key issues that have required development include:

- The development of silicon sensors able to survive 10 years of operation in a high radiation environment;

- The development of front-end and readout electronics able to survive during this period;

- The development of cooling and mechanical structures able to support the silicon modules with extreme accuracy, and at the low operating temperatures previously mentioned.

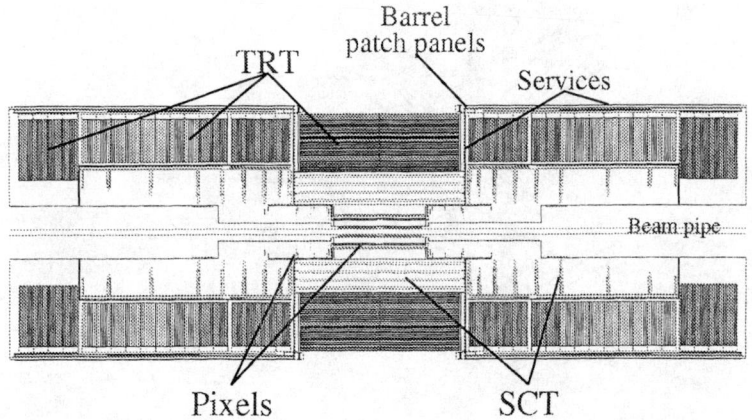

**FIGURE 25.** Layout of the ATLAS silicon tracker.

## Detectors

The total hadron fluence traversing the Pixel+SCT detector has been estimated after an integrated luminosity of 3 years at $10^{40} \mathrm{cm}^{-2} \mathrm{yr}^{-1}$ and 7 years at $10^{41} \mathrm{cm}^{-2} \mathrm{yr}^{-1}$ (the 'standard' 10 year operating period). Normalised to the damage of 1 MeV neutrons in silicon, the integrated fluence reaches $1.4 \times 10^{14} \mathrm{n}_{eq}\ \mathrm{cm}^{-2}$ in the silicon strip detectors.

Both bulk and surface damage occurs during particle irradiation. Surface damage results from the accumulation of charge at the silicon-oxide interface and can be controlled by careful design. Bulk damage follows from the displacement of silicon atoms from the lattice sites, with consequences including:

- The decrease of carrier mobility;

- The reduction of charge collection;

- The increase of leakage current which is directly proportional to the fluence ($\Phi$): $\Delta I_L / V = \alpha \Phi$. This proportionality, $\alpha$, is called the damage constant. $\Delta I_L$ is the change of the leakage current normalised to the depletion volume;

- The change of effective dopant concentration which is directly related to the full depletion voltage. During the irradiation of a n-type substrate material, its effective dopant concentration decreases until the so-called type inversion of the detector bulk. Beyond this inversion, the majority of dopant concentration is of p-type material.

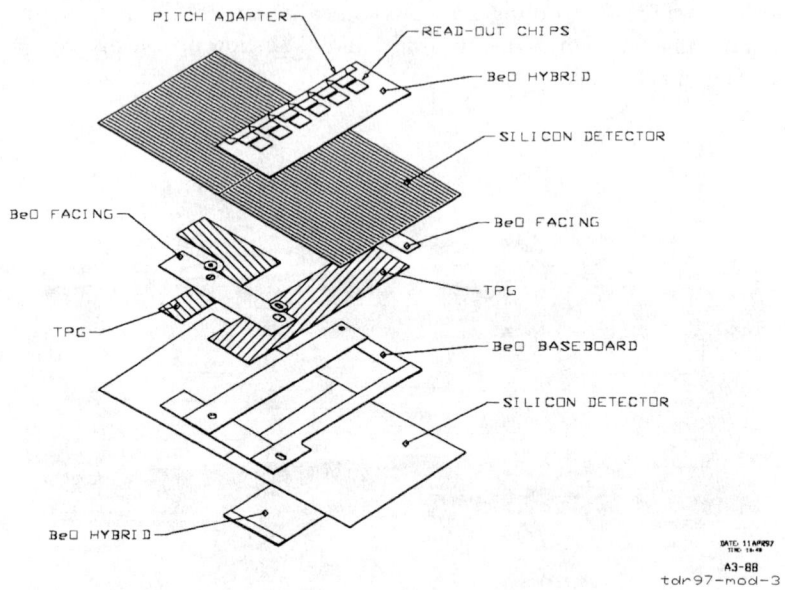

**FIGURE 26.** Expanded view of an ATLAS barrel SCT module.

The removal of donor atoms and the creation of acceptors can be expressed as
$N_{eff} = N_0 \times e^{-c\Phi} + \beta\Phi$
where $N_0$ is the initial donor concentration and $\beta$ and c are constants. Following irradiation, the effective dopant concentration will continue to evolve, depending on the storage temperature. This dependence has been studied intensively by AT-LAS [43] and others. One parametrisation of this evolution is [44]:
$V_d = V_Z + V_S \times \exp(-t/\tau_S) + V_A \times (1-\exp(-t/\tau_L))$
with
$\tau_S[\text{days}] = 70 \times \exp(-0.175\,T)$ and
$\tau_L[\text{days}] = 9140 \times \exp(-0.152\,T)$,
where $T$ is the temperature in [°C]. The time constants $\tau_S$ and $\tau_L$ represent short-term beneficial annealing and the long-term reverse annealing. The constants $V_Z$, $V_S$, $V_A$ are proportional to the fluence and related to the induced damage. The constant $V_Z$ is the fraction of the radiation produced acceptor concentration and gives approximately the minimum value between the beneficial and reverse annealing. The constant $V_S$ is the meta-stable acceptor concentration produced during the irradiation process and $V_A$ is the concentration of damage sites that can become activated acceptor sites due to the anti-annealing. The expected sensor depletion voltage $V_D$ over a 10 year period is then typically as shown in Figure 27.

charge carriers;

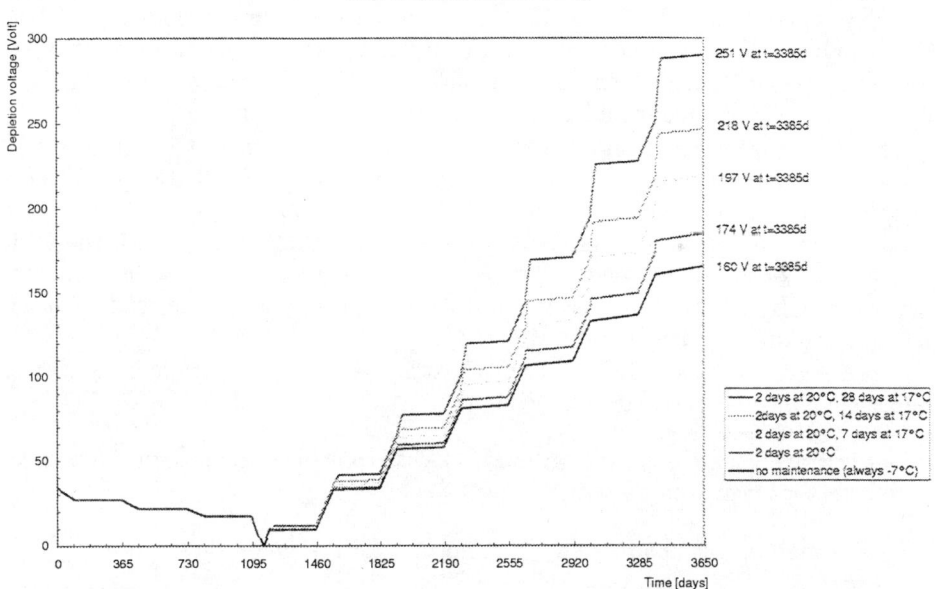

**FIGURE 27.** Expected sensor depletion voltage $V_D$ over a 10 year period assuming an integrated fluence pf $1.4\times10^{14} n_{eq}/\text{cm}^2$ and an operating temperature of -7°C, and different access scenarios. The sensor thickness is assumed to be 300 $\mu$m.

it is proportional to the fluence, and has a temperature dependence given by the Boltzmann equation. To ensure reasonable long-term performance, the detectors will be operated at $-7°C$. The internal radius of the SCT detectors is defined by the required detector signal-to-noise performance over the operating lifetime of the detector.

The original base line sensor design for ATLAS was to use $n^+$-implant strips on an n-type substrate; after inversion, the depletion region starts from the strip side for the $n^+$-strip detectors, while for $p^+$-strip detectors, the depletion starts from the backplane. As for the case of the pixel detectors, this was because of fears that sensors could not be designed to operate reliably at high operating voltages (300-500 Volts).

As a result of intensive efforts by ATLAS and CMS, several manufacturers are now able to supply $p^+$-on-n sensors satisfying the required ATLAS voltage specifications, and for economic reasons, the baseline strip implantation has been changed.

## *Readout Electronics*

The SCT readout electronics is responsible for supplying the strip hit information to the ATLAS level-2 trigger and the ATLAS data acquisition. The front-end preamplifiers and pipeline buffers need to be on the silicon modules to ensure low noise operation, and this implies the use of radiation hard technologies. Other issues, in particular the bunch crossing rate and the track occupancy, force stringent specifications on the signal speed, both the coherent and random noise level, and the signal recovery time including the double pulse resolution.

The chosen front-end readout electronics is based on a one-bit digital (binary) readout architecture. Each channel of a 128-channel chip is bonded to a $\sim 12.4$ cm length strip. The detector design has a detector capacitance of $\sim 1.5$ pF/cm and a total of $\sim 19.2$ pF for an unirradiated detector; for an irradiated detector the value is $\sim 15.5$ pF. Hits above some adjustable threshold recorded by a discriminator are stored in a pipeline. On receipt of a level-1 trigger, the sparse readout of hit patterns from the bunch crossing is made.

The integrated circuits (IC) readout chips are set on electronics hybrids. Those hybrids have two functionalities:

- Electrical: for the power, ground, clocks, control, output data, chip links, detector biasing and filtering;

- Thermal: a high conductivity is important for the power dissipation.

Data links transmit the data off the detector. Other links provide timing and control signals to each module. An optical communication scheme using the VCSEL (Vertical Cavity Surface Emitting Laser) technique is chosen because of its low mass, the lack of electrical pick-up, and an intrinsic component radiation tolerance.

| Parameter | Specification |
|---|---|
| Noise | $\leq 1500$ e$^-$ |
| # Readout channels | 12×128 |
| Efficiency | 99% |
| Noise occupancy | $5 \times 10^{-4}$ |
| Double pulse resolution | 50ns for 3.5fC |
| Large charge recovery (80fC) | 1$\mu$s for 3.5fC |
| Power dissipation | $\leq 3.8$ mW/channel |

**TABLE 6.** Summary of some chip parameters, prior to irradiation.

As noted above, the front-end architecture is based on the amplification, shaping and discrimination of a silicon sensor signal to give binary hit information for each bunch crossing. The following functions must be provided:

- The detection of hits above some noise level for a given beam crossing and the readout of data from 3 bunch crossings centred on the level-1 trigger;

- The storage of hit patterns for the period of the level-1 latency;

- Sparse, data driven readout following a level-1 trigger signal;

- Sundry operations of calibration, synchronisation and redundancy.

The efficiency and occupancy of the full system relies on the signal-to-noise (S:N) ratio provided by the front-end preamplifiers. A target S:N ratio of 12 is expected at the end of the detector lifetime. The behaviour of S:N with the particle fluence has been simulated taking both electronics and detector effects into account: in particular, the electrical parameters of the strip detectors such as resistance and inter-strip capacitance, the diminution of the charge collection efficiency and the ballistic deficit due to charge collection time, and the increase of the detector leakage current.

The detector specification is shown in Table 6. Based on an extensive development program, two 128-channel chip sets exist. In the first, the total functionality resides in a single chip (ABCD) based on the DMILL BiCMOS technology [36]. The second option has the analog part on a single chip (CAFE-M) where the preamplifier, the shaper and the discriminator are realised in a bipolar technology. The digital part, the ABC pipeline, is realised in this scheme as a separate chip in a radiation hard CMOS technology.

Development work on the front-end chip is nearing completion. Figure 28 shows a prototype ATLAS SCT barrel module, and Figure 29 shows the performance of that module. While the performance specifications are close to being met, and studies following irradiation appear promising, the power consumption specifications of both chips has not been achieved (with consequences for the mechanics and cooling). In addition, major system studies must still be demonstrated.

## Mechanics

A total of 2112 barrel and 1976 forward SCT modules are to be mounted on the suppport structure. To maintain the modules at a temperature of $<$-7°C imposes severe constraints on the design of the barrel and forward supports: an excellent dimensional stability over a wide temperature range (-15°C to 25°C) is required, and an efficient evacuation of heat generated by the silicon detectors and their associated electronics and cable plant is needed. The structure must therefore use materials of the lowest possible coefficient of thermal expansion (CTE), and this is achieved using a carbon fibre skin and honeycomb core composite which is tuned to have a low CTE. In the case of the barrel, carbon fibre flanges interlink the 4 barrels, and precision carbon fibre brackets that are glued to the barrel will support individual modules and the (flexible) cooling tubes. A prototype cylinder (Figure 30) has been shown to largely fulfill the ATLAS specifications.

Individual module elements, if correctly constructed, provide a hit reconstruction precision of $\sigma(r - \phi) \sim 18\mu$m. To maintain the design momentum resolution, the four detectors of a module must be known in position relative to each other with a precision of $< 5\mu$m in lateral strip position, $< 25\mu$m in $z$ and $< 25\mu$m in module thickness. After mounting on the structure, the alignment of individual strips must be known with a precision of $\sigma(r - \phi) < 12\mu$m, $\sigma(r) < 50\mu$m and $\sigma(z) < 50\mu$m. This precision can be obtained by a survey using track residuals and

**FIGURE 28.** Photograph of a prototype ATLAS barrel module, showing the front-end ABCD chips, the hybrid, and the detectors.

X-rays. The placement accuracy is rather dominated by the required accuracy for trigger purposes; for this the $r - \phi$ position of any strip must be known without survey or stereo information with a precision $< 200\mu$m.

## Performance of the ATLAS Pixel+SCT tracker

The Pixel and SCT (P+SCT) as described above have on average at least 7 precision space points per track over the full range $|\eta| < 2.5$. It is of interest to understand the stand-alone pattern recognition and track-fitting capability of the P+SCT, its robustness in the case of detector inefficiencies, uncorrelated noise, etc., and the level to which performance specifications are met.

Contrary to (for example) CDF, the use of large semiconductor–based tracking detectors results in a substantial material budget, adversely affecting (due to multiple scattering, bremsstrahlung and photon conversions) both the precision and efficiency of reconstructed tracks. Preceding the outer sensitive region of the P+SCT, the total material reaches almost 0.4 $X_0$, resulting in frequent shower initiation. The material results mainly from electrical and cooling services.

A full simulation of the ATLAS Inner Detector [33] has been made using GEANT [45], including best estimates of the detector, support and service ma-

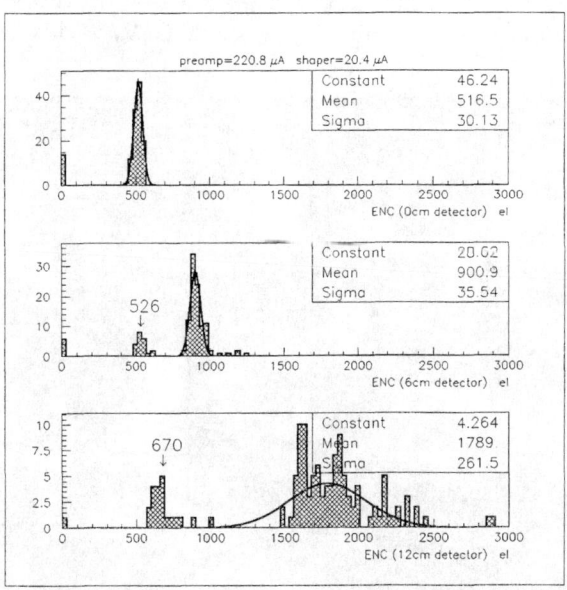

**FIGURE 29.** Performance of the prototype module of Figure 28. Noise (equivalent electrons) with successively no attached detector, a 6 cm strip detector, and a 12 cm strip length. About 20 channels were not connected. A minimum ionising particle signal corresponds to ~22000 electrons.

terial, and parametrisations of detector signal, noise, and resolution from test beam data and design predictions. Studies of the detector performance were then made using isolated tracks ($p$, $\pi^{\pm}$, $\mu^{\pm}$, $e^{\pm}$), and non-isolated tracks generated using PYTHIA [46] for the process $H^0 \to b\bar{b}$ with $m_H = 400$ GeV.

Detailed studies [33,32] have demonstrated the excellent performance of the full Inner Detector. In particular, Figure 31 shows the analytic evaluation of the momentum resolution, including the TRT and a beam constraint of 15 $\mu$m. Beyond $|\eta|=2$, the resolution deteriorates due to the limited tracker length and field non–uniformity. At lower $p_T$, the resolution is dominated by muitple scattering, and at $p_T = 20$ GeV/c, $\sigma(1/p_T)$ varies between 0.8 and 1.2 TeV$^{-1}$. The stand-alone P+SCT deterioration of momentum resolution is a factor $\approx 2$, within specification. The main problem results from the effect of material, and we concentrate only on that point here.

The pattern recognition and measurement precision have been studied [47] for isolated tracks using the iPatRec [48] program which reconstructs P+SCT tracks in a combinatorial search of precision hits. The iPatRec selection requires at least 7 precision hits (from 3 space points in the pixel, $\sim$8 strips in the SCT) with very loose quality cuts on each hit, and at most one missing layer between successive hits.

Using isolated tracks, the direct consequence of material is evident by the number of tracks recording no hit in the TRT. For $p_T = 1$ GeV/c, 10% (30%) of $\pi^{\pm}$ ($e^{\pm}$) tracks have no confirmed TRT hit; for $p_T = 20$ GeV/c, this is reduced to 5%

**FIGURE 30.** Photograph of a prototype barrel for the ATLAS SCT support.

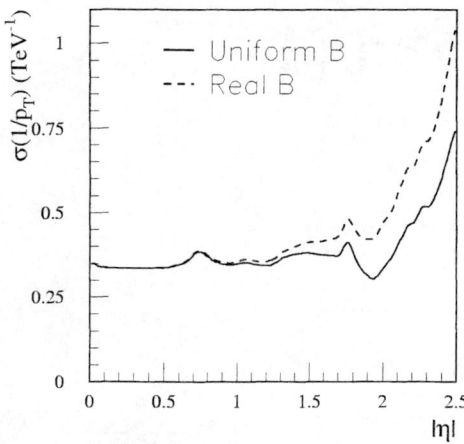

**FIGURE 31.** *Momentum resolution $\pi^+$ tracks of $p_T = 500$ GeV/c as a function of $|\eta|$, for a uniform magnetic field of 2 Tesla, and the actual solenoidal field.*

(14%). The $|\eta|$ dependence of the track efficiency is shown in Figure 32 for isolated $\pi^\pm$ or $e^\pm$. At $p_T = 20$ GeV/c, $\epsilon_{el} \approx 0.83$ in the range $1.4 < |\eta| < 2.2$. Despite the relaxed track selection criteria, the fake rate remains satisfactory ($< 10^{-5}$). Because of material effects, the value of $\epsilon_{rec}$ does not meet specifications, but it remains robust to detector deterioration in the case of isolated tracks, and to the pileup of superimposed minimum bias events.

The impact parameter $d_0$ for $\pi^\pm$ tracks is dependent on the $b$-layer performance; it is not strongly $\eta$ dependent (see below), but is dominated at low $p_T$ by multiple scattering in the pixel layers. The expected performance is:

$$\sigma_T(d_0) = 11 \cdot (1 \oplus \tfrac{5.5 \sin\theta}{p_T})^{1/2} \ \mu m$$

with $p_T$ measured in units of GeV/c.

The specification of track measurement precision assumes Gaussian uncertainties. Again because of material, this is not the case for electrons. While the effects of bremsstrahlung can be corrected on average given a knowledge of the bremsstrahlung position, the measurement accuracy cannot be improved; here we indicate the performance before bremsstrahlung recovery. Figure 33 compares the statistical momentum pulls for $e^\pm$ and $\pi^\pm$ of $p_T = 20$ GeV/c. Both the reconstructed $p_T$ and its uncertainty are sensitively affected by the material, especially for electrons.

More positively, however, it has been shown that the P+SCT track reconstruction efficiency is not strongly affected by the proximity of close tracks such as in a $b$-jet [33,47], and is robust to modest detector deterioration. Despite the problem

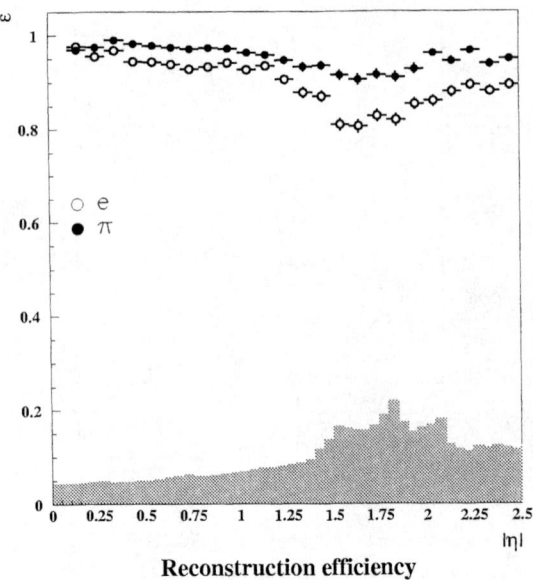

**FIGURE 32.** *Track reconstruction efficiency of isolated electron or $\pi^+$ tracks of $p_T = 20$ GeV/c as a function of $|\eta|$. Superimposed is the traversed P+SCT detector material in $X_0$.*

of material budget, the pixel and SCT detectors are expected to be a powerful tracking tool at the LHC.

# CONCLUSION

In the oral version of these lectures, the author attempted to outline why and how large detector systems have evolved at Hadron Colliders. In that attempt, the author showed the evolution of CDF and D0 with associated Tevatron Collider upgrades, and the requirements placed on the CMS and ATLAS detectors at the CERN LHC.

For reasons of brevity, comments in this written version have concentrated on the CDF and ATLAS experiments; the other experiments could have equally well demonstrated the evolutions discussed.

The developments discussed in this report have resulted initially from a physics motivation. However, many developments (especially in the field of semi-conductor trackers) have only been possible because of fundamental technological developments in micro-electronics.

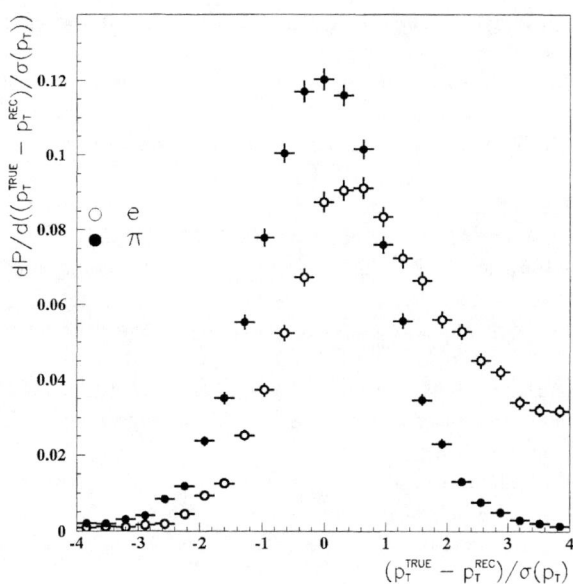

**FIGURE 33.** *Fit pull distribution of $p_T$ for isolated electron or $\pi^+$ tracks of $p_T = 20$ GeV/c, averaged over $|\eta|$.*

## ACKNOWLEDGMENTS

I would like to thank the organizers of the VIII ICFA School on Instrumentation in Elementary Particle Physics, in particular Prof. M. Atac and Prof. S. Kartal, for their hospitality during this extremely successful and friendly Instrumentation School. I would also like to thank all my colleagues who provided material for this presentation and helped me in the preparation of these proceedings. In particular, I acknowledge the help and counsel of F. Strumia, T. Speer, I. Efthymiopoulos, D. Ferrere and X. Wu.

## REFERENCES

1. C. Caso et al., *The Particle Data Group*, The European Physical Journal C3 (1998) 1
2. The Super-Kamiokande Collaboration, *Evidence for oscillation of atmospheric neutrinos*, Phys. Rev. Lett. 81 (1998) 1562-1567
3. D. Boutigny et al., *BABAR Technical Design Report*, SLAC-R-0457, Mar. 1995.
4. The BELLE Collaboration *BELLE Technical Design Report*, KEK-Report 95-1, April 1995.

5. The CLEO Collaboration, *The CLEO III detector : design and physics goals*, CLNS 94-1277, 1994
6. INFN-Laboratori Nazionali di Frascati, *Proposal for a Phi-Factory*, LNF-90/031(R), (1990)
7. The HERA-B Collaboration, *HERA-B Technical Design Report*, DESY-PRC 95/01, 1995.
8. W.W. Armstrong et al., *ATLAS: Technical Proposal for a general-purpose p p experiment at the Large Hadron Collider at CERN*, CERN-LHCC-94-43, Dec 1994.
9. The CMS collaboration, *CMS, the Compact Muon Solenoid: Technical Proposal*, CERN-LHCC-94-38, 1994
10. The MINOS collaboration, *The MINOS Detectors: Technical Proposal*, NuMI-L-337, 1998.
11. H. Shibuya, K. Hoshino, M. Komatsu et al., *Letter of intent : the OPERA emulsion detector for a long-baseline neutrino-oscillation experiment*, CERN-SPSC-97-24 SPSC-I-218 ; LNGS-LOI-8-97
12. F. Arneodo, M. Ambrosio et al., *ICANOE Imaging and CAlorimetric Neutrino Oscillation Experiment : a proposal for a CERN-GS long baseline and atmospheric neutrino oscillation experiment*, LNGS-P21/99, INFN/AE-99-17, CERN/SPSC 99-25, SPSC/P314.
13. K. Nishikawa, *Status of K2K (KEK to Kamioka long baseline neutrino oscillation experiment)*, Nucl. Phys. Proc. Suppl. 77, 198-203 (1999)
14. D. Fournier and L. Serin, *Experimental Techniques*, lecture given at the 1995 European School of High-Energy Physics, CERN 96-04 (1996)
15. A. Savoy-Navarro, *Detector Systems for Future HEP Experiments*, published in: Proceedings of *Instrumentation in Elementary Particle Physics*, 7th ICFA School, (1997)
16. F. Abe et al., Phys. Rev. D50, 2966 (1994).
17. S. Abachi et al., Phys. Rev. Letters 74 2632 (1995)
18. F. Abe et al., Phys. Rev. Lett. 82, 271 (1999); F. Abe et al., Phys. Rev. Lett. 80, 2773 (1998).
19. The CDF II Collaboration *The CDF II Detector: Technical Design Report*, FERMILAB-Pub-96/390-E 1996
20. *Future Electroweak Physics at the Fermilab Tevatron. Report of the Tev_2000 Study Group: Top Physics*, FERMILAB-PUB-96/046, Editors: D. Amidei and R. Brock (1996).
21. Private communication, and Report of LEP experiments at the CERN LEPC meeting, November 1999.
22. J. Valls,*Higgs Searches at the Tevatron in Run I* FERMILAB-CONF-99/263-E. Published in: Proceedings International Europhysics Conference on High-Energy Physics (EPS-HEP 99).
23. See for example the *Higgs Working Group final report, 19-21 Nov 1998 (first draft)*, in http://fnth37.fnal.gov/higgs/draft.html.
24. J. Tseng, The CDF Collaboration, *Rare B Decays, Mixing, and CP Violation at the Fermilab Tevatron* FERMILAB-CONF-99/355-E. Published in: Proceedings 11th Rencontres de Blois: Frontiers of Matter.

25. T. Affolder *et al.*, submitted to Phys. Rev. D, FERMILAB-PUB-99-225-E
26. X. Wu, *Silicon Vertex Tracker: A Fast Precise Tracking Trigger for CDF*, FERMILAB-CONF-99/219-E. Published in: Proceedings 8th International Workshop on Vertex Detectors (VERTEX 99).
27. M. Dell'Orso, L. Ristori, Nucl. Inst. and Meth. A 278 (1989), 436 SVT Technical Design Report, CDF note, CDF/DOC/TRIGGER/PUBLIC/3108
28. CDF II Collaboration, Technical Design Report, FERMILAB-Pub-96/390-E 1996
29. LHC machine
30. LHCb Collaboration. A Large Hadron Collider Beauty experiment for precision measurements of CP-violation and rare decays, CERN/LHCC/98-4, February 1998.
31. ALICE Collaboration. Technical Proposal for a Large Ion Collider Experiment at the CERN LHC, CERN/LHCC/95-71, December 1995.
32. ATLAS Collaboration, *Detector and Physics Performance Design Report*, CERN/LHCC/99-14 and 99-15, May 1999. TDR
33. ATLAS Collaboration, *Inner Detector Technical Design Report*, CERN/LHCC/97-16 and 97-17, April 1997.
34. ATLAS Collaboration, *Central Solenoid Technical Design Report*, CERN/LHCC/97-21, April 1997.
35. ATLAS Collaboration, *Pixel Detector Technical Design Report*, CERN/LHCC/98-13, May 1998.
36. M. Dentan *et al.*, Proc. First Workshop on Electronics for LHC Experiments, CERN/LHCC/95-56, October 1995.
37. ATLAS Collaboration, *Liquid Argon Calorimeter Technical Design Report*, CERN/LHCC/96-41, December 1996.
38. ATLAS Collaboration, *Tile Calorimeter Technical Design Report*, CERN/LHCC/96-42, December 1996.
39. ATLAS Collaboration, *Calorimeter Performance Technical Design Report*, CERN/LHCC/96-40, December 1996.
40. ATLAS Collaboration, *Muon Spectrometer Technical Design Report*, CERN/LHCC/97-22, May 1997.
41. ATLAS Collaboration, *Barrel Toroid Technical Design Report*, CERN/LHCC/97-19, April 1997.
42. ATLAS Collaboration, *End-Cap Toroid Technical Design Report*, CERN/LHCC/97-20, April 1997.
43. P. Allport *et al.*, Nucl. Instr. Methods Phys. Res. A435 (1999) 74.
44. H. J. Ziock *et al.*, Nucl. Instrm. Methods A342 (1994) 96.
45. R. Brun *et al.*, GEANT3, CERN Internal Report CERN DD/EE/84-1 (1986).
46. T. Sjostrand, Computer Physics Communications 82 (1994) 74.
47. A. Clark *et al.*, *Performance of the ATLAS Semiconductor Tracking System*, Proc. 3rd. Hiroshima Symposium on the Application of Semiconductor Tracking Detectors, Melbourne, December 1997. To be published.
48. R. Clifft and A. Poppleton, ATLAS Internal Note SOFT-NO-009 (1994).

# REVIEW TALKS

# Perspectives in High-Energy Physics[1]

## Chris Quigg[2]

*Theoretical Physics Department*
*Fermi National Accelerator Laboratory* [3]
*P.O. Box 500, Batavia, Illinois 60510 USA*

**Abstract.** I sketch some pressing questions in several active areas of particle physics and outline the challenges they present for the design and operation of detectors.

## INTRODUCTION

My assignment at the 1999 ICFA Instrumentation School is to survey some current developments in particle physics, and to describe the kinds of experiments we would like to do in the near future and illustrate the demands our desires place on detectors and data analysis. Like any active science, particle physics is in a state of continual renewal. Many of the subjects that seem most fascinating and most promising today simply did not exist as recently as twenty-five years ago. Other topics that have preoccupied physicists for many years have been reshaped by recent discoveries and insights, and transformed by new techniques in accelerator science and detector technology. To provide some context for the courses and laboratories at this school, I have chosen three topics that are of high scientific interest, and that place very different demands on instrumental techniques. I hope that you will begin to see the breadth of opportunities in particle physics, and that you will also look beyond the domain of particle physics for opportunities to apply the lessons you learn here in Istanbul.

I begin with the remarkable neutrino, a subatomic particle that our instruments must be able to detect both by its presence and by its absence, depending on the circumstances. In particular, I note the interest in observing and characterizing neutrino oscillations in order to determine the properties of the neutrinos, and describe some of the new instruments contemplated to do that. Then I will talk about physics at the high-energy frontier, focusing on what we hope to learn from

---

[1] Review lecture given at the ICFA Instrumentation School, Istanbul, Turkey, June 30, 1999.
[2] Internet address: quigg@fnal.gov.
[3] Fermilab is operated by Universities Research Association Inc. under Contract No. DE-AC02-76CH03000 with the United States Department of Energy.

the top quark in the next set of experiments at Fermilab's Tevatron Collider. Third, I will tell you some of the ways we hope to explore the landscape of spacetime, and explore some of the signs we might find that the three-plus-one dimensions of ordinary experience are not the whole story. My treatment of all of these will be schematic; in a single lecture, my intent is to raise questions and present a wide range of challenges and opportunities.

## NEUTRINO PUZZLES

Neutrinos are tiny subatomic particles that carry no electric charge, have (almost) no mass, move (nearly) at the speed of light, and hardly interact at all. They are among the most abundant particles in the Universe. As you listen to this lecture, inside your body are more than 10 million ($10^7$) relic neutrinos left over from the Big Bang. Each second, some $10^{14}$ neutrinos made in the Sun pass through you. In one tick of the clock, about a thousand neutrinos made by cosmic-ray interactions in Earth's atmosphere traverse your body. Other neutrinos reach us from natural sources, including radioactive decays of elements inside the Earth, and artificial sources, such as nuclear reactors.

Our awareness of neutrinos started with a puzzle in 1914 that led to an idea in 1930 that was confirmed by an experiment in 1956. Today, neutrinos have become an important tool for particle physics and astrophysics, and fascinating objects of study that may yield important new clues about the basic laws of Nature.

### The First Neutrino Puzzle

There was a neutrino puzzle even before physicists knew there was a neutrino. Natural and artificial radioactivity includes nuclear beta ($\beta$) decay, observed as

$$^A Z \to {}^A(Z+1) + \beta^- , \tag{1}$$

where $\beta^-$ is the old-fashioned name for an electron and $^A Z$ stands for the nucleus with $Z$ protons and $A - Z$ neutrons. Examples are tritium $\beta$ decay,

$$^3 H_1 \to {}^3 He_2 + \beta^- , \tag{2}$$

neutron $\beta$ decay,

$$n \to p + \beta^- , \tag{3}$$

and $\beta$ decay of Lead-214,

$$^{214} Pb_{82} \to {}^{214} Bi_{83} + \beta^- . \tag{4}$$

For two-body decays, the Principle of Conservation of Energy & Momentum says that the $\beta$ particle, or electron, should have a definite energy, indicated by the

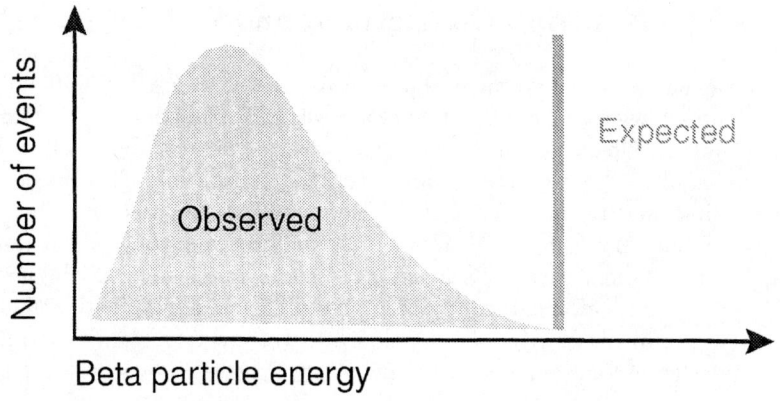

**FIGURE 1.** Expectations and reality for the beta decay spectrum.

spike in Figure 1. What was observed was very different: in 1914, James Chadwick (later to discover the neutron) showed conclusively [1] that in the decay of Radium B and C ($^{214}$Pb and $^{214}$Bi), the $\beta$ energy follows a continuous spectrum, as shown in Figure 1.

How could we account for this completely unexpected behavior? Might it mean that energy and momentum are not uniformly conserved in subatomic events?[4] Although Chadwick did not discover the neutron until 1932, we can use a little chronological license to sharpen our puzzle by considering a cartoon of neutron $\beta$ decay. The continuous $\beta$ spectrum means that, in general, the products of the decay of a stationary neutron will not have balanced momenta (zero net momentum), as shown in the left-hand frame in Figure 2. For the products of a system at rest to drift off in some direction flies in the face of physical intuition, though we have to concede that at the time, physicists' intuition was still largely derived from macroscopic experience.

---

[4] Niels Bohr was willing to consider this possibility.

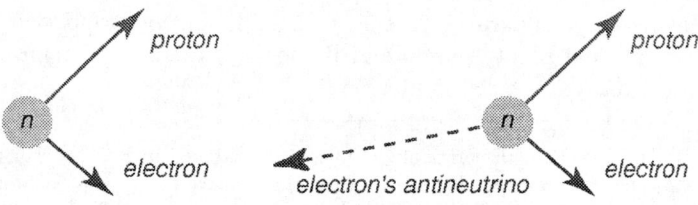

**FIGURE 2.** *(Left)* Apparent nonconservation of energy and momentum in neutron beta decay; *(Right)* Pauli's solution to the nonconservation of energy and momentum: a nearly massless, neutral, penetrating particle that we know as $\bar{\nu}_e$.

# The Neutrino Conjectured and Observed

The $\beta$-decay energy crisis tormented physicists for years. On December 4, 1930, Wolfgang Pauli addressed an open letter[5] to a meeting on radioactivity in Tübingen. Pauli could not attend in person because his presence at a student ball in Zurich was "indispensable." In his letter, Pauli advanced the outlandish idea of a new, very penetrating, neutral particle of vanishingly small mass. Because Pauli's new particle interacted very feebly with matter, it would escape undetected from any known apparatus, taking with it some energy, which would seemingly be lost. The balance of energy and momentum would be restored, as shown in the right-hand frame of Figure 2, by the particle we now know as the electron's antineutrino. The proper decay scheme for the neutron is thus

$$n \to p + \beta^- + \bar{\nu}. \tag{5}$$

Pauli's new particle was indeed a "desperate remedy," but it was, in its way, very conservative, for it preserved the principle of energy and momentum conservation and with it the notion that the laws of physics are invariant under translations in space and time. The hypothesis fit the facts; after the discovery of the neutron in 1932, Fermi named the new particle the neutrino, to distinguish it from the neutron, and constructed his four-fermion theory of the weak interaction. Experimental confirmation of Pauli's neutrino had to wait for dramatic advances in technology.

Detecting a particle as penetrating as the neutrino required a large target and a copious source of neutrinos. In 1953, Clyde Cowan and Fred Reines [4] used the intense beam of antineutrinos from a fission reactor

$$^A Z \to {}^A(Z+1) + \beta^- + \bar{\nu}, \tag{6}$$

and a heavy target (10.7 ft$^3$ of liquid scintillator) containing about $10^{28}$ protons to detect the reaction

$$\bar{\nu} + p \to e^+ + n. \tag{7}$$

Initial runs at the Hanford Engineering Works were suggestive but inconclusive. Moving their apparatus to the stronger fission neutrino source at the Savannah River nuclear plant, Cowan and Reines and their team made the definitive observation of inverse $\beta$ decay in 1956 [5].

---

[5] Pauli's letter (in the original German) is reproduced in Ref. [2]. For an English translation, see pp. 127-8 of Ref. [3]. It begins, "Dear Radioactive Ladies and Gentlemen, I have hit upon a desperate remedy regarding ... the continuous $\beta$-spectrum ..." Pauli concluded, "For the moment I dare not publish anything about this idea and address myself confidentially first to you. ... I admit that my way out may seem rather improbable *a priori* .... Nevertheless, if you don't play you can't win .... Therefore, Dear Radioactives, test and judge." Pauli's neutrino, together with the discovery of the neutron, also resolved a vexing nuclear spin-and-statistics problem.

## Three Families of Leptons

In addition to the electron ($e$) and its neutrino ($\nu_e$), we now recognize two other pairs of pointlike, spin-$\frac{1}{2}$ particles that are not affected by the strong interaction:

$$\begin{pmatrix} \nu_e \\ e^- \end{pmatrix}_L, \quad \begin{pmatrix} \nu_\mu \\ \mu^- \end{pmatrix}_L, \quad \begin{pmatrix} \nu_\tau \\ \tau^- \end{pmatrix}_L. \tag{8}$$

These particles are known as leptons, from the Greek λεπτός = thin, inspired by the small mass of the electron, muon, and their neutrinos compared with the mass of the proton and neutron, the lightest of the baryons (from the Greek βαρύς = heavy).[6]

The muon neutrino is created in charged pion decay,

$$\pi^+ \to \mu^+ \nu_\mu \quad \text{and} \quad \pi^- \to \mu^- \bar{\nu}_\mu, \tag{9}$$

and neutrino beams produced at accelerators are overwhelmingly muon neutrinos. The two-neutrino experiment carried out at Brookhaven National Lab in the early 1960s [6] demonstrated that the neutrinos produced in pion decays do not initiate inverse $\beta$ decay, so that $\nu_\mu$ is distinct from $\nu_e$.

The DONUT (Direct Observation of NU-Tau) Experiment [7] under analysis at Fermilab is a three-neutrino experiment. Using a prompt neutrino beam in which decays of the charmed-strange meson

$$\begin{array}{l} D_s^+ \to \tau^+ \nu_\tau \\ \phantom{D_s^+ \to \tau^+} \hookrightarrow \bar{\nu}_\tau + \text{anything} \end{array} \tag{10}$$

provide the $\nu_\tau$ source, the experimenters aim to observe the reaction

$$\nu_\tau N \to \tau + \text{anything}. \tag{11}$$

I expect to see results from DONUT in the year 2000.

## Neutrino Beams and Detectors

Neutrinos traverse vast amounts of material. The fission antineutrinos detected by Cowan, Reines, and their collaborators have an inverse $\beta$-decay cross section $\sigma(\bar{\nu}_e p \to e^+ n) \approx 10^{-43}$ cm$^2$. Accordingly, their interaction length,

$$\mathcal{L}_{\text{int}} = \frac{1}{\sigma(\bar{\nu}_e p \to e^+ n) N_A (Z/A) \bar{\rho}}, \tag{12}$$

---

[6] We would not choose the same names today, knowing that the tau lepton is twice as massive as the proton ...

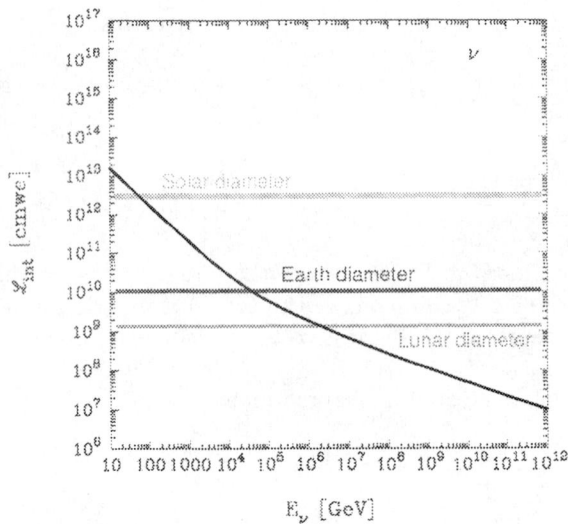

**FIGURE 3.** Interaction lengths for neutrino interactions on an isoscalar nucleon target (from Ref. [8]).

where $N_A = 6.022 \times 10^{23}$ cm$^{-3}$ is Avogadro's number, $\bar{\rho}$ is the specific gravity of the target, and $Z/A$ is the proton fraction of the target nucleons, is very long. It corresponds to about $1.7 \times 10^{19}$ cm of water, or a column density of $1.7 \times 10^{10}$ kilotonnes per square centimeter. On average, a fission neutrino would traverse more than four light-years of lead—the distance from Earth to $\alpha$-Centauri—before interacting. At higher energies, the cross section for the inclusive reaction $\bar{\nu}_e p \to e^+ +$ anything grows—at first $\propto E_\nu$, then more slowly—so the interaction length decreases. The high-energy dependence of the $\nu N$ interaction length is shown in Figure 3. The interaction length of a 100-GeV neutrino is approximately 25 million kilometers of water, or about 230 Earth diameters. If you should stand in Fermilab's neutrino beam, only one neutrino in $10^{11}$ will interact in your body.

What are the consequences of the great interaction length? First, there is the missing-energy signature for neutrinos. Second, there is the difficulty of detecting neutrino interactions. Third—because the other known particles all interact more than neutrinos do—we have the possibility of preparing filtered neutrino beams, which we shall discuss in a moment. Fourth, because neutrinos can penetrate great columns of matter, whereas electromagnetic radiation is blocked by a few hundred grams of material, it is appealing to consider the promise of neutrino astronomy for peering into the hearts of dense structures or looking back at the state of the universe before recombination of ions and electrons into neutral atoms.

At high-energy accelerators, neutrinos are tertiary products of collisions

$$p + \text{Target} \to \text{many } \pi, K \ , \tag{13}$$

followed by the decays

$$\pi \to \mu\nu_\mu, \quad K \to \mu\nu_\mu ,  \tag{14}$$

in an evacuated decay space up to a kilometer long. The decay region is followed by an earthen shield, perhaps made denser by the inclusion of blocks of iron or steel, that absorbs photons, surviving mesons, and other hadrons, and ranges out many of the muons. At Fermilab, the total length of the neutrino beam line is about 3 km.

In the most recent study of deeply inelastic neutrino scattering at Fermilab, 800-GeV protons from the Tevatron deliver $10^{10}$ neutrinos in 5 pings over 2.5 seconds each minute, spread over the 100 ft$^2$ face of the NuTeV Detector. The detector is made up of 690 T of iron, scintillator, and drift chambers. Events are studied from a "fiducial volume" of 390 T. The $10^{10}$ neutrinos per minute produce about 10 – 20 events that are recorded for off-line analysis. Experiments of this kind give us our best look at the interior of the proton, and reveal its quark structure in exquisite detail [9].

## Neutrino Mass

*Neutrinos are very light.* No one has ever weighed a neutrino. The best kinematical determinations set upper bounds [10] on the dominant neutrino species emitted in nuclear beta decay ($m_{\nu_e} \lesssim 15$ eV/$c^2$), charged-pion decay ($m_{\nu_\mu} < 0.19$ MeV/$c^2$ at 90% CL), and $\tau$-lepton decay ($m_{\nu_\tau} < 18.2$ MeV/$c^2$ at 95% CL). Although there are prospects[7] for improving these bounds—and the measurement of a nonzero mass would constitute a real discovery—they are sufficiently large that it is of interest to consider indirect (nonkinematic) constraints from other quarters.

If neutrino lifetimes are greater than the age of the Universe, the requirement that neutrino relics from the Big Bang not overclose the Universe leads to a constraint on the sum of neutrino masses. For relatively light neutrinos ($m_\nu \lesssim$ a few MeV/$c^2$), the total mass in neutrinos,

$$m_{\text{tot}} = \sum_i \tfrac{1}{2} g_i m_{\nu_i} , \tag{15}$$

where $g_i$ is the number of spin degrees of freedom of $\nu_i$ plus $\bar{\nu}_i$, sets the scale of the neutrino contribution to the mass density of the Universe, $\varrho_\nu = m_{\text{tot}} n_\nu \approx 112\, m_{\text{tot}}$ cm$^{-3}$. If we measure $\varrho_\nu$ as a fraction of the critical density to close the Universe, $\varrho_c = 1.05 \times 10^4\, h^2$ eV/$c^2$ cm$^{-3}$, where $h$ is the reduced Hubble parameter, then

$$\Omega_\nu \equiv \frac{\varrho_\nu}{\varrho_c} = \frac{m_{\text{tot}}}{94 h^2 \text{ eV}/c^2} . \tag{16}$$

---

[7] Some of these prospects were reviewed at $\nu$Fact '99 in Lyon by Alvaro De Rújula, Ref. [11].

An assumed bound on $\Omega_\nu h^2$ then implies a bound on $m_{\text{tot}}$. A very conservative bound results from the assumption that $\Omega_\nu h^2 < 1$: it is that $m_{\text{tot}} < 94$ eV/$c^2$.

Recent observations[8] suggest that the total matter density is considerably smaller than the critical density, so that $\Omega_m \approx 0.3$. If we fix $\Omega_\nu < \Omega_m$ and choose the plausible value $h^2 = 0.5 \pm 0.15$, then we arrive at the still generous upper bound $m_{\text{tot}} \lesssim 19$ eV/$c^2$. Taking into account the best (and model-dependent) information about the hot- and cold-dark-matter cocktail [13] needed to reproduce the observed fluctuations in the cosmic microwave background, it seems likely that cosmology limits $m_{\text{tot}} \lesssim$ a few eV/$c^2$. It is worth remarking that the cosmological desire for hot dark matter has been on the wane.

If neutrinos were exactly massless, neutrino physics would be simple. There would be no pattern of masses to explain, no neutrino decays, no mixing among lepton generations, and no neutrino oscillations. The only question would be Why?

If neutrinos do have masses, the electroweak theory has more unexplained parameters: the neutrino masses, three mixing angles, and a CP-violating phase. Neutrinos may decay, and neutrinos can oscillate from one electroweak species to another.

## Neutrino Oscillations[9]

In the quantum world, particles are waves. If neutrinos $\nu_1, \nu_2, \ldots$ have different masses $m_1, m_2, \ldots$, each neutrino flavor may be a mixture of different masses. Let us consider two species for simplicity, and take

$$\begin{pmatrix} \nu_e \\ \nu_\mu \end{pmatrix} = \begin{pmatrix} \cos\theta & \sin\theta \\ -\sin\theta & \cos\theta \end{pmatrix} \begin{pmatrix} \nu_1 \\ \nu_2 \end{pmatrix}. \tag{17}$$

If neutrinos are emitted with a definite momentum $p$, the wave functions corresponding to the two mass eigenstates evolve with different frequencies. As a consequence, a beam born as pure $\nu_\mu$ may evolve a $\nu_e$ component with time. If the neutrino momentum is large compared with the neutrino masses, $p \gg m_i$, then the probability for a $\nu_e$ component to develop in a $\nu_\mu$ beam after a time $t$ is

$$P_{\nu_e \leftarrow \nu_\mu}(t) = \sin^2 2\theta \sin^2\left(\frac{\Delta m^2 t}{4p}\right). \tag{18}$$

Measuring the propagation distance $L = ct$, approximating the neutrino energy as $E \approx pc$, and using the conversion factor $\hbar c \approx 1.97 \times 10^{-13}$ MeV m, we can re-express

$$\sin^2\left(\frac{\Delta m^2 t}{4p}\right) \approx \sin^2\left(1.27 \frac{\Delta m^2}{1 \text{ eV}^2} \cdot \frac{L}{1 \text{ km}} \cdot \frac{1 \text{ GeV}}{E}\right). \tag{19}$$

---

[8] For a review and interpretation of recent observations, see Ref. [12].
[9] For a review of the essentials, see [14].

Arising as it does from a slight frequency mismatch, neutrino flavor oscillation is analogous to the beat-frequency phenomenon observed when two tuning forks with *almost* the same pitch are sounded at the same time. If the forks are sounded individually, producing waves

we might find it difficult to distinguish the pitches. But when we sound them together, the sound intensity

swells and fades periodically because the two sound waves are different, reflecting the physical difference between the two tuning forks.

The probability that a neutrino born as $\nu_\mu$ remain a $\nu_\mu$ at distance $L$ is

$$P_{\nu_\mu \leftarrow \nu_\mu}(L) = 1 - \sin^2 2\theta \sin^2 \left( 1.27 \frac{\Delta m^2}{1 \text{ eV}^2} \cdot \frac{L}{1 \text{ km}} \cdot \frac{1 \text{ GeV}}{E} \right). \tag{20}$$

The probability for a $\nu_\mu$ to metamorphose into a $\nu_e$,

$$P_{\nu_e \leftarrow \nu_\mu} = \sin^2 2\theta \sin^2 \left( 1.27 \frac{\Delta m^2}{1 \text{ eV}^2} \cdot \frac{L}{1 \text{ km}} \cdot \frac{1 \text{ GeV}}{E} \right), \tag{21}$$

depends on two parameters related to experimental conditions: $L$, the distance from the neutrino source to the detector, and $E$, the neutrino energy. It also depends on two fundamental neutrino parameters: the difference of masses squared, $\Delta m^2 = m_1^2 - m_2^2$, and the neutrino mixing parameter, $\sin^2 2\theta$. The amplitude of the probability oscillations is given by $\sin^2 2\theta$, as shown in Figure 4; the wavelength of the oscillations is

$$L_{\text{osc}} = \frac{\pi}{1.27} \cdot \frac{E}{1 \text{ GeV}} \cdot \frac{1 \text{ eV}^2}{\Delta m^2} \text{ km} = 2.48 \frac{E}{1 \text{ GeV}} \cdot \frac{1 \text{ eV}^2}{\Delta m^2} \text{ km}, \tag{22}$$

and the oscillation probability is greatest at a distance $L_{\max} = (k + \frac{1}{2}) L_{\text{osc}}$, where $k$ is an integer.

Many experiments have now used natural sources of neutrinos, neutrino radiation from fission reactors, and neutrino beams generated in particle accelerators to look for evidence of neutrino oscillation.[10]

---

[10] For summaries of the current evidence about neutrino oscillations, see Ref. [15].

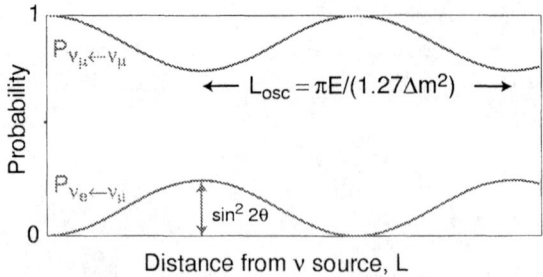

**FIGURE 4.** Evolution of a $\nu_\mu$ beam of energy $E$ depends on the intrinsic parameters $\Delta m^2$ and $\sin^2 2\theta$, and on the experimental conditions characterized by $L$ and $E$.

The nuclear burning that powers the Sun produces neutrinos as well as light and heat [16]. Overall, a network of reactions we may summarize as

$$4p \to {}^4\text{He} + 2e^+ + 2\nu_e + 25 \text{ MeV} \qquad (23)$$

leads to the spectrum of neutrinos[11] shown in Figure 5. Because solar neutrinos interact so feebly, they can only be detected in a very massive target and detector. The Super-Kamiokande Detector in Japan consists of 50 000 tonnes of pure water viewed by 11 000 photomultiplier tubes to detect Cherenkov light. It is sited 1 km under a mountain, under 3 kmwe. The great advantage of the Super-K detector is that it detects neutrino interactions $\nu_e + n \to p + e^-$ in real time, and determines the neutrino direction from the electron direction. Super-K has demonstrated that the brightest object in the neutrino sky is the Sun, which proves that nuclear fusion powers our star. The disadvantage of the water-Cherenkov technique (see Figure 5) is that it is only sensitive to the highest-energy solar neutrinos.

Five solar-neutrino experiments report deficits with respect to the predictions of the standard solar model: Kamiokande and Super-Kamiokande using water-Cherenkov techniques, SAGE and GALLEX using chemical recovery of germanium produced in neutrino interactions with gallium, and Homestake using radiochemical separation of argon produced in neutrino interactions with chlorine. These results suggest the oscillation $\nu_e \to \nu_x$.

Cosmic rays that interact in Earth's atmosphere produce neutrinos in the decays of pions, kaons, and muons, in the approximate proportions $\nu_\mu : \bar{\nu}_\mu : \nu_e : \bar{\nu}_e :: 2 : 2 : 1 : 1$. Five atmospheric-neutrino experiments report anomalies in the arrival of muon neutrinos: Kamiokande, IMB, and Super-Kamiokande using water-Cherenkov techniques, and Soudan II and MACRO using sampling calorimetry. The most striking result is the zenith-angle dependence of the $\nu_\mu$ rate reported last year by Super-K [20,21], which is shown in Figure 6. The electron-like events follow the Monte Carlo simulation, but the muon-like events exhibit a deficit that is

---

[11] The standard solar model is described in Ref. [17]. A wealth of information is available on John Bahcall's home page at http://www.sns.ias.edu/~jnb.

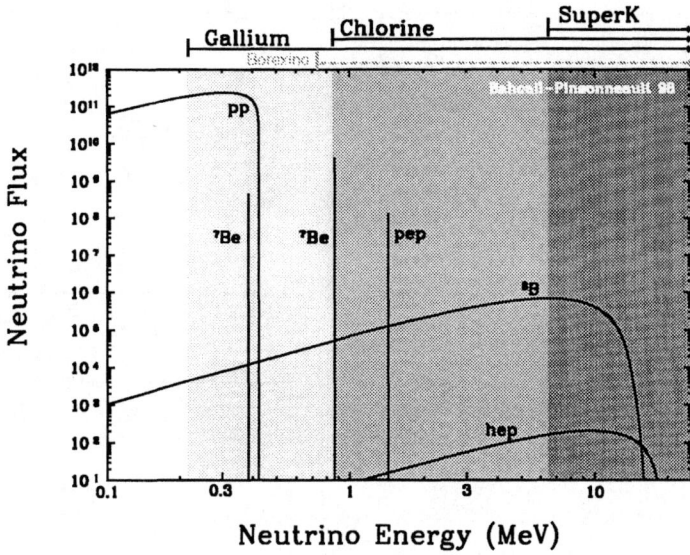

**FIGURE 5.** The spectrum of neutrinos from the *pp* chain predicted by the standard solar model. The neutrino fluxes from continuum sources (*pp* and $^8$B) are given in units of number per cm$^2$ per second per MeV at one astronomical unit. The line fluxes (*pep* and $^7$Be) are given in number per cm$^2$ per second. The detection thresholds for several solar neutrino experiments are indicated at the top of the figure. The figure is from Ref. [18], updated using the data from Ref. [19].

most pronounced for upward-coming neutrinos, *i.e.*, those for which the flight path is longest, up to 13 000 km. These results suggest the oscillation $\nu_\mu \to \nu_\tau$ or $\nu_s$. Auxiliary information disfavors the sterile-neutrino ($\nu_s$) interpretation.

The atmospheric- and solar-neutrino anomalies are both disappearance phenomena. A single experiment reports the appearance of neutrinos that would not be seen in the absence of neutrino oscillations. The LSND experiment [22] reports the observation of $\bar{\nu}_e$-like events is what should be an essentially pure $\bar{\nu}_\mu$ beam produced at the Los Alamos Meson Physics Facility, suggesting the oscillation $\bar{\nu}_\mu \to \bar{\nu}_e$. This result has not yet been reproduced by any other experiment.

A host of experiments have failed to turn up evidence for neutrino oscillations in the regimes of their sensitivity. These results limit neutrino mass-squared differences and mixing angles. In more than a few cases, positive and negative claims are in conflict, or at least face off against each other. Over the next five years, many experiments will seek to verify, further quantify, and extend these claims.

## The Ultimate Neutrino Source?

Our colleagues working to assess the feasibility of very-high-energy muon colliders have given us the courage to think that it may be possible, not too many years in

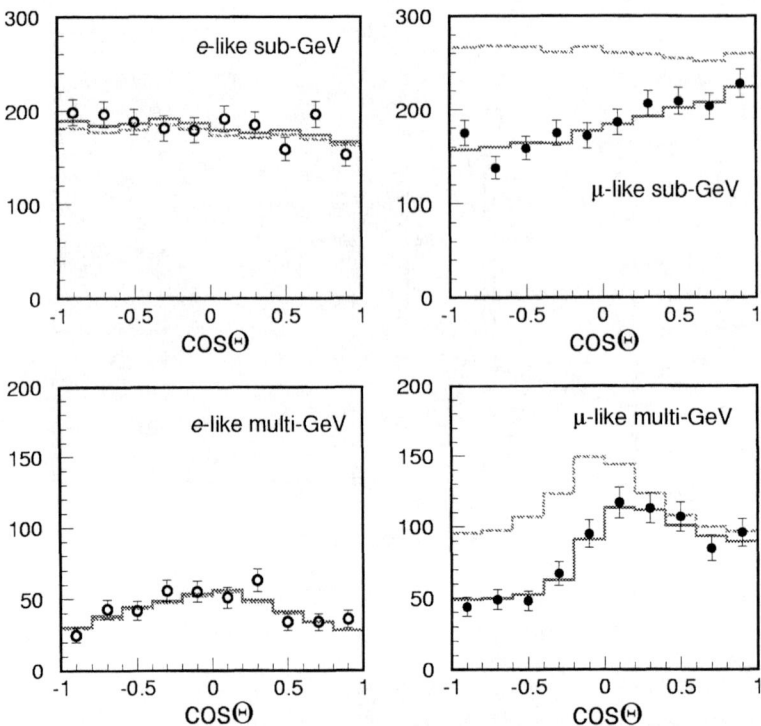

**FIGURE 6.** Comparison of Monte Carlo simulations (dashed curves) and measurements for events initiated in the Super-Kamiokande detector by cosmic-ray $\nu_e + \bar{\nu}_e$ (○) and $\nu_\mu + \bar{\nu}_\mu$ (●). The solid lines show the effects of $\nu_\mu \to \nu_\tau$ oscillations.

the future, to accumulate $10^{20-21}$ (or even $10^{22}$) muons per year. It is very exciting to think of the possibilities that millimoles of muons would raise for studies in fundamental physics.

From the perspective of a muon collider, the 2.2-$\mu$s lifetime of the muon presents a formidable challenge. But if the challenge of producing, capturing, storing, and replenishing many unstable muons can be met, the decays

$$\mu^- \to e^- \nu_\mu \bar{\nu}_e, \qquad \mu^+ \to e^+ \bar{\nu}_\mu \nu_e \qquad (24)$$

offer delicious possibilities for the study of neutrino interactions and neutrino properties [23–25]. We would have at our disposal for the first time not only neutrino beams of unprecedented intensity, but also controlled beams rich in high-energy electron neutrinos. In a *Neutrino Factory*, the composition and spectra of intense neutrino beams will be determined by the charge, momentum, and polarization of the stored muons. The beam from a $\mu^+$ storage ring contains $\bar{\nu}_\mu$ and $\nu_e$, but no $\nu_\mu$, $\bar{\nu}_e$, $\nu_\tau$, or $\bar{\nu}_\tau$. The neutrino spectra are given by

$$\frac{d^2 N_{\bar{\nu}_\mu}}{dx d\Omega} = \frac{x^2}{2\pi}[(3-2x)-(1-2x)\cos\theta]\,,\quad \cos\theta = \hat{p}_\nu \cdot \hat{s}_\mu\,, \qquad (25)$$

and

$$\frac{d^2 N_{\nu_e}}{dx d\Omega} = \frac{3x^2}{\pi}[(1-x)-(1-x)\cos\theta]\,, \qquad (26)$$

where $x = 2E_\nu/E_\mu$ measures the neutrino energy and $\hat{s}_\mu$ specifies the muon's spin direction.

At the energies best suited for the study of neutrino oscillations—tens of GeV, by our current estimates—the muon storage ring is compact. We could build it at one laboratory, pitched at a deep angle, to illuminate a laboratory on the other side of the globe with a neutrino beam whose properties we can control with great precision. By choosing the right combination of energy and destination, we can tune future neutrino-oscillation experiments to the physics questions we will need to answer, by specifying the ratio of path length to neutrino energy and determining the amount of matter the neutrinos traverse. Although we can use each muon decay only once, and we will not be able to select many destinations, we may be able to illuminate two or three well-chosen sites from a muon-storage-ring neutrino source. That possibility—added to the ability to vary the muon charge, polarization, and energy—may give us just the degree of experimental control it will take to resolve the outstanding questions about neutrino oscillations.

*The Detector Challenge.* To distinguish oscillations among $\nu_e$, $\nu_\mu$, $\nu_\tau$, and a possible fourth, "sterile," neutrino $\nu_s$ that does not experience weak interactions, we require a target *cum* detector of several kilotonnes—perhaps several tens of kilotonnes—that can distinguish electrons, muons, and taus, and measure their charges. This is a straightforward requirement for the muons, but the short-lived (0.3 picosecond) $\tau$ and the eager-to-shower electron are more difficult to deal with.

## WHAT CAN WE LEARN FROM THE TOP QUARK?

Top is a most remarkable particle, even for a quark.[12] A single top quark weighs 175 GeV/$c^2$, about as much as an atom of gold. But unlike the gold atom, which can be disassembled into 79 protons, 79 electrons, and 118 neutrons, top seems indivisible, for we discern no structure at a resolution approaching $10^{-18}$ m. Top's expected lifetime of about 0.4 yoctosecond ($0.4 \times 10^{-24}$ s) makes it by far the most ephemeral of the quarks. The compensation for this exceedingly brief life is a measure of freedom: top decays before it experiences the confining influence of the strong interaction. In spite of its fleeting existence, the top quark helps shape the character of the everyday world.

The discovery experiments were carried out at Fermilab's Tevatron, in which a beam of 900-GeV protons collides with a beam of 900-GeV antiprotons. Creating

---

[12] See Ref. [26] for a general introduction.

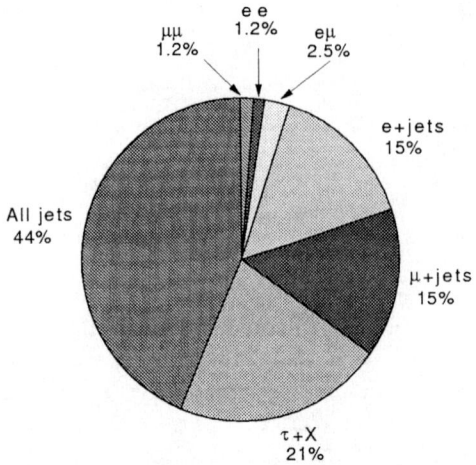

**FIGURE 7.** Branching fractions for prominent decay modes in $t\bar{t}$ production.

top-antitop pairs in sufficient numbers to claim discovery demanded exceptional performance from the Tevatron, for only one interaction in ten billion results in a top-antitop pair, through the reaction

$$\bar{p}p \;\to\; t\,\bar{t} + \cdots$$
$$\phantom{\bar{p}p \to t\,}\hookrightarrow W^-\bar{b} \quad\quad (27)$$
$$\phantom{\bar{p}p \to}\hookrightarrow W^+b$$

Observing traces of the disintegration of top into a $b$-quark and a $W$-boson, the agent of the weak interaction, required highly capable detectors and extraordinary attention to experimental detail. Both the $b$-quark and the $W$-boson are themselves unstable, with many multibody decay modes. The $b$-quark's mean lifetime is about 1.5 ps. It can be identified by a decay vertex displaced by a fraction of a millimeter from the production point, or by the low-momentum electron or muon from the semileptonic decays $b \to ce\nu$, $b \to c\mu\nu$, each with branching fraction about 10%. The $W$ boson decays after only 0.3 ys on average into $e\bar{\nu}_e$, $\mu\bar{\nu}_\mu$, $\tau\bar{\nu}_\tau$, or a quark and antiquark (observed as two jets of hadrons), with probabilities $1/9$, $1/9$, $1/9$, and $2/3$. The characteristic modes in which $t\bar{t}$ production can be sought are shown with their relative weights in Figure 7. Dilepton events ($e\mu$, $ee$, and $\mu\mu$) are produced primarily when both $W$ bosons decay into $e\nu$ or $\mu\nu$. Events in the lepton + jets channels ($e, \mu$ + jets) occur when one $W$ boson decays into leptons and the other decays through quarks into hadrons. Another challenge to experiment is the complexity of events in high-energy $\bar{p}p$ collisions. The top and antitop are typically accompanied by scores of other particles. The discovery experiments scanned $10^6$ events per second.

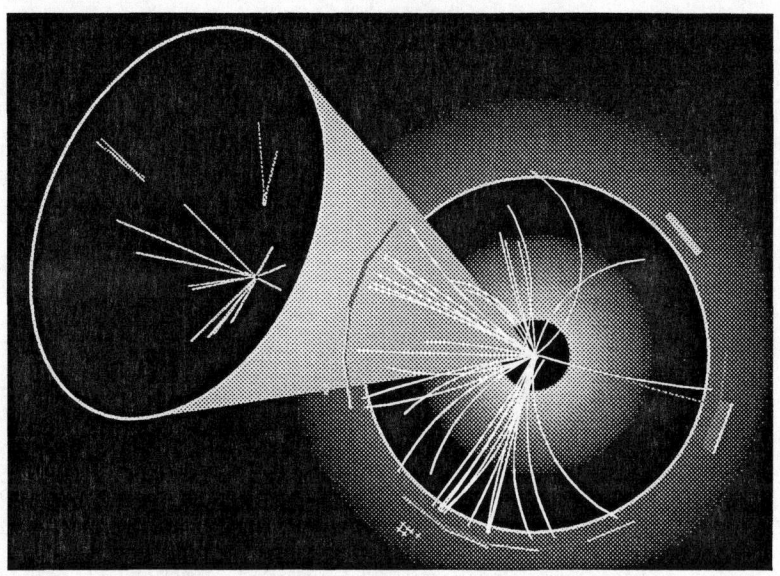

**FIGURE 8.** A pair of top quarks reconstructed in the CDF experiment at Fermilab. Each top decays to a $W$ boson and a $b$ quark. The tower pointing east-southeast in the wide view identifies a positron from $W^+$ decay; the inset shows displaced decays of two $b$ particles.

Each detector is an intricate apparatus operated by an international collaboration of about 450 physicists. The tracking devices, calorimeters, and surrounding iron for muon identification occupy a volume about three stories high and weigh about 5000 tons. The Collider Detector at Fermilab (CDF), a magnetic detector with solenoidal geometry, profited from its high-resolution silicon vertex detector (SVX) to tag $b$-quarks with good efficiency. The DØ Detector (D-Zero) had no central magnetic field, emphasizing instead calorimetric measurement of the energies of produced particles. A top event from the CDF detector, shown in Figure 8, displays the power of the silicon vertex detector to resolve secondary $b$ decays at only a small remove from the production vertex. The DØ event pictured in Figure 9 shows one $W$ boson reconstructed from a muon plus missing energy, and a second $W$ reconstructed from two jets.

*The Detector Challenge.* To select top-quark events with high efficiency, and to measure them effectively, require the ability to decide quickly which few of the $10^7$ events per second in Tevatron Run 2 ($10^8$ per second at the LHC) are interesting; excellent $b$ tagging, resolving secondary vertices within a few tenths of a millimeter of the collision point in a hostile high-radiation environment; efficient lepton identification and measurement of both electron and muon momenta, even for leptons in jets; calorimetry that provides good jet-jet invariant mass resolution

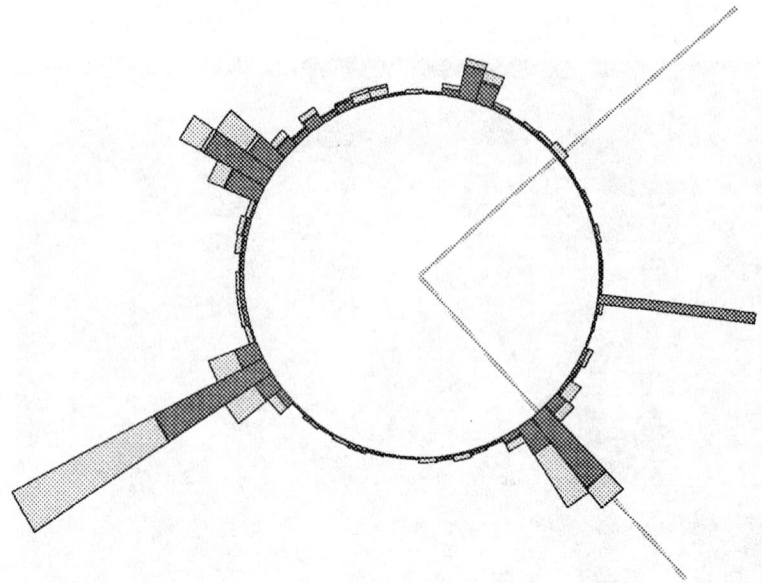

**FIGURE 9.** A pair of top quarks reconstructed in the DØ experiment at Fermilab. This end view shows the final decay products: two muons (thin lines at northeast and southeast), a neutrino (east), and four jets of particles.

and a reliable measurement of missing transverse energy; a hermetic detector.

For the moment, the direct study of the top quark belongs to the Tevatron. Early in the next century, samples twenty times greater than the current samples should be in hand, thanks to the increased event rate made possible by Fermilab's Main Injector and upgrades to CDF and DØ. Boosting the Tevatron's energy to 1 TeV per beam will increase the top yield by nearly 40%. Further enhancements to Fermilab's accelerator complex are under study. A decade from now, the Large Hadron Collider at CERN will produce tops at more than ten thousand times the rate of the discovery experiments. Electron-positron linear colliders or muon colliders may add new opportunities for the study of top-quark properties and dynamics. In the meantime, the network of understanding known as the standard model of particle physics links the properties of top to many phenomena to be explored in other experiments.

How can we expect new experiments to extend our knowledge of the top quark?[13] Tevatron Run 2, to begin in March 2001, is to accumulate 2 fb$^{-1}$ of integrated luminosity. We anticipate a determination of the top mass to $\pm 3$ GeV/$c^2$ in Run 2, $\pm 1$ GeV/$c^2$ with 30 fb$^{-1}$ in Run $2^{bis}$ or in LHC experiments. Combined with a

---

[13] The home page of the Fermilab *thinkshop* on top-quark physics for Run 2 of the Tevatron, at http://lutece.fnal.gov/thinkshop/, contains many useful links. See also the surveys in Refs. [27] and [28].

measurement of the $W$-boson mass to $\pm 40$ MeV/$c^2$, this measurement will make it possible to infer the Higgs-boson mass with increased precision. It is now possible to begin asking how precisely top fits the profile of anticipated properties in its production and decay. It should be possible to determine the $t \to bW$ coupling in single-top production: $\pm 10\%$ in Run 2, $\pm 5\%$ in Run $2^{bis}$. One of the exploratory goals of the new round of experiments will be to search for $t\bar{t}$ resonances, rare decays, and deviations from the expected pattern of top decays. Finally, it may prove possible to begin to probe the ephemeral lifetime of top. On the theoretical front, the large mass of top encourages us to think that the two problems of mass may be linked at the electroweak scale.

## Top Matters!

It is popular to say that top quarks were produced in great numbers in the fiery cauldron of the Big Bang some fifteen billion years ago, disintegrated in the merest fraction of a second, and vanished from the scene until my colleagues learned to create them in the Tevatron. That would be reason enough to care about top: to learn how it helped sow the seeds for the primordial universe that evolved into our world of diversity and change. But it is not the whole story; it invests the top quark with a remoteness that veils its importance for the everyday world.

The real wonder is that here and now, every minute of every day, the top quark affects the world around us. Through the uncertainty principle of quantum mechanics, top quarks and antiquarks wink in and out of an ephemeral presence in our world. Though they appear virtually, fleetingly, on borrowed time, top quarks have real effects.

Quantum effects make the coupling strengths of the fundamental interactions—appropriately normalized analogues of the fine-structure constant $\alpha$—vary with the energy scale on which the coupling is measured. The fine-structure constant itself has the familiar value $1/137$ in the low-energy (or long-wavelength) limit, but grows to about $1/129$ at the mass of the $Z^0$ boson, about 91 GeV/$c^2$. Vacuum-polarization effects make the effective electric charge increase at short distances or high energies.

In unified theories of the strong, weak, and electromagnetic interactions, all the coupling "constants" take on a common value, $\alpha_U$, at some high energy, $M_U$. If we adopt the point of view that $\alpha_U$ is fixed at the unification scale, then the mass of the top quark is encoded in the value of the strong coupling $\alpha_s$ that we experience at low energies.[14] Assuming three generations of quarks and leptons, we evolve $\alpha_s$ downwards in energy from the unification scale.[15] The leading-logarithmic behavior is given by

$$1/\alpha_s(Q) = 1/\alpha_U + \frac{21}{6\pi} \ln(Q/M_U) \ , \tag{28}$$

---

[14] All the important features emerge in an $SU(5)$ unified theory that contains the standard-model gauge group $SU(3)_c \otimes SU(2)_L \otimes U(1)_Y$. The final result is more general.

[15] The strategy is explained in Ref. [29].

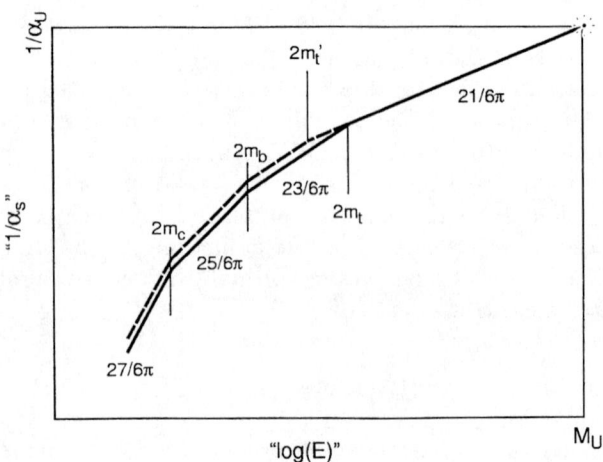

**FIGURE 10.** Two evolutions of the strong coupling constant $\alpha_s$. A smaller value of the top-quark mass leads to a smaller value of $\alpha_s$.

for $M_U > Q > 2m_t$. The positive coefficient $+21/6\pi$ means that the strong coupling constant $\alpha_s$ is smaller at high energies than at low energies. This behavior—opposite to the familiar behavior of the electric charge—is the celebrated property of asymptotic freedom. In the interval between $2m_t$ and $2m_b$, the slope $(33 - 2n_f)/6\pi$ (where $n_f$ is the number of active quark flavors) steepens to $23/6\pi$, and then increases by another $2/6\pi$ at every quark threshold. At the boundary $Q = Q_n$ between effective field theories with $n-1$ and $n$ active flavors, the coupling constants $\alpha_s^{(n-1)}(Q_n)$ and $\alpha_s^{(n)}(Q_n)$ must match. This behavior is shown by the solid line in Figure 10. The dotted line in Figure 10 shows how the evolution of $1/\alpha_s$ changes if the top-quark mass is reduced. A smaller top mass means a larger low-energy value of $1/\alpha_s$, so a smaller value of $\alpha_s$.

To discover the dependence of $\Lambda_{\text{QCD}}$ upon the top-quark mass, we calculate $\alpha_s(2m_t)$ evolving up from low energies and down from the unification scale, and match:

$$1/\alpha_U + \frac{21}{6\pi} \ln(2m_t/M_U) = 1/\alpha_s(2m_c) - \frac{25}{6\pi} \ln(m_c/m_b) - \frac{23}{6\pi} \ln(m_b/m_t). \quad (29)$$

Identifying

$$1/\alpha_s(2m_c) \equiv \frac{27}{6\pi} \ln(2m_c/\Lambda_{\text{QCD}}), \quad (30)$$

we find that

$$\Lambda_{\text{QCD}} = e^{-6\pi/27\alpha_U} \left(\frac{M_U}{1 \text{ GeV}}\right)^{21/27} \left(\frac{2m_t \cdot 2m_b \cdot 2m_c}{1 \text{ GeV}^3}\right)^{2/27} \text{ GeV}. \quad (31)$$

Thanks to QCD, we have learned that the dominant contribution to the light-hadron masses is not the masses of the quarks of which they are constituted, but the energy stored up in confining the quarks in a tiny volume.[16] Our most useful tool in the strong-coupling regime is lattice QCD. Calculating the light hadron spectrum from first principles has been one of the main objectives of the lattice program, and important strides have been made recently. In 1994, the GF11 Collaboration [31]carried out a quenched calculation of the spectrum (no dynamical fermions) that yielded masses that agree with experiment within 5–10%, with good understanding of the residual systematic uncertainties. The CP-PACS Collaboration centered in Tsukuba has embarked on an ambitious program that will soon lead to a full (unquenched) calculation [32].

Neglecting the tiny "current-quark" masses of the up and down quarks, the scale parameter $\Lambda_{\text{QCD}}$ is the only mass parameter in QCD. It determines the scale of the confinement energy that is the dominant contribution to the proton mass. To a good first approximation,

$$M_{\text{proton}} \approx C\Lambda_{\text{QCD}}, \tag{32}$$

where the constant of proportionality $C$ is calculable using techniques of lattice field theory.

We conclude that, in a simple unified theory,

$$\frac{M_{\text{proton}}}{1 \text{ GeV}} \propto \left(\frac{m_t}{1 \text{ GeV}}\right)^{2/27}. \tag{33}$$

This is a wonderful result. Now, we can't use it to compute the mass of the top quark, because we don't know the values of $M_U$ and $\alpha_U$, and haven't yet calculated precisely the constant of proportionality between the proton mass and the QCD scale parameter. Never mind! The important lesson—no surprise to any twentieth-century physicist—is that the microworld does determine the behavior of the quotidian. We will fully understand the origin of one of the most important parameters in the everyday world—the mass of the proton—only by knowing the properties of the top quark.[17]

## WHAT IS THE DIMENSIONALITY OF SPACETIME?

Ordinary experience tells us that spacetime has 3 + 1 dimensions. Could our perceptions be limited? That is the question raised allegorically in *Flatland*, a Victorian fable published in 1880 by a British schoolmaster, Edwin Abbott Abbott (1839 – 1926), and still widely available [34].

---

[16] An accessible essay on our understanding of hadron mass appears in Ref. [30].
[17] For a fuller development of the influence of standard-model parameters on the everyday world, see Ref. [33].

Like any question that tests our preconceptions and unspoken assumptions, "What is the dimensionality of spacetime?" is a legitimate scientific question, to which we should return from time to time. It is given immediacy by recent theoretical work. For its internal consistency, string theory requires an additional six or seven space dimensions, beyond the $3+1$ dimensions of everyday experience. Until recently it has been presumed that the extra dimensions must be compactified on the Planck scale, with a compactification radius

$$R_{\text{unobserved}} \simeq \frac{1}{M_{\text{Planck}}} = \frac{1}{1.22 \times 10^{19} \text{ GeV}/c^2} = 1.6 \times 10^{-35} \text{ m} . \tag{34}$$

Part of the vision of string theory is that what goes on in the small curled-up dimensions does affect the everyday world: excitations of the Calabi–Yau manifolds determine the fermion spectrum, for example.[18]

The great gap between the electroweak scale of about $10^3$ GeV and the Planck scale of about $10^{19}$ GeV gives rise to the hierarchy problem of the electroweak theory [36]. The conventional approach to new physics has been to extend the standard model to understand why the electroweak scale (and the mass of the Higgs boson) is so much smaller than the Planck scale. A novel approach that has evolved over the past two years is instead to *change gravity* to understand why the Planck scale is so much greater than the electroweak scale [37]. Now, experiment tells us that gravitation closely follows the Newtonian force law down to distances on the order of 1 mm. Let us parameterize deviations from a $1/r$ gravitational potential in terms of a relative strength $\varepsilon_G$ and a range $\lambda_G$, so that

$$V(r) = -\int dr_1 \int dr_2 \frac{G_{\text{Newton}} \rho(r_1) \rho(r_2)}{r_{12}} \left[1 + \varepsilon_G \exp(-r_{12}/\lambda_G)\right] , \tag{35}$$

where $\rho(r_i)$ is the mass density of object $i$ and $r_{12}$ is the separation between body 1 and body 2. Elegant experiments that study details of Casimir and Van der Waals forces imply bounds on anomalous gravitational interactions, as shown in Figure 11. Below about a millimeter, the constraints on deviations from Newton's inverse-square force law deteriorate rapidly, so we are free to consider changes to gravity even on a small but macroscopic scale.

That is precisely the possibility raised by the interesting new idea that some extra dimensions of spacetime might be—relatively speaking—large.[19] The new idea is to consider that the $SU(3)_c \otimes SU(2)_L \otimes U(1)_Y$ standard-model gauge fields, plus needed extensions, reside on $3+1$-dimensional branes, not in the extra dimensions, but that gravity can propagate into the extra dimensions. How does this hypothesis change the picture? The dimensional analysis (Gauss's law, if you like) that relates Newton's constant to the Planck scale changes. If gravity propagates not only in

---

[18] For a gentle introduction to the aspirations of string theory, see Ref. [35].
[19] For an enthusiastic overview, see Marcus Chown, "Five and Counting ...," *New Scientist* **160**, No. 2157 (24 October 1998), http://www.newscientist.com/ns/981024/fifth.html.

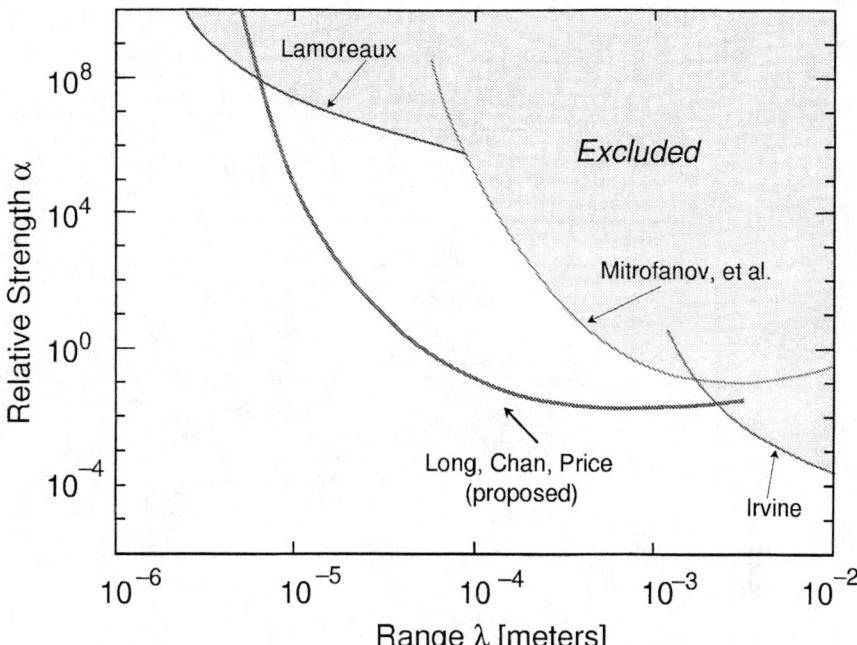

**FIGURE 11.** Experimental limits on the strength $\varepsilon_G$ (relative to gravity) versus the range $\lambda_G$ of a new long-range force, together with the anticipated sensitivity of a new experiment based on small mechanical resonators [38].

the 3+1 dimensions of Minkowski space, but also in $n$ extra dimensions with radius $R$, then

$$G_{\text{Newton}} \sim M_{\text{Planck}}^{-2} \sim M^{\star -n-2} R^{-n}, \qquad (36)$$

where $M^\star$ is gravity's true scale. The correlation between $M^\star$ and the size $R$ of the $n$ large extra dimensions is given in Table 1 for some representative cases. Could the extra dimensions be quasimacroscopic? One large extra dimension seems excluded, since gravity within the solar system obeys Newton's force law in three (not more) spatial dimensions.[20] Notice that if we boldly take $M^\star$ to be as small as 1 TeV/$c^2$, then the scaling law (36) requires the radius of the extra dimensions to be smaller than about 1 mm, for $n \geq 2$.[21] Deviations on the millimeter scale are allowed by current knowledge of gravity on short distances, but will be challenged soon.

If we use the four-dimensional force law to extrapolate the strength of gravity from low energies to high, we find that gravity becomes as strong as the other

---

[20] The semimajor axis of Pluto's orbit is about $6 \times 10^{12}$ m.
[21] Other observational constraints [39] suggest that it is more prudent to choose $M^\star \approx$ 10 - 100 TeV/$c^2$, for which at least three large extra dimensions are required to alter gravity on the millimeter scale.

**TABLE 1.** Radius $R$ of $n$ large extra dimensions for low values of the scale $M^\star$ of gravity.

|  | $n=1$ | $n=2$ | $n=3$ | $n=6$ |
|---|---|---|---|---|
| $M^\star = 1$ TeV/$c^2$ | $10^{13}$ m | $10^{-3}$ m | $10^{-8}$ m | $10^{-14}$ m |
| $M^\star = 10$ TeV/$c^2$ | $10^{10}$ m | $10^{-5}$ m | $10^{-10}$ m | $10^{-15}$ m |
| $M^\star = 100$ TeV/$c^2$ | $10^7$ m | $10^{-7}$ m | $10^{-12}$ m | $10^{-16}$ m |

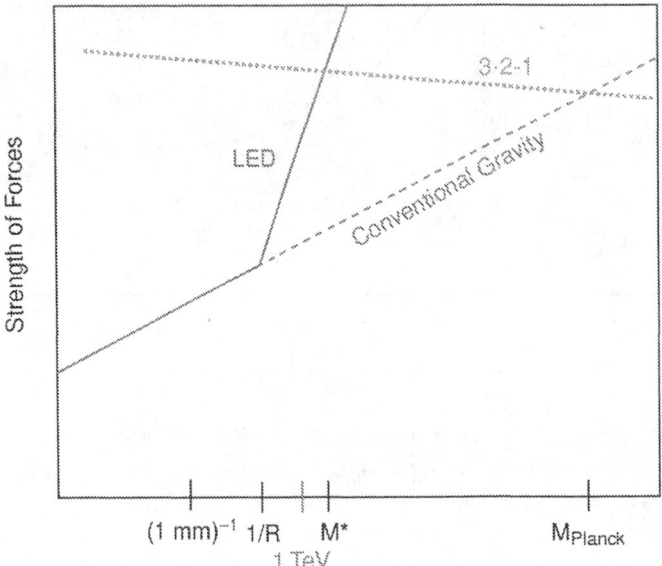

**FIGURE 12.** One of these extrapolations (at least!) is false.

forces on the Planck scale, as shown by the dashed line in Figure 12. If the force law changes at an energy $1/R$, as the large-extra-dimensions scenario suggests, then the forces are unified at an energy $M^\star$, as shown by the solid line in Figure 12. What we know as the Planck scale is then a mirage that results from a false extrapolation: treating gravity as four-dimensional down to arbitrarily small distances, when in fact—or at least in this particular fiction—gravity propagates in $3+n$ spatial dimensions. The Planck mass is an artifact, given by $M_{\text{Planck}} = M^\star (M^\star R)^{n/2}$.

Extra dimensions seen by standard-model particles cannot be larger than about $1$ TeV$^{-1} \approx 10^{-19}$ m, or we would already have probed them at LEP2, HERA, and the Tevatron. We will see in a moment that future collider experiments can examine two or more large extra dimensions with a low gravitational (string) scale $M^\star$.

Although the idea that extra dimensions are just around the corner—either on the submillimeter scale or on the TeV scale—is preposterous, it is not ruled out by observations. For that reason alone, we should entertain ourselves by entertain-

**TABLE 2.** Sensitivities to large extra dimensions, expressed as 95% CL upper limits on the radius $R$ of $n$ extra dimensions (from Ref. [41]).

|  | $n = 2$ | $n = 4$ | $n = 6$ |
|---|---|---|---|
| LEP2 | $4.7 \times 10^{-4}$ m | $1.9 \times 10^{-11}$ m | $6.9 \times 10^{-14}$ m |
| Tevatron Run 1 | $1.1 \times 10^{-3}$ m | $2.4 \times 10^{-11}$ m | $5.8 \times 10^{-14}$ m |
| Tevatron Run 2 | $3.9 \times 10^{-4}$ m | $1.4 \times 10^{-11}$ m | $4.0 \times 10^{-14}$ m |
| Large Hadron Collider | $3.4 \times 10^{-5}$ m | $1.9 \times 10^{-12}$ m | $6.1 \times 10^{-15}$ m |
| 1-TeV polarized linear collider | $1.2 \times 10^{-5}$ m | $1.2 \times 10^{-12}$ m | $6.5 \times 10^{-15}$ m |

ing the consequences. Any particle can radiate a graviton into extra dimensions. An extradimensional graviton is only gravitationally coupled, and so will not interact in the detector: the gravitons go off into extra dimensions and are lost. Their signature is missing energy, $\not{E}_T$. These processes, individually tiny, may be observable because the number of excitable modes is very large. Many authors [39] have considered the gravitational excitation of a tower of Kałuza–Klein modes in the extra dimensions, which would give rise to a missing (transverse) energy signature in collider experiments [40]. We call these excitations *provatons*, after the Greek word πρόβατο for a sheep in a flock. For proton-antiproton collisions, the experimental signatures include $\bar{p}p \to \text{jet} + \not{E}_T$ (parton + graviton), and $\bar{p}p \to \ell^+\ell^- + \not{E}_T$ ($\ell^+\ell^-$ + graviton).

*The Detector Challenge.* To establish a jet + $\not{E}_T$ signature at a hadron collider, we require a hermetic detector with well-controlled $\not{E}_T$ tails to establish and quantify the missing-energy signature; highly efficient rejection of cosmic-ray and accidental triggers; the ability to reject triggers originating from jet mismeasurements; and control over physics backgrounds, including $Z^0$ + jets and $W^\pm$ + jets. The experimental sensitivity to provatons depends on collider energy, luminosity, and species, and the number of large extra dimensions.

A representative analysis of the constraints that may be inferred from anomalous single photon production at $e^+e^-$ colliders and from monojet production at hadron colliders if no signal is seen is shown in Table 2. If a missing-energy signal is found, it will be a challenge to distinguish an extradimensional signal from the classic $R$-parity–conserving signature for supersymmetry. We should be so lucky!

## The Outlook for Extra Dimensions

We are only beginning to explore the possible implications of almost-accessible extra dimensions. Among the fascinating new worlds to explore are the Randall-Sundrum mechanism for localizing gravity in an extra dimension (in the vicinity of a brane or domain wall) [42], and the speculation of Arkani-Hamed and Schmaltz that the fermion mass hierarchy reflects fermion wave packets separated in an extra dimension [43]. The characteristic missing-energy signature of a Kałuza–Klein excitation will be hard to distinguish from other new physics. We need to think in more detail about the backgrounds and optimal search techniques. The search for large extra dimensions reinforces the informative metaphor of a collider and its detectors as an ultramicroscope. Are extra dimensions large enough to see? It is also interesting to consider what might we observe above the threshold for exciting extra dimensions. Although the basic concepts that underlie these speculations have a sound basis in string theory, most of the scenarios put forward so far have more to do with storytelling than with theoretical rigor. However, it is plain that our inability to disprove at once the outlandish idea of large extra dimensions is a measure of the potential for experimental surprises, and an indication of Nature's capacity to amaze us!

## FINAL THOUGHTS

To a great degree, the progress of particle physics has followed from progress in accelerator science and instrumentation. There is no substitute for experiment, and experiment requires inventions in both hardware and software and continuous innovation in analysis. The slogan, "Yesterday's sensation is today's calibration and tomorrow's background,"[22] embodies both the challenge and the opportunity of advances in experimental technique.

In the middle of the revolution we are experiencing—indeed, making—in our conception of Nature, when we deal with fundamental questions about our world, including

> *What are the symmetries of Nature, and how are they hidden from us?*
>
> *Are the quarks and leptons composite?*
>
> *Are there new forms of matter, like the superpartners suggested by supersymmetry?*
>
> *Are there more fundamental forces?*
>
> *What makes an electron an electron, a neutrino a neutrino, and a top quark a top quark?*
>
> *What is the dimensionality of spacetime?*

we cannot advance without new instruments that extend our senses and allow us to create—and understand—new experience far beyond the realm of everyday human experience.

I wish you a productive two weeks in Istanbul, and hope that what you learn about instrumentation, the spirit of experimentation, and the habits of mind necessary to make detectors act as reliable extensions of our senses will open new horizons for you and for science.

---

[22] I believe that this formulation is due to V. L. Telegdi.

# ACKNOWLEDGMENTS

It is a pleasure to thank our Istanbul hosts for their energetic and delightful hospitality. I commend the Panel on Instrumentation Innovation and Development of the International Committee for Future Accelerators for their sponsorship of this series of instrumentation schools, and salute the leaders of the instrumentation courses for their infectious enthusiasm. On behalf of the students, I thank the great laboratories of particle physics, particularly CERN and Fermilab, for providing much of the apparatus that makes possible the hands-on experience that gives this school its special character. Finally, I would like to express my thanks to the students who came to Istanbul from many interesting parts of the globe and whose lively curiosity made my visit memorable. I hope I may have the opportunity to welcome many of you to Fermilab in the future.

# REFERENCES

1. Chadwick, J., *Verh. Deutsch. Phys. Ges.* **16**, 383 (1914).
2. Pauli, W., "Zur älteren und neueren Geschichte des Neutrinos," in *Collected Scientific Papers,* edited by R. Kronig and V. F. Weisskopf (Interscience, New York, 1964), volume 2, p. 1313.
3. Pais, A., "Introducing Atoms and Their Nuclei," in *Twentieth Century Physics,* edited by Brown, L. M., Pais, A., and Pippard, B. (Institute of Physics Publishing, Bristol & Philadelphia; American Institute of Physics, New York, 1955), vol. II, p. 43.
4. Reines, F., and Cowan, C. L., Jr., *Phys. Rev.* **92**, 830 (1953).
5. Cowan, C. L., Jr., Reines, F., Harrison, F. B., Kruse, H. W., and McGuire, A. D., *Science* **124**, 103 (1956).
6. Danby, G., *et al.*, *Phys. Rev. Lett.* **9**, 36 (1962). Schwartz, M. L., *Rev. Mod. Phys.* **61**, 527 (1989).
7. http://fn872.fnal.gov/.
8. Gandhi, R., Quigg, C., Reno, M. H., and Sarcevic, I., *Astropart. Phys.* 5, 81 (1996).
9. Conrad, J. M., Shaevitz, M. H., and Bolton, T., "Precision Measurements with High Energy Neutrino Beams," *Rev. Mod. Phys.* **70**, 1341 (1998), hep-ex/9707015.
10. Caso, C., *et al.* (Particle Data Group), *Euro. Phys. J.* **C3** (1998) 1, and 1999 off-year partial update for the 2000 edition available at http://pdg.lbl.gov/.
11. De Rújula, A., "Puzzles and Challenges in Neutrino Physics," available at http://lyoinfo.in2p3.fr/nufact99/talks/ruj.html.
12. Bahcall, N. A., Ostriker, J. P., Perlmutter, S., and Steinhardt, P. J., *Science* **284** (1999) 1481.
13. Turner, M. S. "Dark Matter and Energy in the Universe" astro-ph/9901109, to be published in *Physica Scripta* (Proceedings of the Nobel Symposium, Particle Physics and the Universe; Enkoping, Sweden, August 20-25, 1998), offers $0.003 \lesssim \Omega_\nu \lesssim 0.15$.
14. Haxton, W. C., and Holstein, B. R., "Neutrino Physics," *Am. J. Phys.* **68**, 13 (2000) hep-ph/9905257.

15. Conrad, J. M., in *Proceedings of the 29th International Conference on High Energy Physics,* Vancouver, edited by A. Astbury, D. Axen, and J. Robinson (World Scientific, Singapore, 1999), p. 25 (hep-ex/9811009), and "Where in the World Is the Oscillating Neutrino?" talk at Inner Space / Outer Space 1999 and PANIC99, available at http://portia.fnal.gov/~jconrad/isos.html; Fisher, P., Kayser, B., and McFarland, K. S., "Neutrino Mass and Oscillation," *Annu. Rev. Nucl. Part. Sci.* **49**, 481 (1999), hep-ph/9906244; DiLella, L., "Accelerator and Reactor Neutrino Experiments," hep-ex/9912010.
16. Bahcall, J. N., *Neutrino Astrophysics* (Cambridge University Press, New York, 1989). Bahcall, J. N. (editor), *Solar Neutrinos: the first thirty years* (Addison–Wesley, Reading, Massachusetts, 1994).
17. Bahcall, J. N., and Pinsonneault, M. H., "Solar models with helium and heavy element diffusion," *Rev. Mod. Phys.* **67**, 781 (1995) hep-ph/9505423.
18. Bahcall, J. N., "Solar Neutrinos: Where We Are, Where We Are Going," *Ap. J.* **467**, 475 (1996) hep-ph/9512285.
19. Bahcall, J. N., Basu, S., and Pinsonneault, M. H., "How uncertain are solar neutrino predictions?" *Phys. Lett.* **B433**, 1 (1998) astro-ph/9805135.
20. Fukuda, Y., *et al.* (Super-Kamiokande Collaboration), *Phys. Rev. Lett.* **81**, 1562 (1998).
21. Mann, W. A. "Atmospheric Neutrinos and the Oscillations Bonanza," Plenary talk at the XIX Int. Symposium on Lepton and Photon Interactions at High Energies, Stanford, Aug. 1999, hep-ex/9912007.
22. Athanassopoulos, C., *et al.* (LSND Collaboration), *Phys. Rev. Lett.* **77**, 3082 (1996); **81**, 1774 (1998); *Phys. Rev.* **C58**, 2489 (1998).
23. Geer, S., *Phys. Rev.* **D57**, 6989 (1998), **59**, 039903E (1999).
24. De Rújula, A., Gavela, M. B., and Hernández, P., *Nucl. Phys.* **B547**, 21 (1999) hep-ph/9811390.
25. Barger, V., Geer, S. and Whisnant, K., "Long Baseline Neutrino Physics with a Muon Storage Ring Neutrino Source" hep-ph/9906487.
26. Quigg, C., "*Top*–ology," *Phys. Today* **50**, 20 (May, 1997); extended version circulated as FERMILAB–PUB–97/091–T hep-ph/9704332.
27. Simmons, E. H., "Thinking about top: Looking outside the standard model," hep-ph/9908511.
28. Willenbrock, S., "Thinking about top within the standard model," hep-ph/9905498.
29. Georgi, H., Quinn, H. R., and Weinberg, S., *Phys. Rev. Lett.* **33**, 451 (1974).
30. Wilczek, F., "Mass without Mass I: Most of Matter," *Phys. Today* **52** (11) 11 (November, 1999).
31. Butler, F., *et al.* (GF11 Collaboration), *Nucl. Phys.* **B430**, 179 (1994).
32. Aoki, S., *et al.* (CP–PACS Collaboration), *Phys. Rev. Lett.* **84**, 238 (2000).
33. Cahn, R. N., *Rev. Mod. Phys.* **68**, 951 (1996).
34. Abbott, E. A., *Flatland: a Romance of Many Dimensions.*
35. Greene, B., *The Elegant Universe* (Norton, New York, 1999).
36. Gildener, E., *Phys. Rev.* **D14**, 1667 (1976); Weinberg, S., *Phys. Lett.* **82B**, 387 (1979).
37. Antoniadis, I., "A Possible New Dimension at a Few TeV," *Phys. Lett.* **B246**, 377 (1990). Lykken, J. D., "Weak Scale Superstrings," *Phys. Rev.* **D54**, 3693 (1996), hep-

th/9603133. Arkani-Hamed, N., Dimopoulos, S., and Dvali, G., *Phys. Lett.* **B429**, 263 (1998), "The Hierarchy Problem and New Dimensions at a Millimeter,"hep-ph/9803315. Ignatios Antoniadis, I., Arkani-Hamed, N., Dimopoulos, S., and Dvali, G., "New Dimensions at a Millimeter to a Fermi and Superstrings at a TeV," *Phys. Lett.* **B436**, 257 (1998), hep-ph/9804398.

38. Long, J. C., Chan, H. W., and Price, J. C., "Experimental status of gravitational-strength forces in the sub-centimeter regime," *Nucl. Phys.* **B539**, 23 (1999).
39. Antoniadis, I., Benakli, K., Quirós, M., "Production of Kaluza–Klein States at Future Colliders," *Phys. Lett.* **B331**, 313 (1994) hep-ph/9403290. Dienes, K. R., Dudas, E., and Gherghetta, T. "Extra Space-Time Dimensions and Unification," *Phys. Lett.* **B436**, 55 (1998) hep-ph/9803466. Giudice, G., Rattazzi, R., and Wells, J., "Quantum gravity and extra dimensions at high-energy colliders," *Nucl. Phys.* **B544**, 3 (1999), hep-ph/9811291. Nussinov, S., and Shrock, R., "Some remarks on theories with large compact dimensions and TeV scale quantum gravity," *Phys. Rev.* **D59**, 105002 (1999) hep-ph/9811323. Hewett, J. L., "Indirect Collider Signals for Extra Dimensions," *Phys. Rev. Lett.* **82**, 4765 (1999) hep-ph/9811356.
40. Spiropulu, M., "Experimental Signals From Higher Dimensions," invited talk at the Aspen Winter Physics Conference (2000), transparencies available at http://www.hepl.harvard.edu/~ms/aspen2000/aspen2000.pdf.
41. Mirabelli, E., Perelstein, M., and Peskin, M. E., "Collider Signatures of New Large Space Dimensions," *Phys. Rev. Lett.* **82**, 2236 (1999) (hep-ph/9811337).
42. Randall, L., and Sundrum, R., "A large mass hierarchy from a small extra dimension," *Phys. Rev. Lett.* **83**, 3370 (1999) hep-ph/9905221; "An alternative to compactification," **83**, 4690 (1999) hep-th/9906064.
43. Arkani-Hamed, N. and Schmaltz, M., "Hierarchies without Symmetries from Extra Dimensions," hep-ph/9903417; Eugene A. Mirabelli, E. A. and Schmaltz, M., "Yukawa Hierarchies from Split Fermions in Extra Dimensions," hep-ph/9912265.

# Confronting New Technological Challenges in HEP

*Aurore SAVOY-NAVARRO*
*LPNHE - UNIVERSITES de PARIS 6&7 - IN2P3/CNRS-France*

## Abstract

*The new technological challenges that will have to be confronted in HEP are mainly due to the new physics issues. What is beyond the standard model? That is the question. This review will first list the physics demands in order to explore this unknown world. It will then show with appropriate examples, how the new physics will require confronting new technological challenges in: designing new accelerators, developing new detectors, designing new front-end readout systems and using the new software and hardware tools for the online readout and DAQ systems.*

## 1 Introduction

It is a very crucial and exciting moment, at the meeting point between two very important periods of time in HEP. The first period covers the last twenty years, where fundamental discoveries and measurements fully established the Standard Model (SM). The second period includes the next twenty years, which should hopefully see the discovery and explanation of the Beyond-Standard-Model (BSM) physics.

In both cases, i.e. SM and BSM, the HEP experimentalists have had and will have to confront new high technological challenges. This was and will be ever more the key of success. This review will therefore show how new physics searches will require confronting new technological challenges, of all kinds, in most of the main experimental aspects: machines, detectors, associated front-end (FE) and readout electronics, data processing and real-time. The first section reviews the physics motivations and requirements in terms of machines and detectors. It also recalls the main physics achievements in the past 20 years as well as the physics goals for the next 20 years. And it gives a global schema of the main technological challenges in HEP.

Each of the following sections will address a specific technological issue.

The first issue concerns what the next machine will be, and it thus describes the next lepton collider, i.e. the $e^+e^-$ or $\mu^+\mu^-$ projects and the VLHC, i.e. the Very Large Hadron Collider project. The second issue is: What detectors? We will especially concentrate on the technological challenges addressed by the tracking systems, and in particular the Si trackers. It includes the integration of the tracking devices in the overall triggering scheme and the problem of radiation hardness. The third issue concerns the front-end readout and points out the growing importance of an all-digital front-end readout and of the introduction of the microelectronics technologies in HEP experiments; going to digital front-end and readout electronics is improving a lot the potentiality of the online processing and monitoring of the data. This is directly connected to the last issue, the technological challenges addressed by the DAQ system and the *"real-time processing and analysis"* of the data.

## 2 High technological challenges: physics motivations

The Standard Model works well but a *"beyond-the-standard-model"* is needed to explain the gap between the electroweak scale and the inaccessible (?) Planck mass, i.e. the so-called *"hierarchy problem"*. Higgs(es), sparticles, extra-dimensions and/or else ??? have to be looked for.

Thus the beyond-the-standard-model implies:

- Higher mass particles, i.e. masses above 100 GeV and up to?

- Higher multiplicity events, let us recall that there will be 1000 particles in average for a standard QCD event at the LHC

- Trickier disintegration modes, and thus:

  - cascade decays
  - production of particles to be tagged besides the *"traditional"* electrons, muons, photons and pions. This includes: heavy flavours, i.e. c-, b- and t-quarks, and also $\tau$'s, $\nu$'s or $\nu$-like particles.

- Rare decays versus high backgrounds

So the BSM requires:

- *High c.m.s. and luminosity machines; it would be even better to have an hadronic and a leptonic machine contemporarily, for questions of complementarity.*

- *Very performing detectors; the tracking system is particularly crucial.*

- *Very sophisticated triggering, including the "tagging" of various "physics objects".*

- *High-rate processing readout and performing DAQ systems, including the capability of "real-time analysis".*

These are the issues that trigger the high technological challenges discussed in this review. Before that, let us summarize the main HEP events of the last 20 years and what the physics prospects are for the next 20 years. As schematized in Fig. 1, in the past 20 years, several crucial steps forward were achieved in HEP. It mainly happened at the two $p\bar{p}$ colliders, namely the S$p\bar{p}$S machine at CERN and the Tevatron at FNAL, and at LEP, at CERN, which is at present the highest c.m. energy $e^+e^-$ machine. The main events that occurred are first the W and $Z^0$ discovery at the S$p\bar{p}$S machine by the UA1 and UA2 experiments in the early 80's. This was the entrance in the $W\&Z^0$ sector. The experiments were working in this mass range until the end of the 90's. In 1997, when LEP reached 183 GeV in c.m.s., the four LEP experiments could penetrate the $W\&Z^0$ pair sector.

The $p\bar{p}$ machines allowed another discovery; the 6th quark, the top, was found at the Tevatron in 1994. The four LEP experiments achieved, from 1989 to 1994, a series of fundamental high-precision measurements at the $Z^0$ peak. This was instrumental in testing with very high accuracy, the SM but also in determining what is left for new physics with the $Z^0$ width measurement.

LEPII, the second round of LEP, allowed the start of the exploration of the W&$Z^0$ pair sector. In 1999, LEPII reached 200 GeV c.m.s.. As outcome, the four LEP experiments [1] were able to determine with a high resolution, of the order of 40 MeV, the W-mass and therefore to compete with the Tevatron experiments, i.e. CDF and D0. LEP was also a crucial and unique probe to explore the Higgs sector.

Moreover, both LEP and the Tevatron, in the last 5 years of the past century, achieved a detailed search for sparticles, covering a wide range in the allowed parameter space. They demonstrated that having, at the same time, a leptonic and a hadronic collider is extremely rewarding in terms of complementarity and also of competitivity.

Now what's next? As sketched in Fig. 1, right at the turn of this new millenium, two types of machines will be running and producing most of the physics data in HEP. On the one hand, there are the B-factories, with Babar at SLAC (USA), Belle at KEK (Japan) and HeraB at DESY (Germany). They should bring more definite answers about CP-violation and a larger coverage

---

[1] *The four LEP experiments are: ALEPH, DELPHI, L3 and OPAL.*

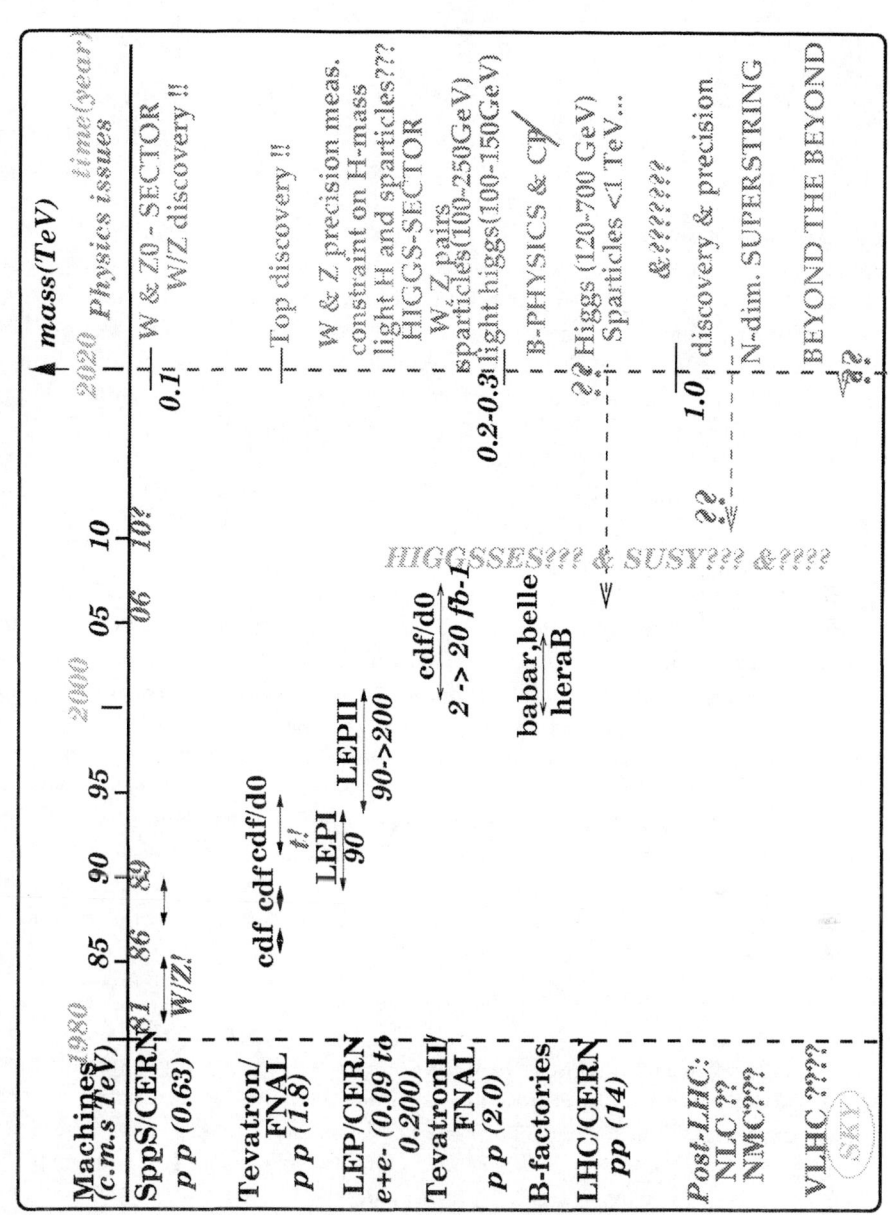

Figure 1: A (The) Future for (of) HEP

of the so-called B-physics. On the other hand and at the same time, the Tevatron at FNAL will run at 2 TeV c.m. energy and an increased luminosity, with two totally renewed experiments: CDF and D0. Important outcomes are expected, in particular in the domains of B-physics and CP-violation [2](somehow in a complementary way to what will be explored with B-factories), as well as in the Higgs and SUSY searches.

Around 2006, the LHC is going to relieve the Tevatron. But is this the end of the story? The HE physicist should, by no means, take a rest at this stage; on the contrary (s)he has to think, already now, of what the next machine will be besides the LHC. This issue is discussed in detail in the next section, but what is roughly sketched in Fig. 1, is that various possibilities are investigated and also the physics needs and benefits from having another machine; among the reasons there are, for instance, the search for Supersymmetry/Higgs(es) at a mass range difficult to access or not accessible with the LHC, and also the need for complementarity and finally some specific issues that are better addressed with leptonic colliders. This is at present an important point of discussion in our community.

To complete this brief panorama it should be pointed out that the past decade has been especially fruitful for the development of new detecting technologies. First CDF undertook, since the very beginning, an almost continuous upgrade of its detector capabilities, and particularly of the tracking system. It was the first $p\bar{p}$ detector to successfully experience a microvertex and to include the tracking information first in the second-level triggering and now even in the first-level; the b-tagging, i.e. the processing of the microvertex information, is also incorporated in the second-level triggering, for Run II. Thanks to these pioneering developments linked to the tracking, CDF demonstrated that $p\bar{p}$ detectors can compete with $e^+e^-$ experiments. For the Run II, both CDF and D0 are being renewed with very advanced detector techniques; This is in order to confront the challenging conditions imposed by the next Tevatron run.

Thus, accompanying and even permitting these important physics advances and discoveries, a lot of progress has been achieved on the technical and technological sides. This concerns both the machines and the detectors. Moreover, the entrance to the *"supercolliders era"*, and especially the LHC project, has triggered an impressive R&D activity during the last decade.

Most of the high technological challenges that will have to be confronted in HEP, are listed in Figs. 2 and 3, and partially discussed in the following sections.

---

[2] *A Btev experiment is also in project at the Tevatron at FNAL, fully dedicated to B-Physics, in much the same way as the LHCb experiment at the LHC*

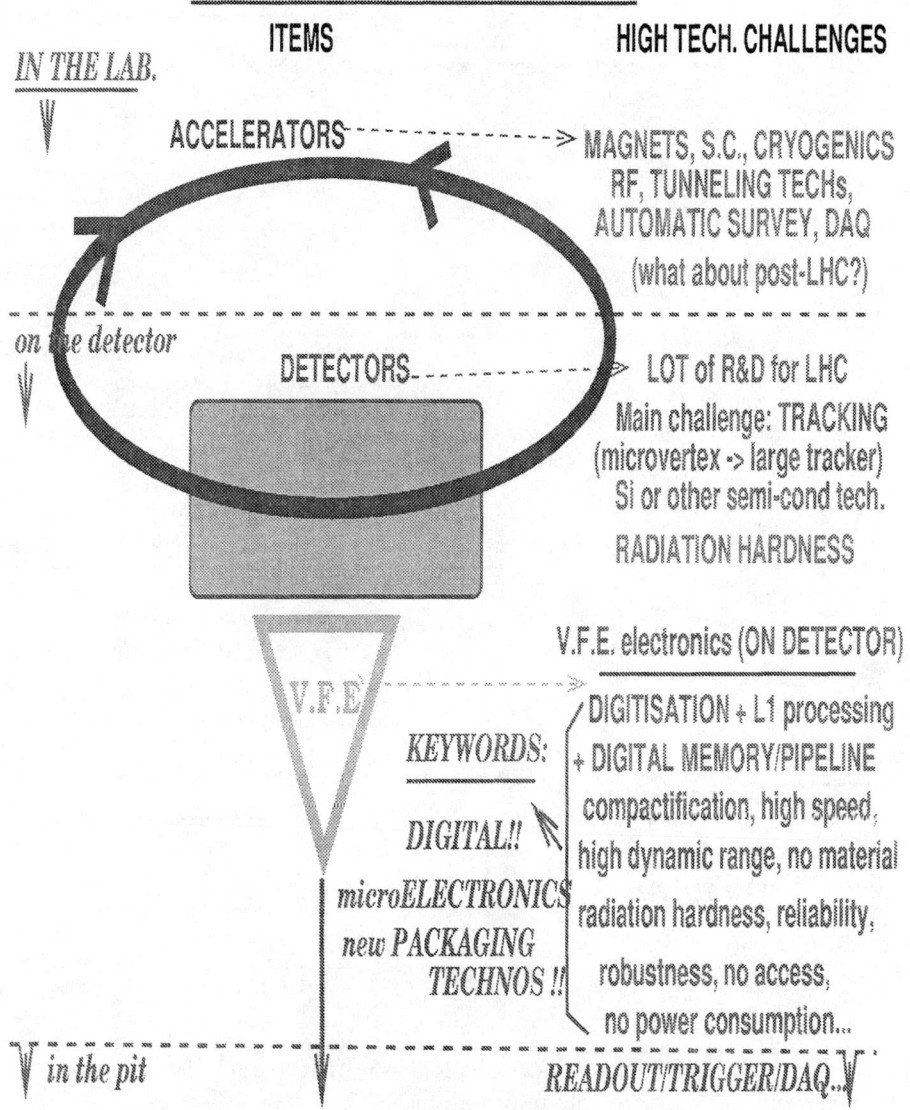

Figure 2: High Technological Challenges in HEP: machines and detectors

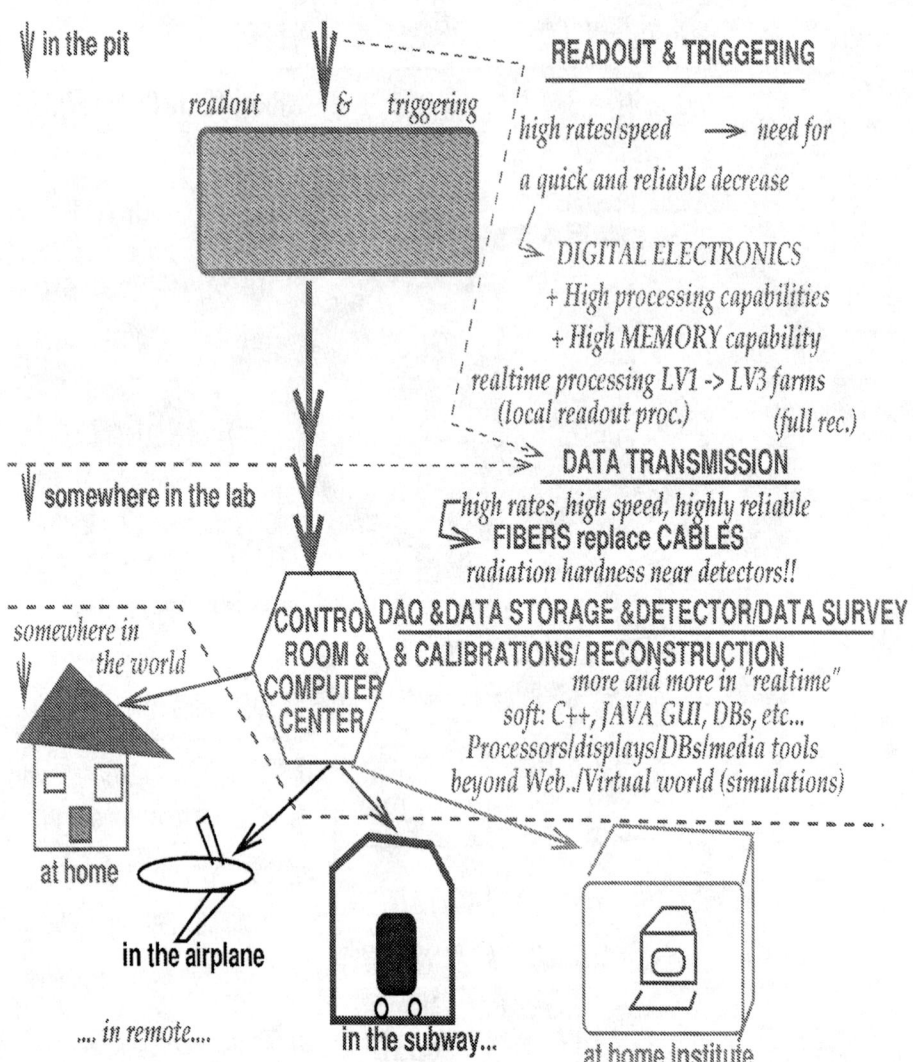

Figure 3: High Technological Challenges in HEP (cont'd): electronics and processing of the informations

This quick overview shows that the HEP field, in the past 20 years, was particularly vivid and successful and produced impressive advances in terms of physics results as well as of detecting techniques. This overview also demonstrates that we are just at the very beginning of a period that may be even more exciting. The next 20 years will hopefully bring crucial discoveries; but to succeed, high technological challenges must be confronted. The aim of this presentation is to discuss some of them, among the most striking and important ones.

## 3 Confronting high technological challenges in HEP: first issue, what is the next machine?

This question relates to another crucial question: *Is there a future or a survival for HEP after LHC?* And the main issue is what machine to build and when?

Indeed the sky does not give all the needed information concerning the source (or the core) of the interaction, as for instance, what the type and the energy of the incident particles are. Therefore, the sky mainly permits *"observations"*. Nothing replaces an accelerator, i.e. a lab experiment.

In the early days of HEP, people started with cosmic-ray experiments, using the sky as a source of accelerated particles. Then, thanks to technological advances, it was possible to re-create, in the laboratory, beams of accelerated particles of different types. This started the *era of accelerators*, which has seen, over the last fifty years, an impressive series of successful developments. The LHC is currently the ultimate step in this quest for re-creating in the laboratory, i.e. with accelerators, beam of particles that reach higher and higher energies with higher and higher luminosities.

The main question now is: *Does o ανθρoπos know how to build post-LHC machines? So what are the new technological challenges to be confronted and Are we able to overcome them?*

For several years, people have been thinking about different possible solutions, among which:

- NLC or the Next Linear Collider, i.e. $e^+e^-$ linear colliders, with an energy per beam going from 0.25 up to 1 or 1.5 TeV. It also includes some alternatives such as an $e\gamma$ or even a $\gamma\gamma$ machine.

- NMC, i.e. the Next Muon Collider, a $\mu^+\mu^-$ collider with an energy per beam going from 0.05 up to 1.5 TeV.

- VLHC, i.e. Very Large Hadron Collider, a pp collider with an energy per beam that could reach 30 and up to 100 TeV.

Is this just science fiction? If not, which of these machines will be the first to be built? The answer to the second question depends on physics issues versus technological challenges and versus money.

For the physics issues, the past experience tends to demonstrate that indeed it is quite important to have, at the same time, a lepton collider and a hadron collider. This is for complementarity, completeness and competition purposes. The scenarios we had so far are: *Petra* and *PEP*, e$^+$e$^-$ machines, versus the ISR pp collider in the 70's and the S$p\bar{p}$S in the 80's. Tevatron I was contemporary to SLC, LEP I and II in the 90's, and Tevatron II will relieve LEP200 until 2006-2008. Then the LHC will take over in 2007–2008 for the physics and at that point it would be very important to have an e$^+$e$^-$ linear collider as companion to the LHC, i.e. starting around 2010. Besides, is it better to have a very high energy muon collider and/or a VLHC pp collider or ???, somewhere around the year 2020. But, at that stage, we are almost science fictioning. However, even science fiction needs R&D already now. And remember that what is called *"science fiction"* now, will become just *"routine"* in ten years'time. All these issues are discussed in the following subsections.

## 3.1 NLC for Next Lepton Collider: some new technological challenges

Different possibilities are currently under study at the R&D level. The first question is: Which lepton collider: e$^+$e$^-$ possibly becoming an e$\gamma$ or even a $\gamma\gamma$ collider ? Or else a $\mu^+\mu^-$ collider? What c.m. energy? We are studying here the main challenges to be confronted, in order to build the next generation of e$^+$e$^-$ machines or a $\mu^+\mu^-$ collider and the compared (dis)advantages.

### 3.1.1 Electron–Positron Linear Colliders

Building high c.m. energy and high luminosity e$^+$e$^-$ colliders [1] implies the three main issues as follow:

⋆ **First issue:** *Synchroton radiation and so Linear Collider technology*

A relativistic particle undergoing centripetal acceleration radiates at a rate given by the Larmor formula multiplied by the fourth power of the Lorentz factor: $P = e^2\alpha^2\gamma^4/6\pi\epsilon_0 c^3$, where $\alpha = v^2/\rho$ is the centripetal acceleration of a particle, with speed $v$, undergoing deflection with a radius of curvature $\rho$. In a synchrotron that has a constant radius of curvature within bending magnets, the energy loss (W), due to synchroton radiation per turn is the above multiplied by the time spent in bending magnets, $2\pi\rho/v$.

So for electrons: $W = 8.85 \times 10^{-5} E^4/\rho$ MeV per turn, where the energy $E$ is in GeV and $\rho$ is in km. For example, at LEPI, $W = 130$ MeV/turn, and at LEPII, 2.8 GeV/turn. So going to 200 GeV per beam in the same tunnel would multiply the value of W by 16, i.e. W would be about 45 GeV/turn.

*This is the reason why the circular collider technique for an $e^+e^-$ machine must be abandoned at this stage and replaced by the linear collider technique.*

The principle of a linear collider is sketched in Fig. 4. As shown there,

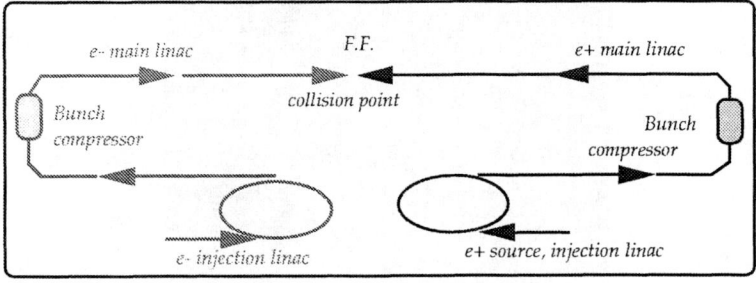

Figure 4: Principle of an $e^+e^-$ linear collider

the electrons and positrons are first accelerated by the *injector linacs* up to 2 GeV; they are then cooled in *damping rings* to make ultra-low emittance and led through pre-accelerators to the *main linacs*, reaching, at their ends, the maximum beam energy. They are finally squeezed by a *Final Focus System* to a nanometre level.

The SLC at SLAC, which is the highest c.m. energy $e^+e^-$ linear collider, at present, is 4 km long. As we will see in the next subsection, NLC (the Next Linear Collider) would be of the order of 25 km.

⋆ **Second issue:** *High accelerating fields and so gigawatt technology*

The accelerating fields in the traditional linear colliders are 10 MV/m; Therefore, 75 km long accelerating tubes would be needed to get 750 GeV/beam. This would mean a total length of more than 150 km. What about the price of such a collider!? In order to reduce the collider length to less than or at most 25 km, accelerating fields of more than 120 MV/m are required. This means that *High Power Klystrons (HPK)* are needed. By HPK we mean klystrons capable of producing peak power of 2 GW per 10 m, and an accelerating structure that can put up with such a high power.

Figure 5 shows a photograph of a klystron. Klystrons are microwave tubes that use the same technology as airport radar and home microwave ovens. But, the high technological challenges here, i.e. the gigawatt technology, is to provide the necessary level of power to the NLC, the HPKs must provide a million times the power of an ordinary light bulb or about 75 thousand times the power of a powerful kitchen microwave oven.

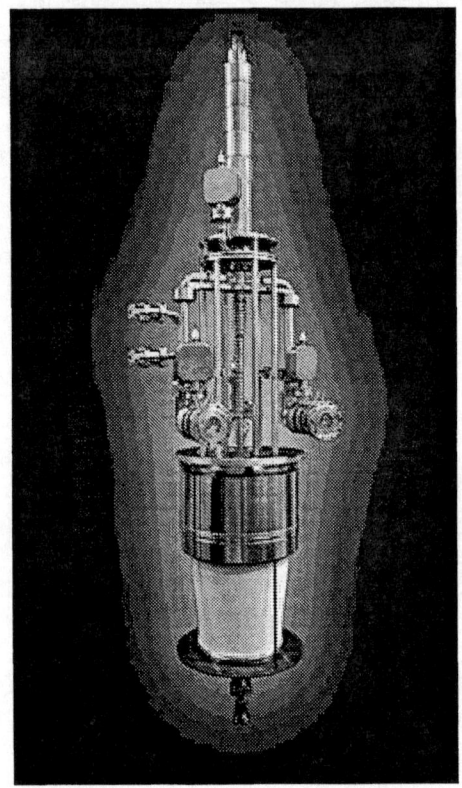

Figure 5: Photograph of a klystron

⋆ **Third issue:** *Ultra-low emittance/final focus and so nanometre technology*

In order to increase by a factor 100 the currently available luminosity, both beams must be squeezed down to the nanometre level, and they thus must be ultra-low emittance beams. Therefore, after full acceleration in the main linacs, the *Final Focus System* will have to focus the electron and positron beams down to the required nanometre level and make them collide stably. This requires strong lenses (i.e. focusing magnets) and an active feedback system to remove vibrations. The nanometre technology, which is used to make the nanometre-sized beams collide stably head-on, is the technology developed for VLSI fabrication or ultra-high precision machining. This is experienced at the FFTB (Final Focus Test Beam) at SLAC. For several years, the FFTB project has been developing the *"nanometre technology"*. Figure 6 shows a quadrupole magnet used to focus the beam at the Final Focus section. It shows the support structure that allows fine adjustments in the magnet orientation for precise beam focusing.

Figure 6: Quadrupole magnet in the FFTB at SLAC

In conclusion, there are three key-issues for the next generation of $e^+e^-$ machines, namely: 1) to use the *linear collider technique*, which will keep a reasonable machine length and be more cost saving; 2) however, this option requires high accelerating beams, HPKs, and thus quite a challenging *gigawatt technology*, and 3) an ultra-low emittance and stable precise final focus, thus a *nanometre technology*.

Several projects are studied in the world: *CLIC* at CERN, which would be able to reach up to 3 TeV c.m. energy; *TESLA* at DESY, with 1 TeV c.m.s., the NLC at SLAC and JLC at KEK, Japan, which are both being considered to start at a relatively low energy in c.m., i.e. 500 to 600 GeV, or even 350 GeV in the case of JLC. The most sophisticated of these projects is certainly CLIC. Figure 7 shows the overall general layout. However for both economical and physics reasons, the best would be a machine that starts with a c.m.

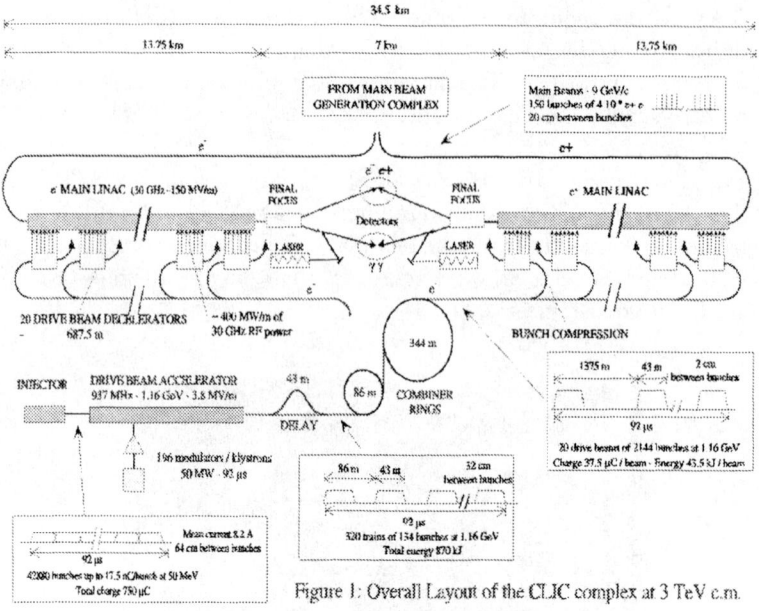

Figure 7: General layout of the CLIC $e^+e^-$ project at CERN

energy around 500 GeV and is upgraded up to 1 and then 1.5 TeV. Another possibility with an $e^+e^-$ facility is to transform it into an $e\gamma$ or even $\gamma\gamma$ facility. The advantages of the $\gamma$ options have been discussed for a long time, mainly by theoreticians, that find them especially attractive for the Higgs searches. But $\gamma$-beams mean additional technical problems to be confronted; these are not discussed here.

### 3.1.2 Muon colliders

The Next Muon Collider (NMC) project is another alternative for a lepton collider. It is under study at the R&D level at FNAL [2]. Figure 8 shows the general schematics of a muon collider. The main components are first a proton accelerator that can also be used for intense K-physics; The proton beam hits a target to produce $\pi$'s, which are then focused with a superconducting magnet. The $\pi$'s are then travelling in the *"Pion decay channel"*, where $\mu$'s are produced by the decay of $\pi \longrightarrow \mu\nu_\mu$. The muons pass through the *"muon ionization cooling channel"*, where they are kept and focused before they decay (remember the $\mu$ lifetime of 2.2 $\mu$s). The surviving $\mu$'s are then accelerated

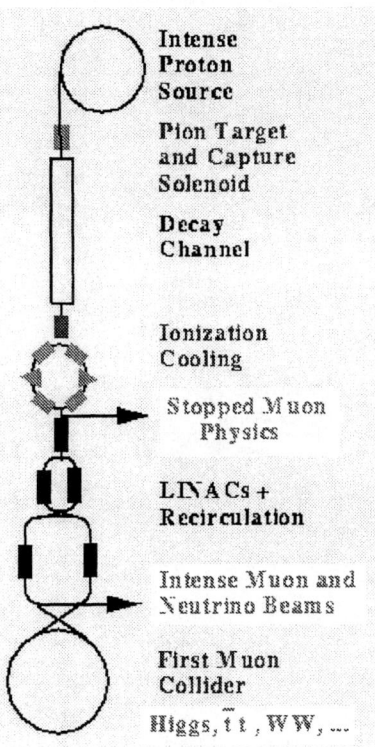

Figure 8: General layout of the NMC project at FNAL

in two stages, allowing also the production at each stage of neutrino beams of respectively low and high energy. Finally the $\mu^+$ and $\mu^-$ that are produced, are accelerated and collide in the final $\mu$-collider.

Three beam energies have been considered in some detail so far: 50×50, 200×200 and 1500×1500 GeV/beam. Table 1 gives the main parameters of these three alternatives.

The advantages of the $\mu$ versus the electron is that $m_\mu/m_e = 207$, thus enormous reductions in radiative losses. Moreover, it permits higher energy to be reached, smaller collider rings to be used, and, the possibility of Higgs production in the s-channel. Figure 9 shows how, for instance, a 0.5 TeV or even 4 TeV c.m.s. $\mu^+\mu^-$ collider would fit in the FNAL site, and how these compare with the LHC at CERN or an $e^+e^-$ linear collider with 0.5 up to 1 TeV c.m.s.. The *compactness* of such a machine gives a good reason to suspect

Table 1: Muon Collider Parameters for three collider energies

| Energy/beam in GeV | 50×50 | 50×50 | 200×200 | 1500×1500 |
|---|---|---|---|---|
| | Broadband | Narrowband | | |
| Rate (Hz) | 15 | 15 | 15 | 15 |
| Muons/bunch | $4 \times 10^{12}$ | $4 \times 10^{12}$ | $2 \times 10^{12}$ | $2 \times 10^{12}$ |
| Bunches | $1 \times 1$ | $1 \times 1$ | $2 \times 2$ | $2 \times 2$ |
| Circumference (m) | 300 | 300 | 1000 | 6000 |
| Bunch $\sigma_Z$ (cm) | 9 | 13 | 2.3 | 0.3 |
| Spot $\sigma_r$ ($\mu$m) | 187 | 270 | 24 | 3.2 |
| $\beta^*$ (cm) | 9 | 13 | 2.3 | 0.3 |
| $\Delta E/E$ (%) | 0.007 | 0.002 | 0.08 | 0.08 |
| Luminosity ($cm^{-2}s^{-1}$) | $2 \times 10^{31}$ | $1 \times 10^{31}$ | $1 \times 10^{33}$ | $5 \times 10^{34}$ |

that, if feasible, a muon collider will be significantly cheaper than alternative, futuristic high energy colliders.

However there are significant technological challenges in the designing of an accelerator that would be capable of making, accelerating and colliding $\mu^+$ and $\mu^-$ bunches before nearly all of the $\mu$'s have decayed, which they do with a rest lifetime of 2.2 $\mu$s.

This is why, although the muon collider is an old idea (60's), it has not yet become a concrete project. With modern technology now is thinkable to make $\mu$ bunches of a few $10^{12}$ $\mu$'s and to accelerate to acceptable levels the electron background from the $\mu$ decay ($\mu \longrightarrow e\nu_e\nu_\mu$). Among several challenges still to be overcome in order to build an NMC, the main ones are:

- *The pion production target and the decay channel*

- *The $\mu$-cooling*

- *The background environment for detectors*

Again the aim of this review talk is not to describe each point in detail, but more to give a general overview of the questions, pointing out some of the main issues.

⋆ **First issue:** *The pion production target and the decay channel*

A dedicated R&D project is devoted to the study of the pion source and decay channel. The main steps in the pion production target are:

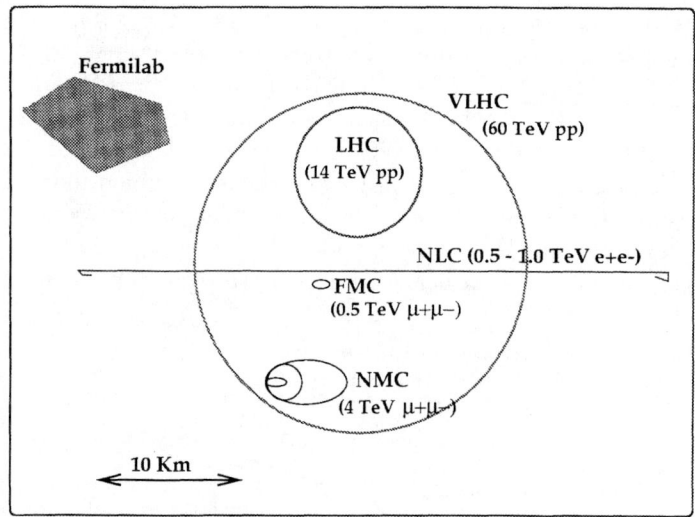

Figure 9: The NMC project at FNAL: comparison in size with other machines

- Of the order of $10^{15}$ protons per second hit a high-Z target, producing 4 MW

- The pions with $P_T \leq 200$ MeV must be captured in a 20 T solenoid

- Pions must then be transferred to a 1.25 T solenoidal decay channel

- The energy spread of the $\pi/\mu$ bunch must be compressed with RF cavities

Three main technological problems are related to this:

- *As 400 kW are deposited in the target, it is necessary to move the target material away from the beam and to cool it remotely; a baseline solution is to use a liquid metal jet.*

- *The first RF cavity should be about 3 m away from the target, thus the question is: Will it operate in this high radiation environment?*

- *High-power low-frequency RF will be needed*

⋆ **Second issue:** *The muon ionization cooling*

To "cool" the muon bunch means to turn the diffuse muon cloud into a bunch of small longitudinal and transverse dimensions, suitable to be accelerated and injected into a collider. The difficulty is that $\mu$-cooling must be fast, i.e. $\leq 2.2$ $\mu$s, which means that the electron or stochastic cooling techniques do not work. Figure 10 shows the schematic of the ionization cooling; the muons lose energy by $dE/dx$ and the longitudinal momentum is replaced by RF (transverse cooling). Minimizing the heating from Coulomb scattering implies small $\beta_\perp$, i.e. strong focusing, using high field solenoids or lithium lenses, and low-Z absorber, using liquid hydrogen. The energy "cooling" uses a

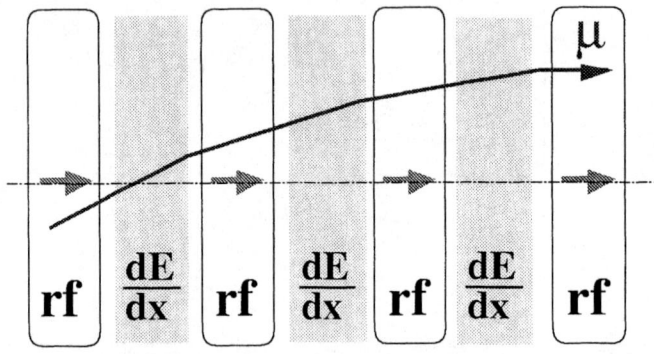

Figure 10: Transverse muon decay by dE/dx

wedge plus dispersion, and an alternating gradient to focus the beams. Figure 11 shows the basic concept of a complete cooling channel for a high luminosity muon collider. It would consist of about 20 to 30 stages, each about 25 m long, and each reducing the 6-D phase space by a factor of order 2. To confront the muon cooling technological challenges, an R&D collaboration, *MUCOOL* was set with the following goals:

- To develop special RF modules giving high peak accelerating gradients

- To design, build and test an alternating solenoid transverse cooling section

- To develop long liquid lithium lenses with a high surface field

- To build short cooling sections and test their performances in a low energy muon beam

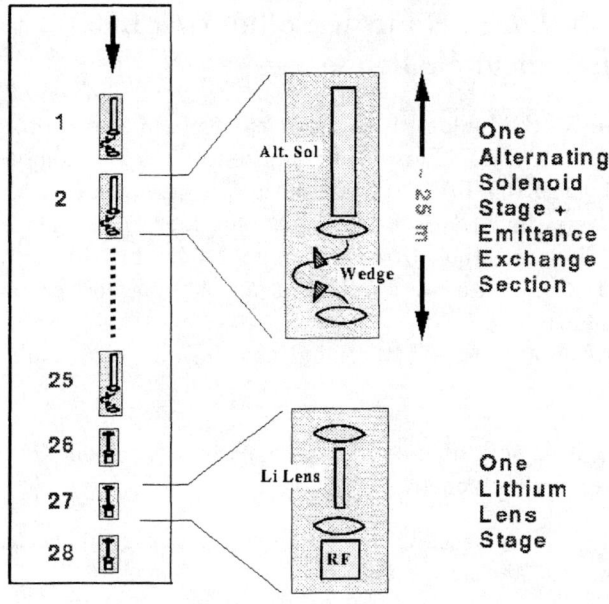

Figure 11: Cooling channel concept for the muons in an NMC collider.

To conclude, the muon collider option is attractive from several points of view: compactness, huge reduction in radiative losses with respect to $e^+e^-$ colliders, allowing higher c.m. energy to be reached, and the possibility of an s-channel production of the Higgs boson. Indeed there is still a long way to go before making it a feasible option, as can be seen with the brief description of some of the technological challenges to be confronted.

However, this important R&D program, which has been undertaken by the NMC collaboration, proposes appealing *by-products*, such as intense K-physics beams and intense high energy muon and neutrino beams. These by-products are certainly interesting for K- or $\nu$-physics and do not necessarily imply to go to very high c.m. energy muon colliders.

## 3.2 A Very Large Hadron Collider (VLHC), some new technological challenges

The goal of the VLHC builders is to make feasible a proton–proton collider of $50 \times 50$ or $100 \times 100$ TeV. This would require to dig a very long tunnel of up to 1160 km (for the 100 TeV per beam option) at as low a cost as possible.

If at the time the LHC is being built, this looks like science-fiction for most of us, it is interesting to recall that in 1954, at the time of table-top accelerators, Enrico Fermi was thinking of an orbiting accelerator, even one encircling the Earth.

In order to build a VLHC, the main new technological challenges to be confronted are:

- *The magnetic path:* it must be made of robust, reliable, fully equipped and inexpensive magnets.

- *The tunnelling techniques*: new technologies to dig tunnels that are very long, low cost and not accessible to human beings.

- *The instrumentation*: it must be robust to environment, remotely surveyed, reliable. This includes: electrical connections, power distribution, fibre optic network links, DAQ and monitoring systems.

A VLHC project is under study at FERMILAB [3]; indeed it pursues the basic concept of a precursor in the field, namely the FNAL funding Director, R.R. Wilson. [3]

### 3.2.1 The long magnet path of a VLHC machine

It is based on a *double-C transmission-line magnet* proposed by B. Forster. It is a two-in-one warm iron superferric magnet, built around a 75 kA superconducting transmission line. The overall design of this transmission-line magnet is shown in Fig.12. Its key features are: a simple cryogenic system, making little use of superconductors, a small cold mass, a low heat leakage, with continuity in length, no quads or spool pieces, a warm bore vacuum system and standard construction methods. Concerning the magnetic design, as shown in Fig. 13, the current is returned in the cryogenic distribution lines located in the structure tube underneath the magnet. The steel yokes above and below the transmission line concentrate the magnetic flux in pair gaps which provide

---

[3] " *Whether the next large proton accelerator (20 TeV ?) is built ..., to be affordable, innovations in construction must be made. The design of a superferric magnet ring buried in a pipe in the ground is explored to see what reductions in cost might result*", R.R. Wilson, Snowmass, July 1982, at the first Summer Study for the SSC.

Figure 12: Photograph of the 2m long model of the double-C transmission-line magnet.

opposite bend fields in the twin apertures needed for pp collisions. Pole tips are shaped to provide the alternating gradient necessary to focus quadrupoles and allows the magnet to be continuous over long distances. The field lines circulating along the transmission lines are concentrated into the beam gaps by the iron pole tips.

The length of the magnet assembly is about 250 m long, and the vacuum hardware, instrumentation and cryogenics are preassembled and tested in factory.

For the superconductors, high temperature superconductors (HTS) are tentatively rejected because of their cost. The most important development is the order-of-magnitude increase in current density per dollar of conventional NbTi for low field applications, such as MRI magnets for instance. This is the currently preferred option.

### 3.2.2 The tunnelling technologies

To make true the old idea of a *"low cost superferric magnet ring"*, another key-issue is the long tunnel. It may become feasible in the next two decades, with the dramatic recent advances in:

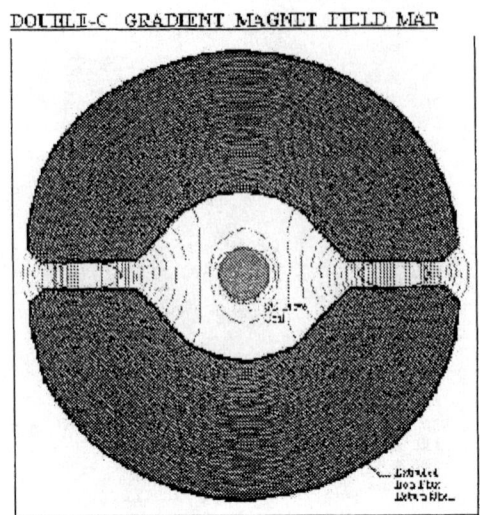

Figure 13: Magnetic field of the transmission line magnet.

- Advance tunnelling technology for small diameter non-human-accessible tunnels.
- Accurate remote-guidance systems for boring machine steering.
- Industrial applications of remote manipulation and robotics.
- Digitally multiplexed electronics to minimize cables.

The trenchless technology has a rapidly growing importance as a practical solution to expansion and repair of underground utilities. It is motivated in part to build new and rebuild old infrastructure with minimum surface disturbance. *The VLHC can be a catalyst to this environmental crucial industry.* There are two competing technologies: the *microtunnelling* developed in Europe and Japan, and the *horizontal directional drilling* developed in the USA.

⋆*Microtunnelling technology*

Microtunnelling is a *trenchless technology* for constructing pipelines to very small tolerance, i.e. ± 1 inch. It is a remotely controlled, laser-guided system and so personnel entry is not required. The typical production rates are 30 to 60 feet/day, but rates above 200 feet/day could be achieved. It can be used in a variety of ground conditions (soft clay to rock; above or up to 100 feet below the water table). The microtunnelling costs are continuously dropping.

It should be noted that most of the microtunnels built up to now are straight or gently curved. Laser ring gyro systems, with accuracies of 2 inches, are under development.

*⋆Horizontal directional drilling technology*

It is a US invention, developed primarily for oil and gas exploration. The biggest problem is the accuracy: currently it is $\pm$ 1 or 2 feet accurate. Moreover the density variations in the rock cause the drill to reer from the desired direction. An active research is under way to improve the accuracy in guiding the drill head.

*⋆Other related technological issues*

As corollary of microtunnelling are two main issues: *robotics* and *surface penetrations*. Robotics or remote manipulation is a sine qua non condition for these microtunnelling techniques. It is now used for the repair of sewer pipes of around 30 inches in diameter, with access every 300–400 feet via manhole. The robots cut holes, put in patches, cut roots out, install new lateral connections etc. It is a rapidly expanding industry. Besides, virtual reality consoles on the surface will aid complex work deep underground.

Cryogenics and power supply/quench protection require *access shafts and surface buildings* about every 80 km. Not all of them need to have human access, and the fewer the better for cost considerations.

### 3.2.3 Instrumentation

The needed instrumentation mainly consists of the electrical connections at the end of the magnets for power distributions, the fibre optic network links for communication network, and ring-wide clock and control distribution, slow controls and the readout electronics of the beam position and beam loss monitors.

The magnet electronics must be: radiation hard, redundant (fault-tolerant), multiplexed to reduce the cabling, hermetically sealed, modular and pluggable for replacement by robots. Thus digital and multiplexed electronics allow most of these requirements, if not all of them, to be fulfilled.

To conclude this section, it is important to point out the direct connections between the construction of such a machine and a very interesting aspect of environmental engineering. In this sense, HEP can serve as *catalyst for developing new technologies that will be helpful for the everyday life and environmental issues.* This is why issues such as microtunnelling, robotics, low-field magnets such as those used for MRI applications, have been discussed in some detail.

Besides these applications to everyday life, the physics motivations of the VLHC still have to be well established. It will certainly allow us to go *"beyond-the-beyond"*, but work has yet to be done to justify the need for such a high c.m. energy. However, meanwhile and in parallel, it is important to pursue the R&D work, because of all the technological challenges that have to be confronted to build such a machine. But, because of the recent dramatic advances in the various technological issues discussed here, the old idea of a low-cost, low-field, iron-dominated magnet in a small-diameter pipe may become feasible, and also even Fermi's dream.

## 3.3 A post-LHC machine: present concluding remarks

The points to be retained from this section are:

- *The HEP community is studying the next supercollider.* Indeed this is crucial not only for the survival of this research field, but more importantly for allowing us to push ahead the quest of our understanding of the elementary constituents of matter and of the associated elementary forces. **Accelerators are unique probes to penetrate this unknown world**.

- There are several active R&D projects on accelerators, especially the NLC, NMC and VLHC options, concentrated in the largest HEP laboratories all over the world (Europe, Japan and USA).

- Constructing these new machines means to confront new high technological challenges of various types. Some of these developments could serve as catalysts for everyday-life applications and also profit from them.

- Because of the time needed for developing these new techniques, designing, building and starting to operate such machines, the already started R&D projects must be actively pursued in parallel with the related physics studies. These studies, together with the results obtained with the currently running machines, should help in deciding what the best option is, both what type of particle ($e, \mu, p$, or $\gamma$) and what c.m.

energy have to be chosen, not to mention of course the choice of the machine location (but this is politics).

Such decisions should not stop pursuing the R&D projects on the other options: these may be useful in the longer term.

- It would be important to have in an almost **contemporary way**, an $e^+e^-$ machine with the LHC; this is for *complementarity, completeness and competition issues*.

# 4 Confronting high technological challenges in HEP: which detectors and related electronics, and DAQ systems?

The past decade led to impressive advances in all aspects of detecting techniques, i.e. detectors, electronics and data processing. They were driven by the striking evolution of HEP and also of the more and more demanding environmental conditions. High c.m. energy hadronic colliders, i.e. the Tevatron at FNAL and the LHC at CERN, pushed at the frontiers of what detectors were expected to be able to do. LEP was also instrumental in developing, in particular, very performant tracking systems such as the TPCs of ALEPH or DELPHI and the Si-microvertex detectors.

These crucial progresses were made possible by the impressive and fast advances in many aspects of high technologies in the industry, such as microelectronics, hardware and software computing techniques, civil engineering, solid state, new materials, cryogenics, micro-mechanics, etc...

This section explains, with some chosen examples, the high technological challenges that will have to be faced for building the detectors, their associated front-end and readout electronics, and the related DAQ systems for the forthcoming and far-future HEP experiments [4].

## 4.1 New technological challenges for building HEP detectors

This section will concentrate on the tracking systems that will serve as an example here. Tracking is indeed the detecting technique that evolved the most in the past 20 years.

### 4.1.1 The growing importance of the tracking

Let me first recall the main revolutionary steps in the development of the tracking technique. The development of the proportional multiwire chambers, in the sixties and early seventies, marked the entrance of this detecting technique in the *"gotha" of the electronically read-out detectors*. The evolution to the drift chamber technique allowed enough input signal to be obtained and thus the front-end and readout electronics to be better attacked. The development of the microvertex devices with CCDs or Si-pads and then pixels, gave access to the region very near the interaction point. In parallel, the striking advances in (micro)electronics allowed us to accompany and even to push ahead the striking evolution of various tracking techniques.

For a long time, the information provided by tracking devices was *too slow* to be part of the real-time data processing. They were typically part of the so-called *offline processing and analysis*.

Especially important in the case of the hadronic colliders, is the online processing, including the triggering system. The trigger is more and more sophisticated and includes various levels of decision. For instance what was totally unthinkable some years ago is becoming true in the tracking system of the CDF experiment for Run II, namely to have the tracking information included in the full trigger system and therefore available already at the Level-1 trigger. Moreover, in this experiment, the microvertex information will be available at Level-2 [5].

*Introducing the tracking data in the online system and expanding the detecting capabilities of the tracking devices made the role of these detectors evolve dramatically.* Indeed they are the key-detectors for most of the physics issues.

### 4.1.2 From multiwire chambers to solid-state tracking devices

The evolution from multiwire chamber detectors using a gas mixture as ionization medium, to the *"hybrid"* detectors, i.e. detectors mixing multiwire and solid state, namely a microstrip structure, and then to the *"all-silicium"* trackers, made of silicium pads or pixels, demonstrates the growing importance of the solid-state technology for building the new tracking devices [4]. The aim here is to make you aware of this evolution.

The final choice by CMS [6] of an *"all-silicium" tracking system*, instead of the previous MSGC baseline, is an example of this evolution. Both techniques are very challenging, in particular in the LHC environment and at the

---

[4] *For details about the functioning and performances of each of these tracking techniques, I refer to the dedicated lectures at this school.*

demanded LHC scale. Both R&D projects were pushed very far and were quite successful. Several high tech industries demonstrated to be interested in constructing MSGCs for the overall CMS tracking system, and able to realize it.

Now the all-silicium option in CMS has no alternative to success. For this, it must overcome new technological challenges, in terms of intrinsic performances, stability, mechanical structure, pads dimensions, industrialized construction, quality control, radiation hardness, maintenance and robustness, cost. It will be extremely instructive and exciting to experience how such a tracking system, using 6" Si pads, covering about 160 m$^2$, and representing a total of about 5 million channels, will operate at the LHC.

For the new central tracking, the CDF experiment has chosen a completely opposite approach. CDF stays with a "traditional" drift chamber technology, similar to the one used in the tilted wires drift Central Tracking Chamber used in Run I, but this technique is pushed to its extremes in order to allow this device to be fast enough to cope with the ultimate bunch crossing time of Run II at the Tevatron, i.e. 132 ns. This detector is being mounted in the experimental hall and should begin operation in Summer 2000 for the engineering run of CDF and the Tevatron. This has some consequence, as pointed out in the next section.

### 4.1.3 Tracking information included in the triggering system

The very advanced drift chamber technology of the central outer tracking (COT) in CDF, makes it a *fast device*. It thus permits the introduction of the information from the COT already at the Level-1 trigger. This is indeed a revolutionary approach, which concludes the series of upgrades performed by CDF, all along, in order to gradually introduce the central tracking information within the overall triggering system and at all the levels of decision.

Moreover, CDF has also built, at the Level-2 trigger, a processing system combining the COT and the Si-microvertex detector information. It will allow a refined online reconstruction of the whole track from the interaction point up to the end of the available tracking space. It will permit in addition to perform an *online b-tagging* and to introduce it at an early stage of the trigger decision chain, namely at Level-2. This is particularly crucial in the case of pp colliders (see Fig. 14). Building this fast and performing tracking device and achieving the associated performant online processing of its information is made possible by the dramatic advances in high tech microelectronics. This progress occurred over the last five years in various aspects of microelectronics such as the development of fast RAMs, FPGAs and of the semi-conductor technology.

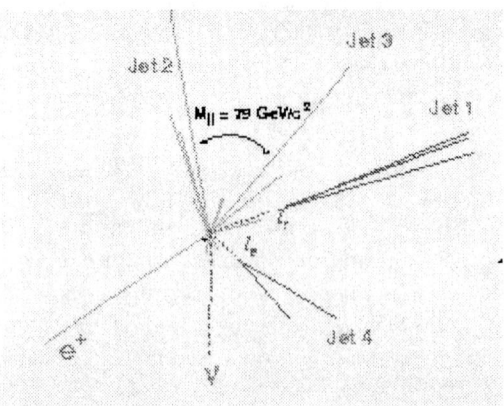

Figure 14: An example of the importance of the tracking: use of the microvertex for the b-tagging in CDF experiment at Run I at the Tevatron. The figure shows one of the first top event discovered by CDF with the double b-tagging.

Figure 15 shows the block diagram of the overall CDF II trigger scheme [5]. A consequence of all this progress can be seen from the scheme in Fig. 15. Indeed it shows that the tracking system is on an equal footing with the other two detecting systems, namely the calorimetry and the muon detector, and this, in spite of its much more dangerous location. This detecting system is the one located nearest the interaction point. It is therefore on the *"front-line"* or at the *"hard core region"* of the interaction. So the role and importance of this device have increased as the physics needs were demanding more and more sophisticated processed information. As examples, let me recall that the Higgs(es) searches are dominated by b-tagging or $\tau$-tagging or even $c\bar{s}$-tagging (in the case of charged Higgs). The SUSY searches also require, very often, to tag heavy flavours or charged leptons or to have a refined information about a jet, including its charged content. QCD measurements also rely more and more on the tracking information as well as on heavy flavours tagging, for a general study of bottom and top physics. The jet recognition and measurements, in particular for multijet topologies, are much more accurate using the tracking information.

---

[5] *For more details on all these advanced tracking issues, see the lectures given at the ICFA Instrumentation School in 1997 at Leon by ASN, on "Detector Systems for Future High Energy Experiments", section on Tracking, and references therein.*

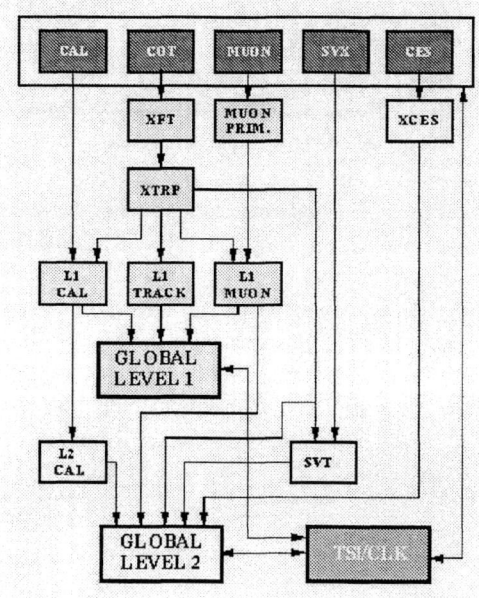

Figure 15: General block diagram of the trigger system of CDF II at the Tevatron. Note that the tracking information COT is included in Level-1 and the microvertex information from the SVX (SVT) is part of Level-2.

*So for many different reasons, the tracking is certainly becoming the "key-detector" of future experiments. Its role has increased and become predominant, because of the physics issues and also the technological advances that allow these issues to be confronted by the new tracking devices.*

## 4.2 New technological challenges for building the front-end and readout electronics

The electronic chain associated to a detector must first collect the input signal from the detectors. This is the role of the *"front-end electronics"*, which amplifies and shapes the input analogue signal from the detecting device. The digitization may occur at the very beginning, i.e. in the front-end, or else the second stage called *"readout"*. At the readout stage, after digitization, the detector information is preprocessed and sent to the DAQ system [7]. The trigger system receives the signal from the readout front-end, before or after digitization. If it takes the signals before they are digitized, the trigger logic is done by combining analogue signals. Otherwise, the trigger logic will be all digital. Then the information is further processed in the two processing chains, i.e. the trigger and the DAQ chains, and combined with the preprocessed information from the other detectors. In order to face the constraints imposed, on the one hand by the new and future H.E. accelerators and their harsh environmental conditions and, on the other hand, by the new physics requirements, the electronics associated to the detectors must demonstrate very high performances and satisfy a long list of conditions, among which:

- *Be fast (for instance at the LHC it must work at 40 MHz, whereas at Tevatron II the rates will ultimately be of order 10 MHz).*

- *Have a very low noise (therefore an optimized connectivity).*

- *Have a very large dynamic range(as the c.m. energy of the machines are increasing, high mass objects are looked for, therefore increasing the necessary dynamic range).*

- *Have a very low power consumption and therefore be easy to cool down.*

- *Be a high-density device (in terms of the content in electronics component), and thus compact.*

- *Be easy to produce in large numbers, and thus easy to standardize.*

- *Be realible.*

- *Be well shielded.*

- *Be easy to repair and maintain.*

- *Be radiation-hard or at least radiation-tolerant.*

- *Permit sophisticated (pre-)processing of the information at the different stages of the electronic chain, including the trigger.*

- *Be of low cost*

All these conditions and requirements motivate the use of

- *DIGITAL ELECTRONICS rather than analogue, and*

- *MICROELECTRONICS HIGH TECH RESSOURCES*

### 4.2.1 Digital front-end and readout electronics

In the last ten years, one of the main breakthroughs in this field was to introduce digitization at the earliest stage of the electronic chain, which reads out and processes the signals from the detectors [7]. What is meant by earliest stage is the front-end. Right after being amplified and shaped, the analogue signal provided by the detector is thus digitized, before it is sent to the trigger and the DAQ chains. This was made possible by the dramatic advances in some aspects of the high tech electronics. In particular the industry of cellular phones triggered an impressive progress in the domain of fast ADCs of large dynamic range. As an outcome, these ADCs are used in HEP experiments to equip the front-end of the calorimeters and allow the signals of these detectors to be immediately digitized. Several HEP experiments have already adopted or will adopt this option. This is the case for instance of Belle at KEK, Babar at SLAC, CDF at the Tevatron II and CMS at the LHC.

Adopting this way of processing the input signals has very important consequences. The treatment of the signal is much more reliable and easy to transport, and it allows, at a very early stage in the readout chain, sophisticated processing of the information and reduction of the data to be transferred; this is also true, of course, ofr the trigger system, which becomes quite refined and flexible, with an increased potentiality. Moreover, the overall readout, DAQ and trigger chains are *programmable*. They are easily and *remotely* monitored and calibrated as well as adjusted (as for instance thresholds etc). Going digital gives a large potential of capabilities and flexibility as well as freedom. Moreover, it makes all the important advances in this high tech field available, with constantly increasing ressources. This certainly is the way to go!

### 4.2.2 Towards the use of microelectronics and related high tech ressources

A corollary of the fast digitization of the signal is the use of microelectronics and its related ressources. As an example, the use of *new packaging technology* is briefly discussed here.

In the industrial world, already for some time, the increasing complexity of the VLSI circuits and the always strongest demands on better performances, as for instance the speed and the density, have led to an impressive progress in the *interconnects and packaging* (IP) technologies [8]. This gave rise to a plethora of different solutions. Among the available packaging technologies, there are: Dual-In-Line(DIL) packaging, the Quad Flat Pack (QFP), the Pin Grid Array (PGA) and the Ball Grid Array (BGA) in either ceramic or plastic base. There are various technologies to mount integrated circuits, among which: the microbonding, the flip-chip and the Tape Automated Bonding

(TAB). Moreover, there are various MultiChip Module (MCM) technologies for mounting electronics systems with a complex architecture (see Fig. 16).

It is important to note that, when designing electronics circuits or systems,

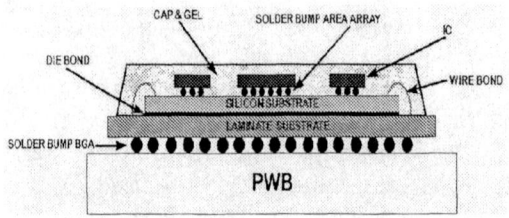

Figure 16: Schematic view of a MCM-D packaging technology as developed by the SUMMIT European ESPRIT project. For further details, see [9].

interconnects and packaging are part of the problem, from the initial phase of the conception, and play an essential role for:

- Best solving the noise issues,

- Optimizing the requested performances,

- Optimizing the overall system architecture (best routing and interconnexions as well as arrangement of the components),

- Best solving the space (i.e. density), volume and material issues,

- Best solving the cooling problem,

- Best addressing the testability (both during production and for maintenance), repairability and reliability issues.

As can be seen from this list, IP technology helps in fulfilling the requirements on the electronics for future HEP experiments. It is also important to note, when designing a circuit, that there exists an IP solution for each case. Indeed, it must take into account all the environmental constraints and the required performances. Besides, the thermal issues are also part of the design, and they enter the choice of IP solution from the initial phase of conception of the complete electronic system. Therefore the choice of the interconnects and packaging technologies plays an essential role when designing the complete electronic front-end, readout, trigger and DAQ systems.

As an example, Fig. 17 shows a prototype of a digital front-end readout for calorimetry, including a low-noise and fast preamplifier and a fast and

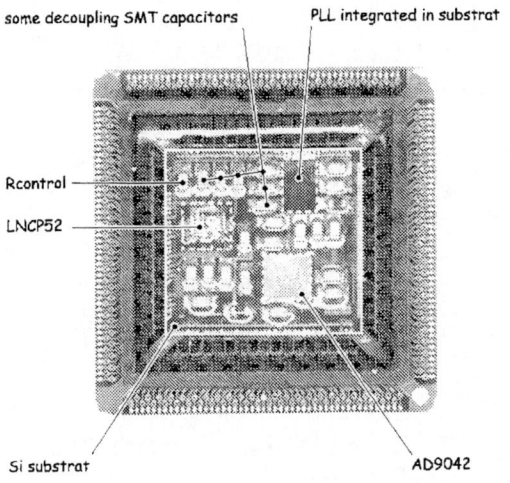

Figure 17: View of a MCM-D packaging including a prototype of a full digital front-end and readout chain for LHC calorimetry [9]. The main components are a fast and low-noise preamplifier LNCP52 and a fast ADC AD9042.

large dynamic range ADC. This circuit includes a PLL and all the necessary surroundings for this electronic chain. It is a prototype of a complete digital front-end and readout chain, *inserted in an active MCM-D packaging.*

*The dramatic advances in the treatment of the signal delivered by the detectors are based on the impressive progress of the high tech microelectronics and digital electronics.* Surely these are the major high technological challenges to which HEP experiment will be confronted, for building their readout electronics. The challenges are mandatory in order to allow these experiments to confront both the new physics requirements and the new environmental conditions imposed by the future machines.

## 4.3 New technological challenges for building the DAQ systems

The evolution of the DAQ system is associated, on the one hand, to the technical advances on the readout electronic chain mentioned above. On the other hand, this evolution is strongly influenced by the development of highly performing processors, of high capability and fast transmission systems for high fluxes of data (fast and high-capacity switches and optic fibres), as well as of new software tools. For instance, *the Internet and the Java revolutions* [6], and the performing and high capability object-oriented databases are instrumental.

All these impressive and challenging high technological developments are leading to the growing importance and even the key-role of the online and real-time tasks and therefore of the DAQ system. This is important to have in mind when designing and setting-up forthcoming HEP experiments. These online tasks include, apart from the data-taking operations, the constant monitoring of the DAQ and of the detector operations, all the calibration tasks and even the full reconstruction of the complicated events by *online processor farms*. All these operations are accessible in *remote mode*, which makes the online operations more flexible and friendly-access. The data taking will be available from any Institute in the world, which takes part in an experiment. The extended use of high capacity and fast online databases will allow us more and more, to *keep the long term history* of the data taking and will be of important use for the monitoring and also for the analysis of the data. Moreover, the online database is becoming the *"bank of updated parameters"* and therefore the key-place for processing the *corrected physics quantities* that are used to fully reconstruct the event.

*More and more, and thanks to the striking progress in computing technologies and associated hardware, the online and real-time processing of the data will replace the off-line analysis. Moreover these progresses are also permitting to work in a remote way, thus facilitating the life of the experimentalist and giving her/him easy access to the information at any time and from anywhere.*

## 5 Concluding remarks

To explore the "beyond-the-standard-model", new accelerators and new detectors must be built; the LHC is on the rails and we must already actively

---

[6] *"First proposed as a mechanism for enhancing Web content for selecting tasks, the Java language has taken off as a serious general-purpose programming language. Industry and academia alike have expressed interest in using the Java language as a programming language for scientific and engineering computations. Applications in these domains are characterized by intensive numerical computing and often have very high performance requirements. Programming techniques are developed that lead to Java numerical codes with performances comparable to FORTRAN and C, the more traditional languages for this field. The techniques are centered around the use of a high performance numerical library, written entirely in the Java language, and on compiler technology". IBM SYSTEMS JOURNAL, Vol. 39, No 1, 2000, article on "Java programming for high-performance numerical computing" by J.E.Moreira et al., from IBM.*

work towards the next step. A linear collider, possibly almost contemporary to the LHC, is certainly a first very appealing project. In addition we must investigate what the post-LHC will be: CLIC, NMC or VLHC or ??

The quest for beyond-the-standard-model, initiated at LEP II and the Tevatron I will be pursued at Tevatron II, LHC and beyond. To undertake this long exploration, new technological challenges must be confronted in an incredibly large spectrum of technological fields, such as: civil engineering, robotics, cryogenics, supraconductors, computing (software and hardware), solid state, electronics and microelectronics, telecom, mechanics and micromechanics, new materials.

It is interesting to note that HEP can serve as a catalyst and also as a large-scale demonstrator for these new technological developments. Thus H.E. experimentalists must work in closer and closer contact with industry and must follow the evolution of technologies.

The aim of this review talk was to make you aware of these developments and changes, and to convince you that HEP is at the frontier of high technologies, and therefore more challenging and exciting than ever.

# References

[1] See: *http://cern.web.cern.ch/CERN/Divisions/PS/CLIC/Welcome.html*, and
*http://www-lc.fnal.gov/*, and references therein

[2] See: *www.fnal.gov/projects/muon–collider/*, in particular:
S. Geer, *Muon Colliders: Overview and Technical Challenges*", Talk to UEC Exec. Committee, FNAL, January 1999.

[3] See: *http://www-ap.fnal.gov/VLHC/* and references therein.

[4] A. Savoy-Navarro, *Detector Systems for future HEP experiments*, AIP Proceedings of the ICFA 97 School on Instrumentation,Leon (Mexico), July 1997, P 47–116.

[5] R.E. Blair et al., CDF Collaboration, *The CDF II Detector Technical Design Report*, FERMILAB-Pub-96/390E, Nov. 1996 and references therein.

[6] CMS Collaboration, New Tracking TDR document in preparation.

[7] A. Savoy-Navarro, *Realization and test of a fast digital readout for LHC calorimeters: present performances*, Proceedings of the 6th International

Conference on Calorimetry in High Energy Physics, Frascati, June 8–14 1996.

[8] A. Savoy-Navarro, *On the use of new interconnects and packaging technologies in high energy physics experiments*, Proceedings of the 6th International Conference on Advanced Technology and Particle Physics, Como, Italy, Oct. 5–9 1998, Nuclear Physics B (Proc. Suppl.)78(1999)August 1999.

[9] J. Lopez Contreras, Silicon Substrate Multichip Modules for Innovative Products (SUMMIT), Proceedings of EUPAC 98, 3rd European Conference on Electronic Packaging, Nuremberg, June 15–17, 1998, and references therein.

# Advanced Photodetectors for High Energy Physics Particle Astrophysics and Medical Imaging

## Muzaffer Atac

*Fermilab, Batavia, Illinois and University of California at Los Angeles*

**Abstract.** This article is a survey of advanced major photodetectors for High Energy Particle Physics, Particle Astrophysics and Medical Imaging. It is intended to give a broad background to **ICFA** level students. Description, operation and some applications of such photodetectors will be described.

## INTRODUCTION

There have been major advances in photodetectors in the last two decades due to advancements in High Energy Particle Physics, Particle Astrophysics and Medical Imaging. Advancements in Solid State technology, rapid advances in electronic circuitry and computer technology helped in this advancement. Photon Detectors are so many types that we will select some of them due to their specific applications.

A human eye is the most incredible photon detector in image forming in color to a resolution of 2 x 2 $\mu m^2$, about 60 times a second (Encyclopedia Britannica 1999). A human eye may have about $1.3 \times 10^4$ by $1.3 \times 10^4$ pixels (sensors). Detection threshold of an eye is about 2500 photons per $cm^2$. Such a threshold can only be achieved by staying in a very dark room for at least half an hour. Then we need to slowly increase exposing our eyes to light, not to damage the retina.

## Photon Detectors

A photon detector is a device that converts photons to electronic signals. We can classify them as vacuum photon detectors, solid-state photon detectors and hybrid photon detectors. Some of them convert photons into electrons and holes without internal multiplication (photodiodes), and some of them have internal multiplication of electrons by cascade processes (photomultipliers, PMTs). They can measure time from picoseconds to seconds, energy from eV to multi GeV, and positions from $\mu m$ to meters. Commercially available photodetectors can be sensitive to wavelengths, $\lambda$ from UV (200-300nm) to far infrared (IR up to 50 $\mu m$).

# Vacuum Photodetectors

All vacuum photodetectors have windows inner surface coated with low work function materials (e.g. bialkali coating). When a photon enters through the window it may free an electron from the low work function material (photocathode). This process is formulated well by Einstein. Albert Einstein wrote a paper about the quantum nature of the photoelectric effect in 1905 and awarded his Nobel Prize in 1921 for the photoelectric effect,

$$E = h\nu - \varphi$$

Where, E is the energy of the exiting electron from the photocathode, h is the Planck Constant, $\nu$ is the frequency of the incoming photon and $\varphi$ is work function of the photocathode material. It is clear from the formula that the photon has to carry more energy than the work function of the photocathode. The photon may not produce a detectable electron due to the reason that photocathode material has to be very thin, otherwise the electron may not get through the material. This thickness is very precisely adjusted. The efficiency of producing detectable electron is expressed in quantum efficiency (QE),

QE ($\lambda$) = number of detected photoelectrons / number of incident h$\nu$

QE in percent can be expressed as a function of radiant sensitivity $S_R$ and wavelength $\lambda$ as,

$$QE = S_R (1239.5 / \lambda) \cdot 100 \%$$

A typical QE for a bialkali photocathode and for $\lambda = 420$ nm wavelength is about 20 %. The QE drops off rapidly for larger $\lambda$, and the window has to be made of quartz for the UV photons to penetrate and produce photoelctrons from the photocathode. Among them are single anode PMTs, multi-anode PMTs ( MAPMTs ), multi channel plate PMTs ( MCP-PMTs ). They will be briefly described here. A most commonly used PMT is shown in Figure 1.

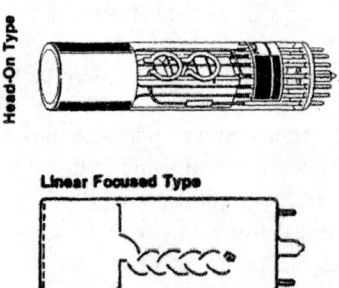

**FIGURE 1.** The most commonly used type of photomultiplier tube.

In a vacuum PMT the photoelectron is accelerated to the dynodes and cascade multiply and collected by the anode. Depending on the electric fields between the dynodes and the number of dynodes one photoelectron may become up to few times $10^6$. This is an exponential multiplication process. There are variety of PMTs , Venetian–blind, metal dynode, wire mesh dynode and multianode (MAPMTs).They come in various shapes and sizes.

MAPMTs have been mainly manufactured by Hamamatsu Company, Japan. They may be produced with multi-wire anode or multi-anode pads. They are also produced having wire-mesh dynodes or with Venetian-blind dynodes. Figure 2 shows a multi wire anode MAPMT that was assembled with amplifiers close to the base. Some of these tubes were used for a double Compton Scattering Camera [1].

Scattering

**FIGURE 2.** An assembled multi wire anode MAPMT that was used for a double Compton Scattering Camera. It had 16 X and 16 Y cross-wire anodes. 32-channel preamplifiers are mounted close to the base.

Figure 3A illustrates a cutaway view of a MCP (Multi Channel Plate) PMT together with one of the channels. Lead Glass channels are highly resistive, therefore high electric field gradient is developed in the channels, when a high voltage is applied between the two ends of the MCP. When a photoelectron is entered in one the channels, it is cascade multiplied, forming an avalanche of electrons. A gain of $10^3$-$10^6$ can be obtainable depending on the channel length and the applied high voltage. This feature of the MCP can be used for boosting photo-image as illustrated in Figure 3B.

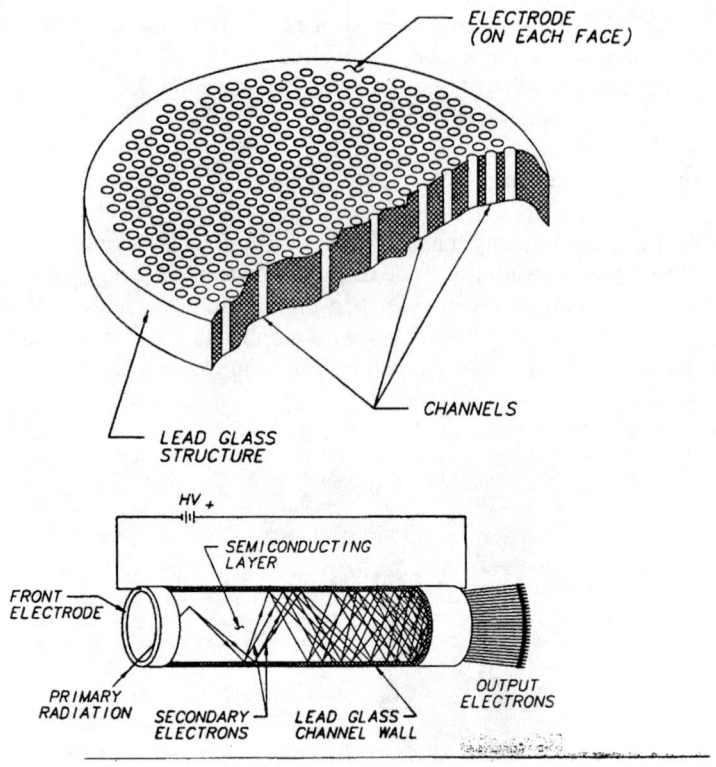

**FIGURE 3A.** A cut schematic view of a MCP-PMT together with a view of a micro channel showing the avalanche multiplication process.

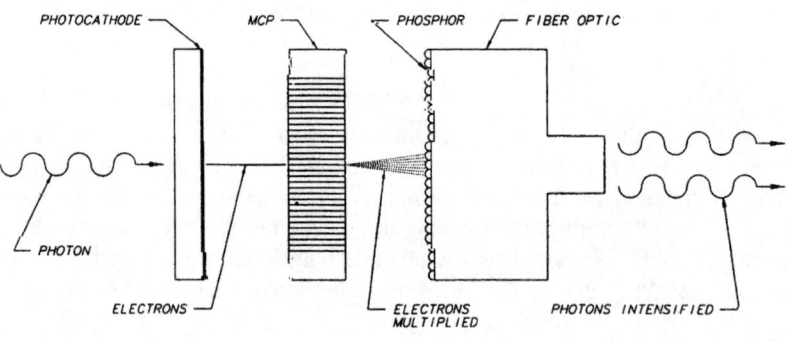

**FIGURE 3B.** Illustrates how a MCP-PMT can be used for boosting an optical image.

# Solid State Photodetectors

Among them are PIN-junction diodes, Charge Coupled Devices (CCD), Avalanche Photodiodes (APD), and Visible Light Photon Counters (VLPC). The first two do not have electronic charge gain, but the APD and the VLPC can produce substantial internal electronic gain. The first three of them are manufactured by several companies, but the VLPC is made exclusively by a branch of Boeing Co. in Los Angeles, California.

All of these devices above are made of semiconductors. There are two types of semiconductors: intrinsic semiconductors and extrinsic semiconductors, and there are two types of photo-effects, intrinsic photo-effect and extrinsic photo-effect. In the case of intrinsic photo-effect, a photon creates an electron hole pair from a bound semiconductor lattice. For the case of extrinsic photo-effect, a photon creates a free electron and a bound hole by interacting with an impurity atom in a semiconductor lattice. The effects are described in the Figures 4-a and b. An electron-hole pair in a semiconductor can be created if,

$$h\nu \supseteq E_g, \text{ another way of writing } hc/\lambda \supseteq E_g$$

where, **h** is the Planck constant, $\nu$ is the frequency of the photon, $\lambda$ is the wavelength of the photon, **c** is the speed of light, and $E_g$ is the bandgap of the semiconductor.

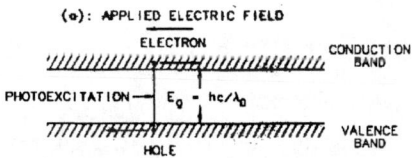

**FIGURE 4A.** A schematic diagram for the intrinsic photoeffect.

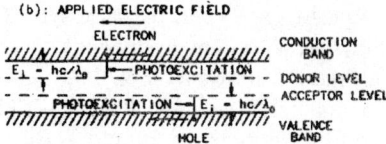

**FIGURE 4B.** A schematic diagram for the extrinsic photoeffect

## PIN-JUNCTION DIODES

These photodiodes converts the photons into electrical currents with no internal gain. They can be used for detecting burst of photons of sufficient intensity possibly from scintillating crystals in detecting gamma-rays. They can have good quantum efficiencies around 70 % to visible photons with silicon oxide antireflective coating. A cross section sketch of a PIN diode is shown in Figure 5-a and b. It is a n-type silicon wafer with p-layer doping as shown in the Figure 5a. Figure 5b shows the energy level diagram. The photons are absorbed in the valence band giving rise the electrons into the conduction band.

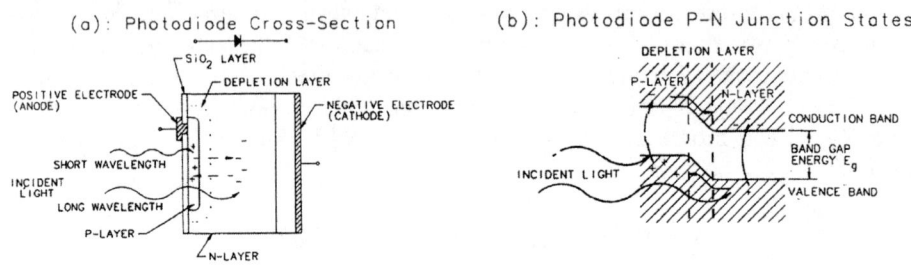

**FIGURES 5A AND 5B.** (A) A cross-section view of an APD structure. (B) Energy level diagram of p and n-layers together with the energy band-gap.

## CHARGE COUPLED DEVICES (CCD)

Charge coupled devices are imaging devices that are very widely used in video cameras, as charge particle tracking devices for High Energy Particle Physics, for medical imaging and Astronomy. They are mainly Silicon based devices. GaAs – CCD is being developed for X-ray imaging. Pixel size and thickness of a CCD may vary depending on the applications. As small as 10 micron pixel size CCDs have been manufactured. They are they are slow devices. Electronic charges from the photo conversions are collected in the pixels and read out typically 50-60 times a second to form images. The readout speed very much depends on the size of the CCD and the clock rate of the readout. Figure 6 shows a block diagram of a relatively old type CCD with two-phase clock operational Fairchild CCD202. It shows pixels(charge storage wells), vertical and analog transport registers, clock lines, amplifier, gate and video output. The charges accumulated in the storage wells are first moved to vertical analog transport registers, and then shifted out from pixel to pixel to the horizontal analog registers and from there shifted out by using the two-phase clock system. Typical clock rate is around 15 MHz. More information about the operation of CCD can be found in [2].

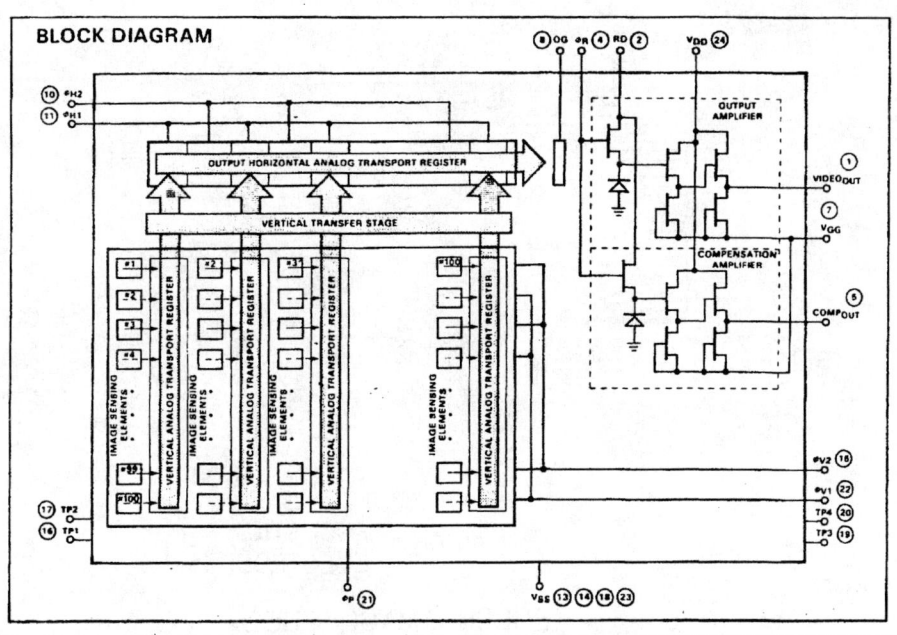

**FIGURE 6.** A block diagram of two-phase clock operational CCD.

## ELECTRON BOMBARDED CCD (EBCCD)

A cross-section view of an EBCCD is shown in Figure 7. In this case, photoelectrons produced from the photocathode can be electrostatically focused under a substantial electric field. About 8 to 12kV potential may be applied between the cathode and anode depended on the required gain. These electrons with substantial kinetic energy hit the CCD and produce gain by ionizing the silicon of the CCD. It is an image intensifier where charge gain is required [3]. The fiber optic window reduces the image distortion.

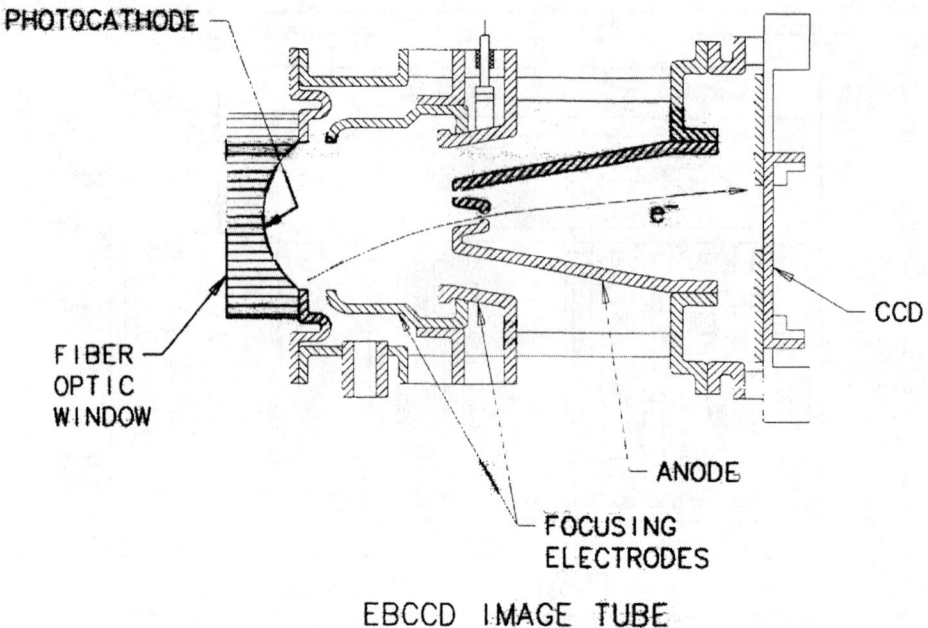

**FIGURE 7.** A schematic diagram of an electrostatically focused EBCCD that can boost an optical image.

## AVALANCHE PHOTO DIODES

Avalanche photo diodes (APD) are silicon base devices that can run in the avalanche mode and the Geiger (breakdown) mode. A cross-section view of an APD together with bias-voltage gain characteristics are shown in Figures 8a and b. Typical gain in the avalanche mode is 50-100 and the gain may reach $10^7$ in the Geiger mode of operation. They are used mainly in the avalanche mode for optical coupling of fibers in communication and in High Energy Particle Physics experiments for detecting photons from scintillating crystals with electromagnetic calorimeters [4].

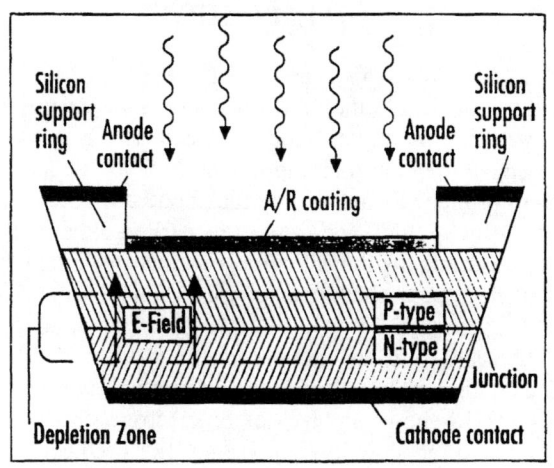

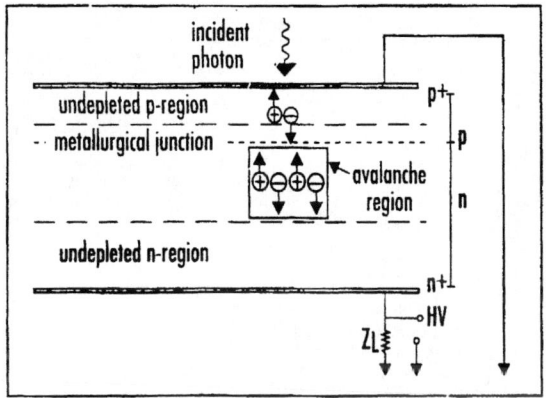

**FIGURES 8A AND 8B.** (A) Cross section of the API APD. Note the beveled edge which prevents early breakdown. (B) Photon absorption, charge drift, and multiplication under bias in the APD.

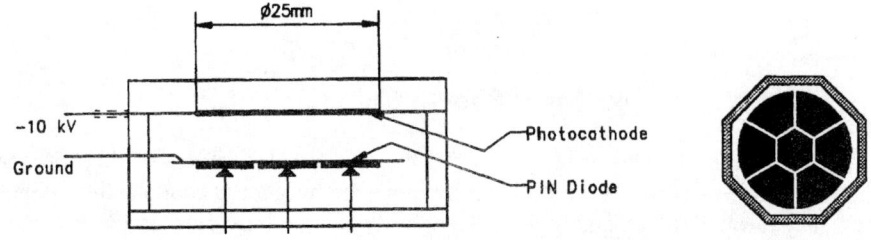

**FIGURE 9.** A cross section view of a 7 silicon pad Hybrid PMT.

## HYBRID PMT, HPMT

Hybrid Photomultiplier Tubes are very much needed when there is a strong magnetic field in the region where they are used as photodetectors [5]. DEP Co. in Holland has been developing these with multi-anode structure as proximity focusing devices which makes them operate in strong magnetic fields without large gain drop. Photoelectrons that are produced from the photocathode are accelerated under a high voltage (around 10kV) hit a PIN-Diode (see the earlier pages) and produce large number of electron-hole pairs in the silicon. An electronic gain of 2000-3000 is obtainable in such a device due to the kinetic energy gained by the electrons under the applied electric field. Approximately 3.6eV energy is required for producing an electron-hole pair. Cross-section and multi-pad anode view is shown in Figure 9. HPMT can be used as a photon counter due to the internal gain and long integration time preamplifier. Figure 10 shows a pulse height spectrum obtained by G. Anzivino et.al. [6] at low rates due to the slow low noise amplifier used for obtaining the pulse height spectrum.

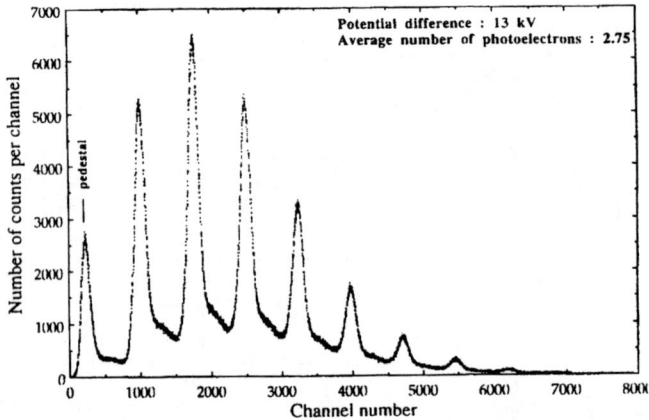

**FIGURE 10.** Pulse-height spectrum of an electrostatically focused HPD exposed to a small number of photoelectrons per light pulse.

## Visible Light Photon Counters (VLPC)

The Visible Light Photon Counters (VLPC) are the derivatives of the Solid State Photo Multipliers (SSPM) which have high avalanche gain capability, high quantum efficiency (around 80%), and high rate capability Silicon base devices [Refs.7,8,9]. M. Atac has pioneered the development of the VLPC's, at UCLA working together with M. Petroff of Rockwell International Science Center, now is a subsidiary of Boeing Co. The devices can be used for scintillating fiber tracking and for medical imaging due to their above characteristics.

Operation principles of the VLPCs are given in Reference 8, therefore we will discussed them briefly here. They are Impurity Band Conduction (IBC) devices that are minimized in quantum efficiency in the Infrared (IR) region while maximized in quantum efficiency for the wavelengths around 550 nm relative to the original device, SSPM which was discovered by Rockwell International Science Center. The VLPC's and the SSPMs are silicon based devices with some levels of donor and acceptor concentrations in silicon that are formed by molecular epitaxy technique.

A schematic diagram of the VLPC cross-section is shown in Figure 11.

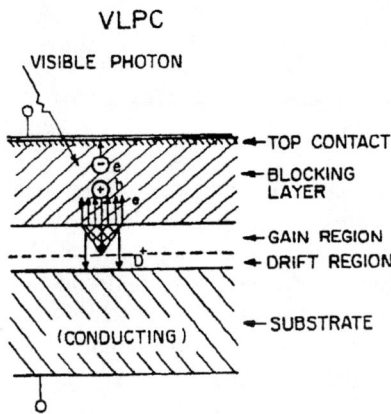

**FIGURE 11.** Schematic of the operational principles of the VLPC.

In a VLPC, a neutral donor is a substitutional ion with an electron bound to it in a hydrogen-like orbit with an ionization potential of about 0.05eV. Because of this very small energy (band-gap) required for creating electron-hole pair the devices need to run at cryogenic temperatures. Nominally they run at temperatures around 7K. The gain plus the drift region is less than 10μm. When the concentration of impurities is sufficiently high, they form an energy band separated from the conduction band by the ionization potential. When the applied electric field between the substrate and the top contact is sufficiently high, about $2 \times 10^3$-$10^4$ V/cm, each electron in the gain region starts an avalanche of free electron-hole pairs within 1nsec. The avalanche gain could reach up to $5 \times 10^4$ when applied voltage reaches 7volts. The avalanche may occupy about 10micron diameter area for about few microseconds while the rest of the area of 1mm$^2$ is continuously available for detecting photons. The VLPC's are now produced in 2x4 chips having 8 pads for a convenient connection to a ribbon cable. Characteristics of the VLPC's are given in Table I.

## Table I

| | |
|---|---|
| Quantum efficiency optimized for 530nm | 80% |
| Avalanche gain | $3\text{-}5\times 10^4$ |
| Thermal electron pulse rate at 6.5K | $\sim 5\times 10^3/\text{sec.mm}^2$ |
| Saturation pulse rate | $5\times 10^7/\text{sec.mm}^2$ |
| Pulse risetime | < 3nsec |
| Average power per pixel | 1.5 microwatt |
| Optimum operating voltage | ~6.5V |
| Optimum operating temperature | 6-7K |
| Dynamic-range (linear) | 3000 photoelectrons |
| Effect by magnetic field | No effect up to 12kG |

Quantum efficiency, avalanche gain and dark count pulse rate as functions of temperature and bias voltage are given in Figure12. The nominal bias voltage and the temperature settings that are given in the Table I can easily be found from the curves. Neither the quantum efficiency nor the gain is a strong function of the bias voltage and the temperature. We will see in the following paper that achieving 0.1K stability of the temperature is relatively easy by making use of Enthalpy of the boil-off He gas of a liquid Helium dewar.

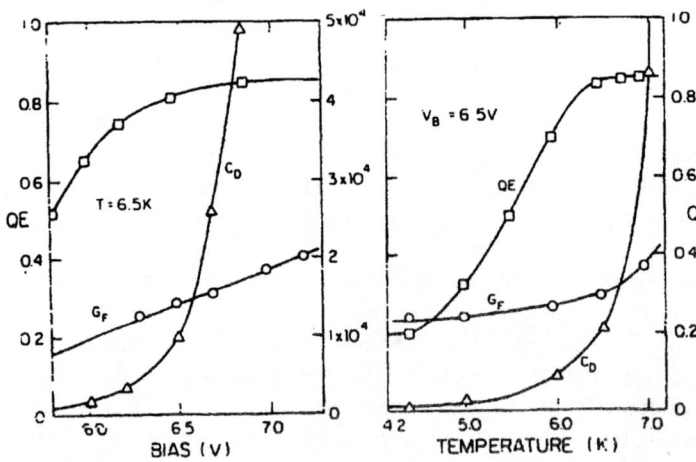

**FIGURE 12.** Quantum efficiency, avalanche gain and dark count pulse rate as functions of temperature and bias voltage.

# REFERENCES

1. Atac, M. et al.; Nuclear Physics B, 10B (1989) 139-142.
2. Damerell, C.J.S.; Rutherford Appleton Laboratory Preprint, RAL-P-95-008.
3. Hamamatsu Co., Japan.
4. McIntyre, R.J.; IEEE Trans. Electron Devices ED-20 (1973) 637-641.
5. DEP Co. Catalog, Holland.
6. Anzivino G. et al.; Nucl. Instr. And Meth. Phys. Res. A (1995) 76-82.
7. Petroff, M. D. and Atac, M.; IEEE Trans. On Nucl. Sci. NS-36 (1989) 163.
8. Atac, M. et al.; Nucl. Instr. And Meth. A314 (1992) 56.
9. Atac, M. et al.; Nucl. Instr. And Meth. A320 (1992) 155.

# Visible Light Photon Counters For High Rate Tracking Medical Imaging and Particle Astrophysics

Muzaffer Atac

*Fermi National Accelerator Laboratory*
*Batavia, Illinois 60510, U.S.A.*
*University of California at Los Angeles*
*Los Angeles, California 90095, U.S.A.*

**Abstract.** This paper is about the applications of the Visible Light Photon Counters (VLPCs) to High Energy Particle Physics Experiments using scintillating fibers, Particle Astrophysics, and Medical Imaging. The VLPC is a derivative of the Solid State Photomultiplier (SSPM) that can detect single photons at a high avalanche gain with very high quantum efficiency with excellent time resolution.

## INTRODUCTION

Photodetectors with high quantum efficiency had long been searched for in High Energy Particle Physics research, Particle Astrophysics and Medical Imaging. M. Atac has pioneered the development of the VLPCs for these applications, working together with M. Petroff of Rockwell International Science Center [1]. The VLPCs can provide quantum efficiency around 80% for wavelength about 630nm with high gain and excellent time resolution. The Rockwell Group accidentally discovered the SSPMs while they were working on Blocked Impurity Band devices. They had high quantum efficiency for the infrared (IR) photons. What we needed was a device that could deliver high quantum efficiency for the visible photons in the range of 420-600 nm wavelengths. It took about five iterations and six years to reduce the infrared sensitivity of the VLPCs to a few percent, while increasing the quantum efficiency to around 70-80 % for the visible photons.

## Tracking With Scintillating Fibers And VLPCs

High Energy Physics Collider Experiments (CDF and D0 at Fermilab) are designed to run at high luminosities. They require fine segmentation, good spatial resolutions, fast timing and good time resolutions due to the high luminosities and large number of secondary charged particles that are produced by proton-antiproton or proton-proton collisions at very high energies. Silicon Microstrip detectors are recommended for the Large Hadron Collider (LHC) Experiments, CMS and ATLAS, but relatively moderate luminosity collider experiments, CDF and D0, at Fermilab could use scintillating fiber tracking. In fact, the D0 experiment [2] is now constructing a scintillating fiber tracking system that will use 80,000 fibers and the same number of VLPCs. E835, Charmonium Experiment at Fermilab has used about 1,000 VLPCs very successfully [3]. After about one year of running time, the group found no measurable change in the performance of the devices. This was the first High Energy Physics experiment with scintillating fiber tracking using these photodetectors.

Scintillating-fiber ribbons can be put on carbon fiber composite cylinders as axial and stereo layers for collider experiments due their flexibility. Figure 1 shows such a configuration.

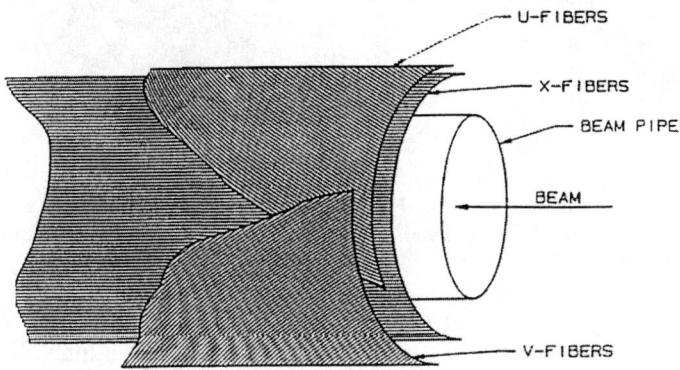

**FIGURE 1.** Scintillating Fiber Configuration of a Proposed Vertex Tracking for Collider Experiment

For a scintillating fiber tracker, the fibers are connected to clear optical fibers using specially designed optical connectors to transmit the photons that are produced by the passage of charge particle tracks to the VLPCs. The optical fibers are connected to a cassette that houses the VLPCs. A 32 channel experimental cassette is shown in Figure 2. In this case, 1mm sensitive area VLPCs and 835 micron scintillating fibers were used. The optical connector between the scintillating fibers and the clear fibers, and the connector on top of the cassette are very precisely made not to lose photons as they are transmitted to the VLPCs. Our experiments showed that the losses are less than 5%. A 32-channel amplifier [4] board is also shown in the figure. The electronic signals are carried to the amplifiers using Teflon coated 75 micron thick gold plated Stainless steel 304 wire ribbons in the cassette.

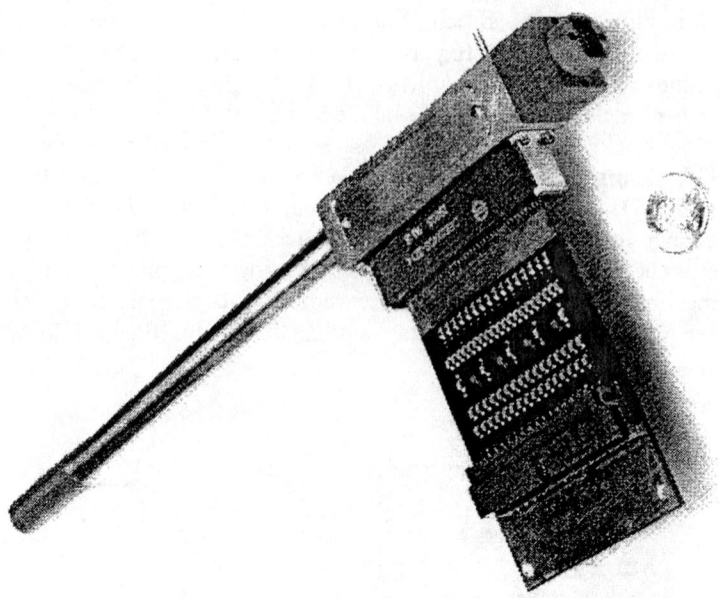

**FIGURE 2.** A photograph of a 32 channel VLPC cassette. The penny in the picture shows the compactness of the unit.

The cassette has a thin wall Stainless steel 304 tubing of 12mm diameter for reducing heat conduction to the dewar. The tube is attached to a copper housing that contains the VLPCs. The details of the 8 liter liquid helium cryostat is shown in Figure 3. An oxygen free high conductivity (OFHC) Cu tubing that is immersed into the liquid He, surrounds the VLPC Cu housing to bring the low temperature up to that level. The Cu tubing plays an important role in keeping the temperature at the optimum level of 6.5K independent of the liquid He level. The cryostat with the cassette lasts for about five days before replenishing the liquid helium. In this case we use the full Enthalpy of the boil off gas (in the dewar) goes through the cassette making it very economical in the usage of the liquid helium. Inner part of the cassette is shown in Figure 4. Optical fibers that transmit the photons into the VLPCs are aligned within 25μm relative to the 1mm diameter sensitive areas.

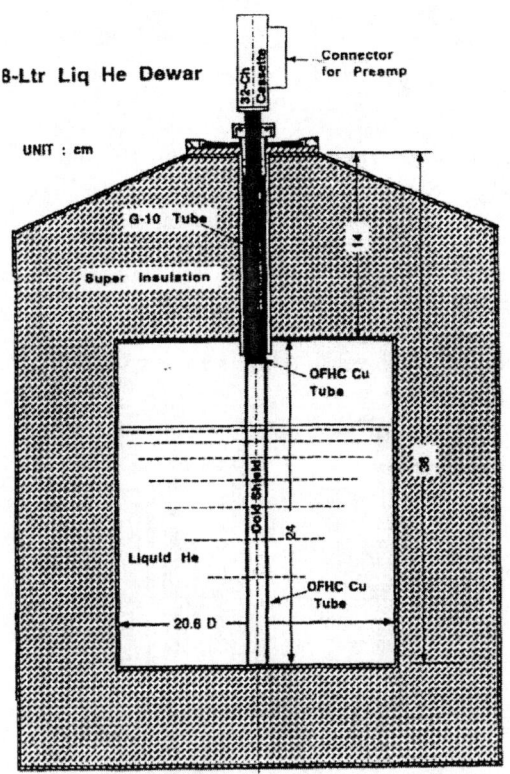

**FIGURE 3.** The structure of the 8-liter cryostat.

## Scintillating Fiber Tracking Tests

Some scintillating fiber tracking experiments were carried out using 500μm overall diameter multiclad fibers (produced by Kuraray Co., Japan), the above described cassette and the cryostat assembly. Six photoelectrons in the average were detected per minimum ionizing track passing through the fibers at the middle of a 280cm length of the scintillating fibers (were mirrorized at the far end by aluminum sputtering) which were coupled with 500cm of multiclad clear fibers. The scintillating fibers were polystyrene base with 1500ppm 3HF+ 1% PTP doping [5]. An ADC count versus number of photoelectron (pe) calibration was taken before every measurement as shown in Figure 5. An LED was used for this, illuminating the optical fibers from the top of the cassette. As seen in the figure, after 6pe a small saturation appears. This is due to the amplifier [4] saturation. The calibration was obtained by making a cut at N>2 and fitting to a convolution of Poisson and Gaussian functions:

$$f(x) = N \sum_n \frac{1}{\sigma_n \sqrt{2\pi}} \exp\left[-\frac{1}{2}\left(\frac{x-n}{\sigma_n}\right)^2\right] \frac{\exp(-n_{pe})}{n!} \quad (1)$$

where, the free parameters were normalization factor (N) and the mean value of the Poisson distribution ($\sigma_{pe}$). Sigma of each Gaussian ($\sigma_n$) was fixed by determining the peaks with the LED runs for the bias voltage and temperature.

$$\sigma_n^2 = \sigma_0^2 + \sigma_c^2 \cdot n \quad (2)$$

where, $\sigma_c$ is the sigma of the n-th peak and $\sigma_0$ is a value of the pedestal.

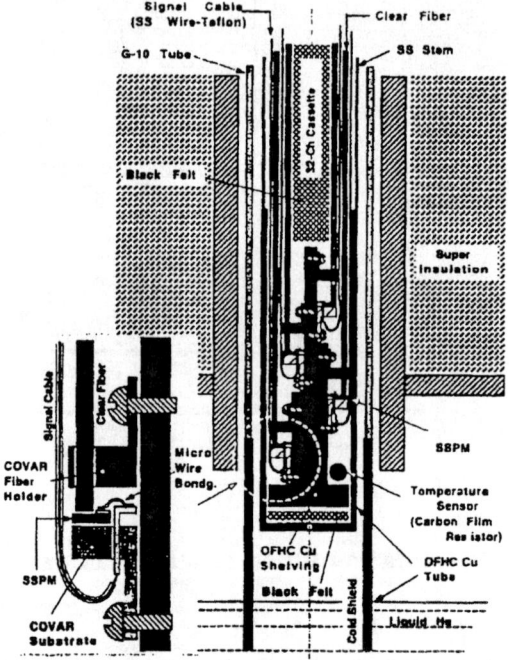

**FIGURE 4.** Enlarged view of the cassette.

Some beam tests were carried out at the Meson 6-west test beam at Fermilab using 830μm (scintillating core 0f 725μm) multiclad 3HF using two of the 32 channel cassettes. Four staggered doublets of scintillating fiber ribbons were used for the tests. Figure 6 shows a 15 GeV hadron (most likely pion) track with 15 photoelectrons obtainable in the average. This would give us a doublet tracking efficiency of better than 99.7%.

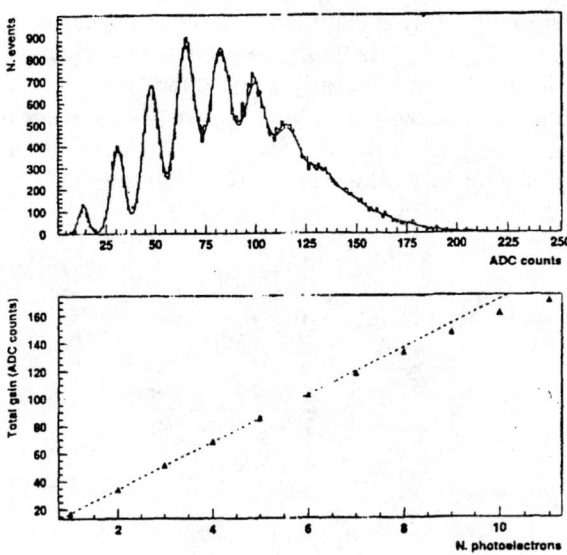

**FIGURE 5.** Multi-photoelectron peaks and a calibration plot. As we see in the figure, multi-photon peaks are well resolved. For this reason, we call the devices "Visible Light Photon Counters" (VLPC).

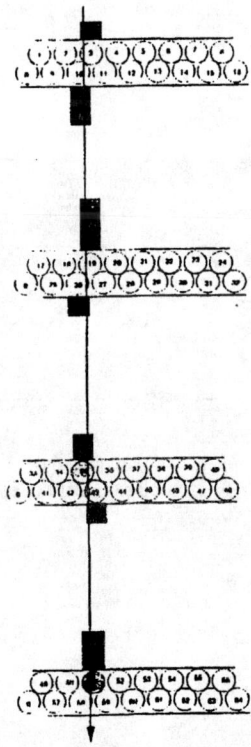

**FIGURE 6.** A minimizing ionizing track. The number of detected photons is indicated in the bars.

The Collider Detector Facility at Fermilab (CDF) has considered using scintillating fiber tracker with VLPC system and some track reconstructions were carried out. Figures 7A and 7B show how efficiently a top quark event can be reconstructed. Figure 7A shows the R-φ and the Figure7B shows the R-Z view of one of the earlier detected top-antitop quark pair event. The tracks, which were very efficiently found, were the tracks detected by the Central Tracker (CTC) of the CDF.

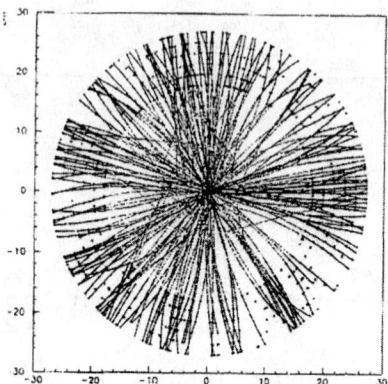

**FIGURE 7A.** A typical top + 6 MB event in the R-φ view. Crosses are axial hits, solid lines connect hits used, and dotted curves are extrapolation inward of the final parameters.

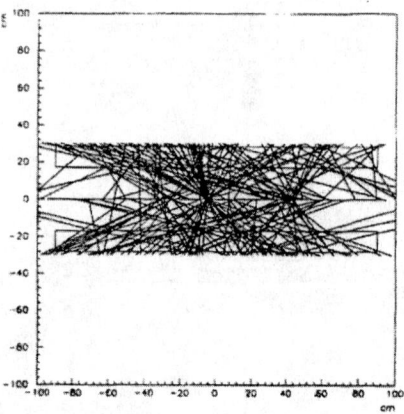

**FIGURE 7B.** R-Z view of a top + 6 MB event. Crosses show locations of associated stereo-hit/axial-segment points. Lines show extrapolation of fitted segment to beam line.

# Use of VLPCs for Medical Application

*Single fiber Tracker for Stereotactic Biopsy and Intraoperative Lumpectomy of Breast Cancer*

Breast cancer in women of child bearing age is the second leading cause of death in the USA [6]. Early detection has allowed for less extensive surgical procedures and/or decreased need for chemotherapy since a substantial majority of questionable lesions detected by mammography are benign. There is a growing interest among the health care professionals and patients in finding alternatives to surgical biopsy for diagnosing such lesions. State-of-the-art stereotactic breast biopsy is comparable in sensitivity to surgical biopsy, and the procedure is quicker, cheaper and easier than the standard practice of preoperative, mammographically guided localization followed by surgical biopsy.

The problems mentioned above can be ameliorated by a nuclear medicine procedure using a beta detector at the end of a 0.8mm diameter and 3mm long plastic scintillator and a fiber optic cable [7]. By positioning the detector within a few millimeters of the suspected area, small lesion, usually not detectable using gamma radiation detectors, can be identified for activity. The fiber optic cable with a small scintillating fiber attached (fused) to the tip can either be inserted into a core biopsy, or can be used during ductogram to identify the duct system containing microcalcified clusters. When inserted into a surgical wand, it could be used to ensure that all residual tumor was removed during lumpectomy. This diagnostics alone is very much needed to prevent recurrence and spread of malignant tissues.

We have developed a prototype suitable probe that uses a rather small diameter biopsy needle (in the current study an 18 gauge needle with an external diameter of 1.25mm) containing a 0.83mm diameter and 3mm length of 3HF (above mentioned) multiclad scintillating fiber, which is fused to the same diameter multiclad clear optical fiber of 200cm length.

Photons emitted from the scintillating fiber by the passage of betas are transmitted through the optical fiber (attenuation length of the photons in the optical fiber is 900cm) and are detected by a VLPC. For the set up, a cassette and the cryogenics mentioned in the tracking section above were used. The probe assembly and rather inexpensive data acquisition system are shown in Figure 8.

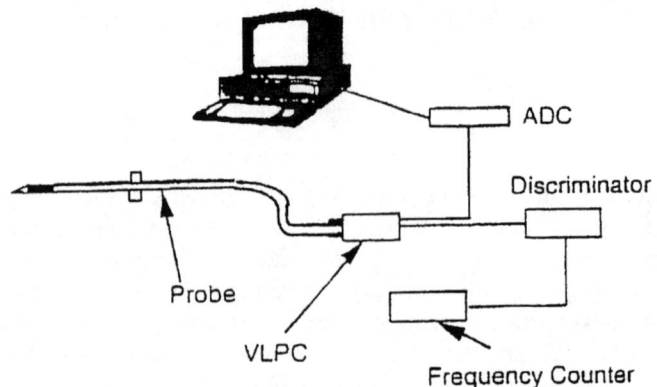

**FIGURE 8.** A schematic view of the biopsy needle probe together with a simple data acquisition system.

The signals from the VLPC were amplified by a transimpedance amplifier (TIA), fed into a discriminator and counted by a commercial scaler. The VLPC produces an avalanche gain around 30,000 per photoelectron. We obtained less than 2 counts per minute as background rate when the threshold was set above three photoelectrons. We measured experimentally that the average number of photoelectrons produced by the VLPC was more than 40, by the passage of the betas going through the thin scintillating fiber. Pulse height spectrum obtained using a $Bi^{207}$ beta source is shown in Figure 9. Only a small fraction of the 1MeV beta particle energy is left in the thin scintillating fiber, giving rise to the pulse height spectrum.

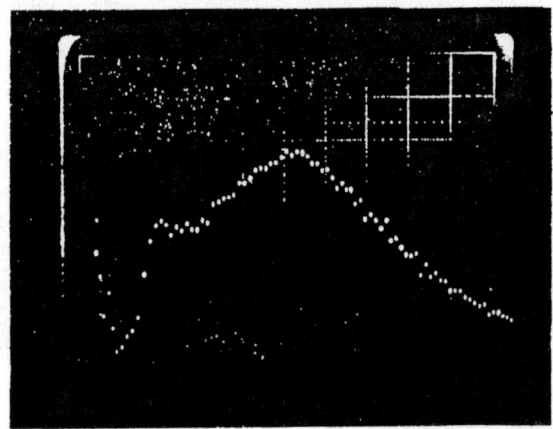

**FIGURE 9.** Pulse height spectrum obtained using $Bi^{207}$ beta source. The average energy released in the scintillator fiber is about 60KeV. The peak value corresponds to 40 photoelectrons detected by the VLPC.

# Experimental Results

In order to determine point spread function, we moved the probe linearly relative to the 1μC $Bi^{207}$ source without and with 1.5mm thick Lucite sheets (mimicking tissue equivalent density) in between the source and the probe, and recorded the counts per second. The source diameter was approximately 4mm and it was not collimated. The results plotted, in Figure 10 show that the 1MeV betas from the source were very much attenuated after one sheet of Lucite, but we can find the source position after 4.5mm thickness. We expect that the intrinsic resolution of the probe be better than 1mm. The curves also show that the probe is sensitive to betas, and not to the gammas, although only 8% of the decays of the source produce betas and the rest being the gamma activity. The probe like this can be cooled by liquid Helium vapor for safety. A photograph of the probe is shown in Figure 11.

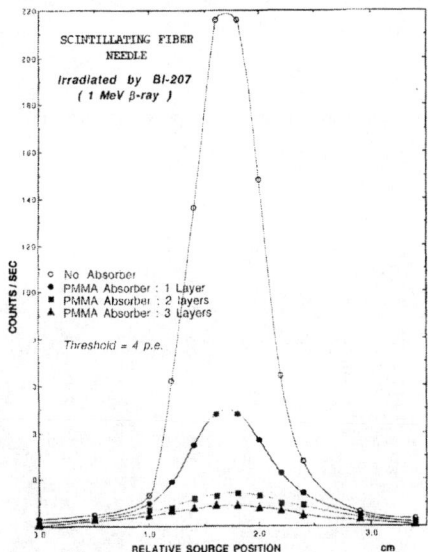

**FIGURE 10.** Results from the $Bi^{207}$ source test. The curves clearly indicated that the 1MeV betas are rapidly absorbed by the 1.5mm thick Lucite sheets, and there are not many counts from gamma conversions in the scintillator although 90% of the decays from the source are gamma rays in this case.

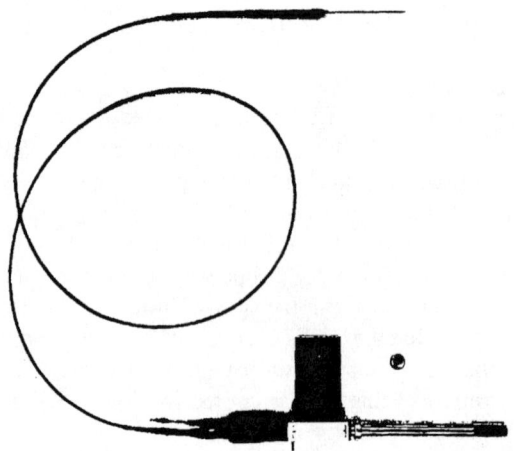

**FIGURE 11.** Photograph of the probe with the 2 meter long optical fiber between the biopsy needle and the VLPC unit.

As a first experiment, a preliminary test was done using a rat bearing R3230 adenocarcinoma. The experimental arrangement is shown in Figure 12. As shown in Figure 13, the biopsy needle was moved in an x,y matrix points and counts were recorded. Even from outside of the skin, the probe indicates where the radionucleide concentration is located.

**FIGURE 12.** Test done with a rat having an R3230 AC in the hind leg. The rat was administered 432 microcurie FDG i.v.

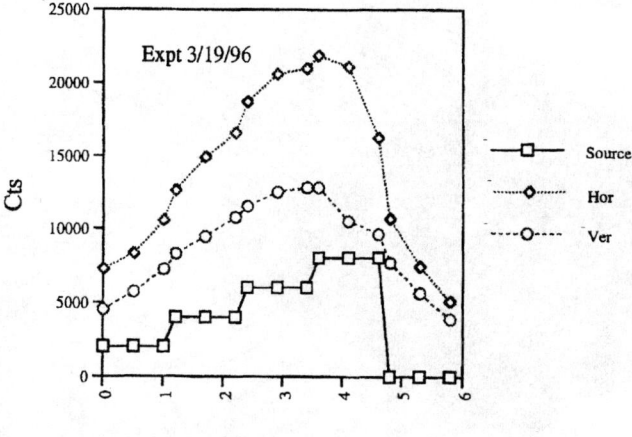

**FIGURE 13.** Two dimensional scan, even from outside of the skin, indicates where the radionucleide concentration is.

## USE OF VLPCS FOR PARTICLE ASTROPHYSICS

A possible way of using VLPCs as photodetectors for detecting relatively low energy gammas is shown in Figure 14. In this case, wavelength shifting fibers are attached to a matrix array of scintillating crystals in an x,y array. The crystal size can be sufficiently large to detect multi-MeV gammas from outer space. Scintillating crystal like BGO can be used in this case. Rubrene doped multiclad waveshifter fiber could be a good match for the VLPCs. Optical fiber that is coupled to wavelength shifter fibers carry the photons to the VLPCs. This idea was proposed earlier by M. Petroff for medical imaging, but it did not work well due to low energy gammas producing very few photoelectrons. I am convinced that it will work here due to multi-MeV gammas.

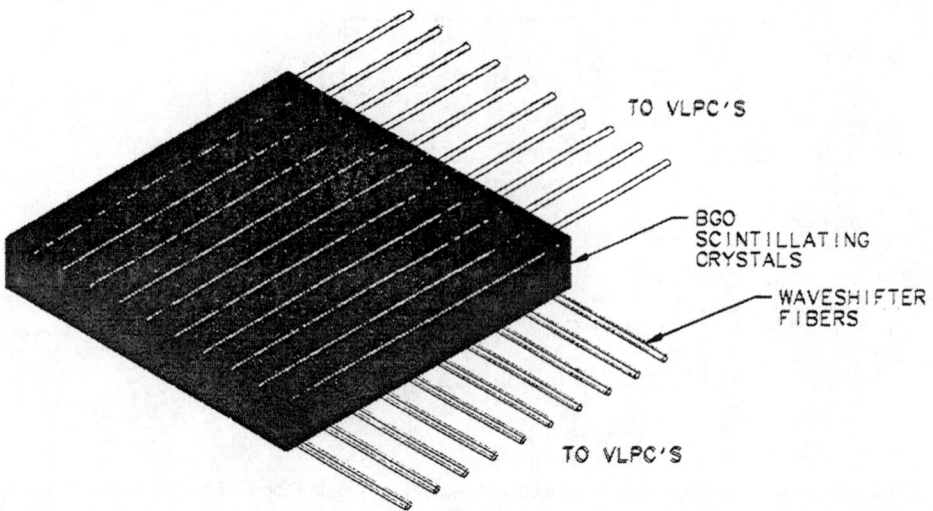

**FIGURE 14.** The principal scheme for detecting gamma rays in a two dimensional readout. More layers can be added depending on the energy of the gammas to be detected.

# REFERENCES

1. Anderson, J.; Atac, M. and Cihangir, S., "Lifetime of Cosmic Ray Muons," ICFA 99, VIIIth HEP Instrumentation School, Istanbul, Turkey.
2. Gutierrez, Gaston, et al; Contribution to Proceedings of Conference on Scintillating and Fiber Detectors, 1997, p. 221.
3. Musso, Roberto, et al; Contribution to Proceedings of Conference on Scintillating and Fiber Detectors, 1997, p. 335.
4. Zimmerman, Tom, *IEEE Trans. Nucl. Sci.* **NS-37(2)**, 439 (1992).
5. Atac, M., et al; CDF/ANAL/Tracking/Public/3569, July 1997.
6. Titcomb, C., *Hawaii Medical Journal* **49**, 18 (1990).
7. Atac, M. et al; Unpublished Report.

# Detecting Dark Matter
### By
### Roger L. Dixon

## 1.) Abstract

Dark matter is one of the most pressing problems in modern cosmology and particle physic research. This talk will motivate the existence of dark matter by reviewing the main experimental evidence for its existence, the rotation curves of galaxies and the motions of galaxies about one another. It will then go on to review the corroborating theoretical motivations before combining all the supporting evidence to explore some of the possibilities for dark matter along with its expected properties. This will lay the ground work for dark matter detection. A number of differing techniques are being developed and used to detect dark matter. These will be briefly discussed before the focus turns to cryogenic detection techniques. Finally, some preliminary results and expectations will be given for the Cryogenic Dark Matter Search (CDMS) experiment.

## 2.) Preliminary Definitions

The following set of equations will provide the basic definitions for the quantities to be discussed during the talk:

$\rho \equiv$ mass density

$\rho_{critical} \equiv$ Density Required to close the Universe

$\Omega_x \equiv \rho_x / \rho_{critical}$  **For any system. For example, x might be luminous stars or baryons**

$\Omega_0 \equiv \rho_0 / \rho_{critical}$  **Where $\rho_0$ is the total Density of the Universe now**

$\Omega_0 \equiv 1$  **For a closed Universe**

$H_0 \equiv$ Hubble Constant $\equiv 100 h$ km sec$^{-1}$ Mpc$^{-1}$

Where $.4 \leq h \leq 1.0$

## 3.) Rotation Curves

The evidence for dark matter comes primarily from its apparent gravitational interaction with the visible universe. To see this we look first at the solar system to remind ourselves of the simplicity and effectiveness of Newton's laws in explaining the motions we observe there. The planets move around the sun in perfect congruence with the mathematics which describe Newton's general principles of motion as shown in figure 1. There is no evidence that anything might be amiss. The velocity of the planets falls as

$$v \sim \frac{1}{\sqrt{r}}$$

just as predicted by Newton.

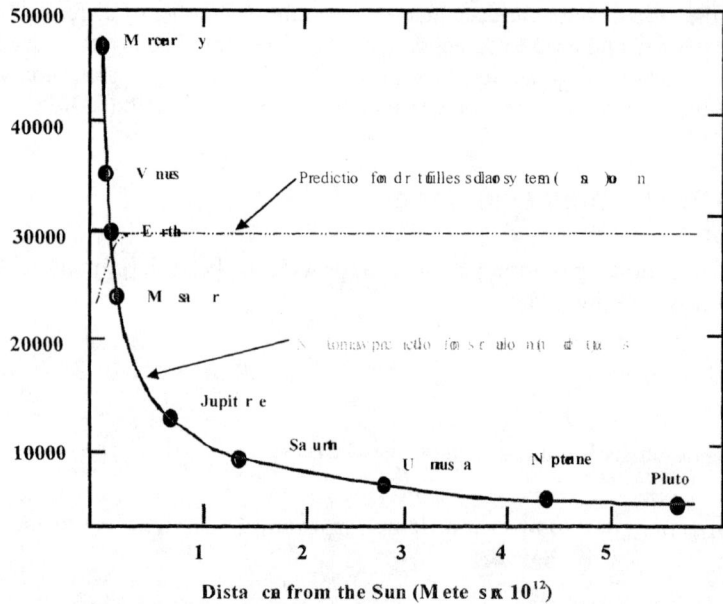

Figure 1

However, if we make the same kind of observations for the motions of the stars within our own galaxy we run into immediate problems. The velocity function appears to be better described by a constant rather than the inverse root distance behavior expected. This can best be seen in figure 2 where we have plotted the velocity distribution for a distant galaxy that is much easier to see than our own. This perplexity leads us to ask what we might expect of the solar system if it were filled with a large amount of dust out to great distances. Note that the curve for this possibility is shown in figure 1. In that case we get a constant distribution of velocities in spacewith distance from the center much as is observed for the galaxy. Could our galaxy be filled with some invisible material which behaves gravitationally as a dust cloud? That is the evidence.

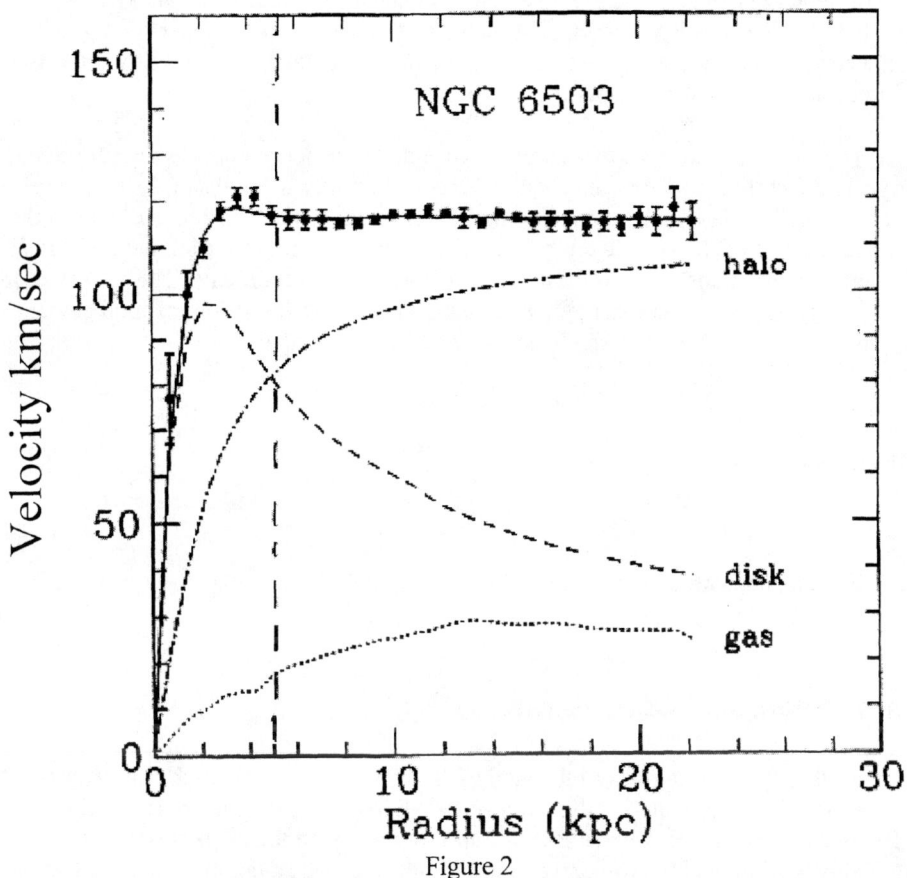

Figure 2

Surprisingly we see no evidence that the velocity ever begins to decrease with distance. The implication is that the invisible halo is much larger than the visible part of the galaxy, and the amount of matter making up the halo is more than 10 times the visible matter.

It is an easy matter to compute the energy density of the halo from basic principles. We find that:

$$\rho \sim \frac{1}{r^2}$$

By using the best determinations we have for the luminous mass and the rotation curves, we determine that :

$$\rho = .3 \rightarrow .6 \quad \text{GeV} \cdot \text{cm}^{-3}$$

As evidence from larger scales is considered the amount of missing matter needed to explain the discrepancies becomes even greater. For example, one can measure the velocities of the member galaxies in a cluster. The maximum velocities observed give a measure of the depth of the gravitational well they are trapped in my means of the virial theorem. These observations indicate that more than 90% of the matter in the universe is not visible. The alternative is that we don't understand gravity. When theoretical considerations are added to the experimental evidence the case for dark matter becomes even more compelling as we shall see.

To begin our transition to theoretical considerations let us consider the Big Bang theory and its implications for dark matter. One of the major triumphs of the theory is that it predicts the relative abundance of the lightest of elements based on the rates of the nuclear reactions which produce the helium, deuterium, and lithium from the primordial hydrogen. Observations confirm these predicted ratios very accurately. This allows us to use the theory and measurements of abundances to place constraints on the total amount of baryonic matter in the universe. When we do this we find that:

$$\Omega_{baryon} \leq .07$$

While the fraction of visible matter is:

$$\Omega_{visible} \leq .007$$

Hence, we can conclude that many of the baryons in the universe may, in fact, be invisible to us.

## 4.) Theoretical Considerations

Further impetus is given to the idea of dark matter if we look at the universe on the very largest of scales and combine our observations with some theoretical ideas. In the 1960's the Cosmic Microwave Background was observed for the first time. It was surprisingly uniform and isotropic, which caused the theoretical community considerable concern. Why should the entire universe be so smooth when much of it had not been in causal contact before the last scattering of the radiation at recombination time. This problem was resolved by inflation theory, which suggests that the universe went through a very rapid period of expansion after the big bang. This would allow all of the visible universe to have been in casual contact at an early time, thus providing an explanation for the smooth microwave background.

There are also implications of inflation for the dark matter problem. One of the questions that arises is why the universe is so close to $\Omega = 1$ after all this time. This is called the naturalness problem. To make the argument clear imagine that you have thrown a stone into the air with a velocity as close to escape velocity as you can. The stone would not be expected to by anywhere near escape velocity after 14 billion years unless you managed to give it a very precise set of initial conditions. Otherwise, the velocity will deviate very rapidly from the escape velocity. The same argument can be applied to the universe. The initial conditions must have been determined very precisely in order for the universe to still be within an order of magnitude of $\Omega = 1$ corresponding to a flat universe.

Another argument for $\Omega = 1$ comes from the inflationary dynamics themselves. Inflation can be thought of as rapidly blowing up a balloon by many orders of magnitude. The result is that the surface of the balloon becomes arbitrarily flat very quickly. If we replace the balloon with the universe, we expect that the universe becomes arbitrarily close to perfectly flat during the inflationary period. Of course, the argument that the universe is flat implies that there is either lots of matter that is not seen, or some other effect is at play in the cosmos. This second possibility will be addressed now.

By observing near and far supernovae and using them as distance candles, two groups have shown that there is yet another factor in the universe which must be taken into account. The observers had initially hoped to measure the deceleration parameter of the universe, which would have allowed them to determine the mass that was slowing the expansion. Instead, their data indicates that expansion of the universe is accelerating. Another way to say this is that $\Lambda \neq 0$ where $\Lambda$ is Einstein's famous fudge factor invented to allow a static universe. If $\Lambda$ is not zero, it may indicate that there is not as much matter in the universe as inflation would lead us to believe. Nevertheless, the supernova data combined with the other density measurements already described here indicates that there is still a lot more matter than observed. A rough tabulation of the makeup of the universe is given in Table 1.

The final theoretical consideration to be addressed here will be that of the structure of the universe. The theorist is given to ask how all the structure came about. What mechanism formed the stars, planets, galaxies, clusters of galaxies, and super clusters? This is particularly perplexing given the very smooth, structureless beginnings of the universe. Of course, there was some structure present in the early universe. More accurate measurements of the CMB beginning with the COBE satellite results in the 1990's have shown that there is indeed structure in the microwave background. This means there was structure before recombination time, and we are led to ask how any structure that was present at those very early times survived. Normal matter was subjected to a tremendous radiation pressure which prevented it from clumping up. As a result, no structure seeded by normal baryonic matter could have begun forming until after recombination time. This does not allow enough time for observed structure.

One possible solution to the problem would be the existence of a form of matter not affected by the radiation pressure. Neutrinos are one candidate particle with this property. It is very likely that at least some neutrino species have some mass.

- **Visible Matter** .01
  - Evidence
    - Telescope observations
  - Composition
    - Ordinary matter-- protons and neutrons
- **Baryonic Dark Matter** .05
  - Evidence
    - BBN
  - Composition
    - Matter too dim to see
- **Nonbaryonic Dark Matter** .3
  - Evidence
    - Gravity, Motion
  - Composition
    - WIMPs, Axions, Neutrinos
- **Cosmological Dark Matter** .6
  - Evidence
    - CMB, Supernova Data
- **Total** ~1

Table 1

To this end Davis, Efstathiou, Frenk, and White (DEFW) simulated neutrinos in the early universe and tried to reproduce the structure that is observed today. They found that neutrinos tend to form structure from the top down and produce a much lumpier universe on the larger scales than we observe. Also, the structure at the smaller scales does not begin to form in time to produce the observed small scale structure. What is needed to solve this problem is a kind of matter that would have been nonrelativistic before recombination time, so that the structure could begin forming early. Hence, the idea of cold dark matter was born. The simulations of DEFW indicate qualitative agreement, but quantitatively, there are still problems. If one looks in detail at the power spectrum of the structure, which is shown in figure 3, one finds that the agreement is not satisfactory without adjusting some of the parameters of the theory. Perhaps, both neutrinos and CDM are needed to produce the Universe we observe.

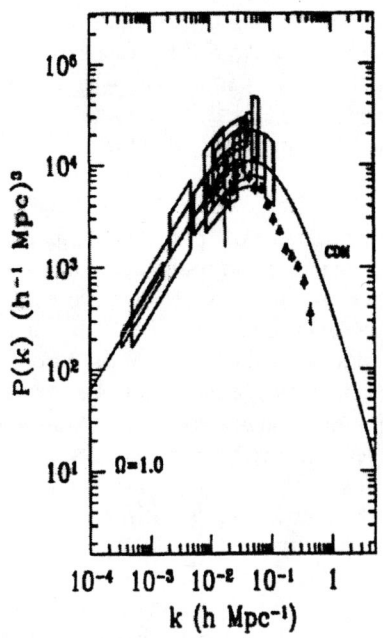 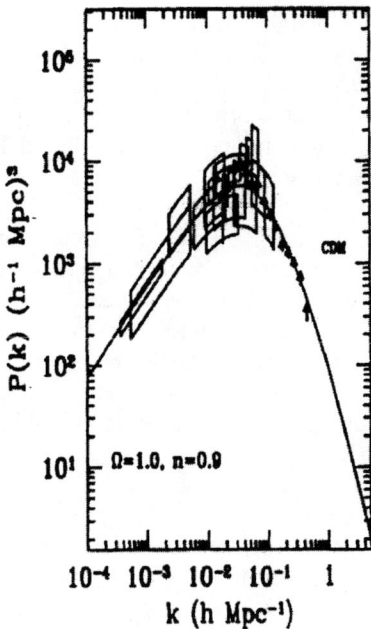

Figure 3

Among the candidates for cold dark matter are the supersymmetric particles which are well motivated in particle theory. Nevertheless, any weakly interacting massive particle (WIMP) will do. The idea is that they are produced in the big bang and annihilate with one another until the temperature of the universe falls to a point where they "freeze out". Their density and energy become low enough so that the annihilations no longer occur often enough to deplete the supply and the universe is left with a relic abundance of the particles. If nothing else, it is amusing that the annihilation cross-section which is calculated for the WIMP particles in order to give an abundance which would produce an $\Omega = 1$ universe agrees to within an order of magnitude of the weak coupling strength.

There are many good candidates for the WIMPs including massive neutrinos and supersymmetric particles. This does not exhaust the list. In the rest of this talk we will discuss detecting WIMPs without giving too much consideration to just which candidate they might be. All we assume is that they interact weakly and that they are massive.

## 5.) Detection

Detection of WIMPs can be accomplished in two manners. The first is called indirect detection and relies on detecting the products of the WIMP annihilation process. This method may be enhanced if the WIMPs tend to cluster in the center of the sun or the earth. One looks for the neutrinos from the annihilations of the particles and antiparticles. Several groups are employing this strategy, but we will not focus on it here. Instead, we will discuss direct detection of WIMPs. Since the particles are weakly interacting they are expected to scatter off of a nucleus with a weak scattering cross-section. It is these nuclear recoil events we hope to detect in our direct detection experiments.

There are many possible backgrounds to nuclear recoils. Any particle coming into a detector, whether it be from a radioactive decay in some of the surrounding material or a cosmic particle from deep space, can cause an event to occur in our detectors. It will help, but is not absolutely necessary if some of these events can be eliminated by means of discrimination in the detectors. NaI crystals provide a good example of the sort of discrimination we might expect. The decay time probabilities of the light pulse in NaI are quite different for events produced by gamma interactions in the crystal than for those resulting from nuclear recoil events produced by neutron scatters. This allows a distinction between the two kinds of events to be made in the detectors, which can help to eliminate some of the background. Of course, it will always be very difficult to eliminate the neutron scattering background. This difficulty must be attacked by other means, which will be discussed later.

Another important feature of nuclear recoils is demonstrated in the equation for the counting rates. The rate for nuclear recoils is given by:

$$\frac{dR}{dQ} = \frac{\sigma_0 \rho_0}{\sqrt{\pi} m_\chi m_{eff}} F^2(Q) T(Q)$$

And:

$$F(Q) \sim f(A, Q, m_\chi) e^{-kQ^2}$$

Where R is the rate of the interactions, Q is the nuclear recoil energy, and F(Q) is the nuclear form factor. Since F(Q) falls off rapidly with $Q^2$ where Q is the energy of the recoiling nucleus it is evident that it is very important to have a low threshold for detection in our detectors as that is where most events occur.

## 6.) Cryogenic Detectors

For the remainder of the talk we will focus on cryogenic detectors which have the ability to go to very low thresholds, but at a disadvantage for producing large mass experiments. These detectors are typically made of germanium, but are not limited to that material. For example, silicon, sapphire, and tungsten crystals are also used. Typically, the cryogenic detector measures the energy of the nuclear recoil by detecting the phonons produced in the crystal. Discrimination can be achieved by also measuring ionization or scintillation light and making the ration between phonon energy and the ionization or scintillation energy.

An excellent example of a cryogenic detector with very good discrimination properties is shown in figure 4. This detector has been produced by the CRESST collaboration. The plots show that the ratio of scintillation energy to phonon energy is much lower for nuclear recoils than it is for the electromagnetic events. The events are cleanly separated into two distinct bands allowing discrimination to be made on an event by event basis.

Figure 4

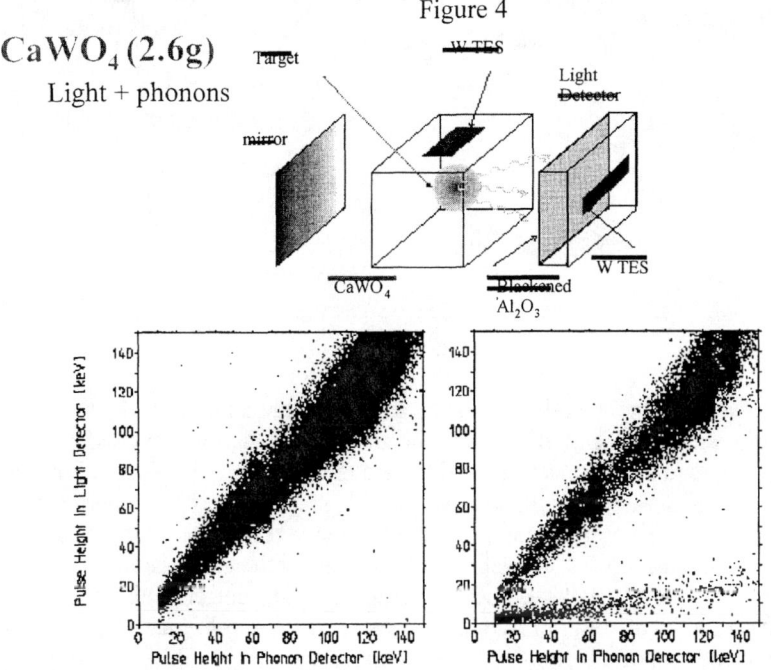

The Cryogenic Dark Matter Search (CDMS) uses detectors that measure both phonon energy and ionization. Several detector designs have been investigated by the CDMS collaboration and one of these will be described here.

Figure 5 shows the CDMS ZIP detectors which empty the tungsten transition edge sensors to measure the phonon energy in addition to the ionization energy which is measured by putting a small bias voltage across the crystal. The detectors are made to operate at 20 mk crystal temperature, but the phonon sensors on the surface are maintained at 70 to 90 mk.

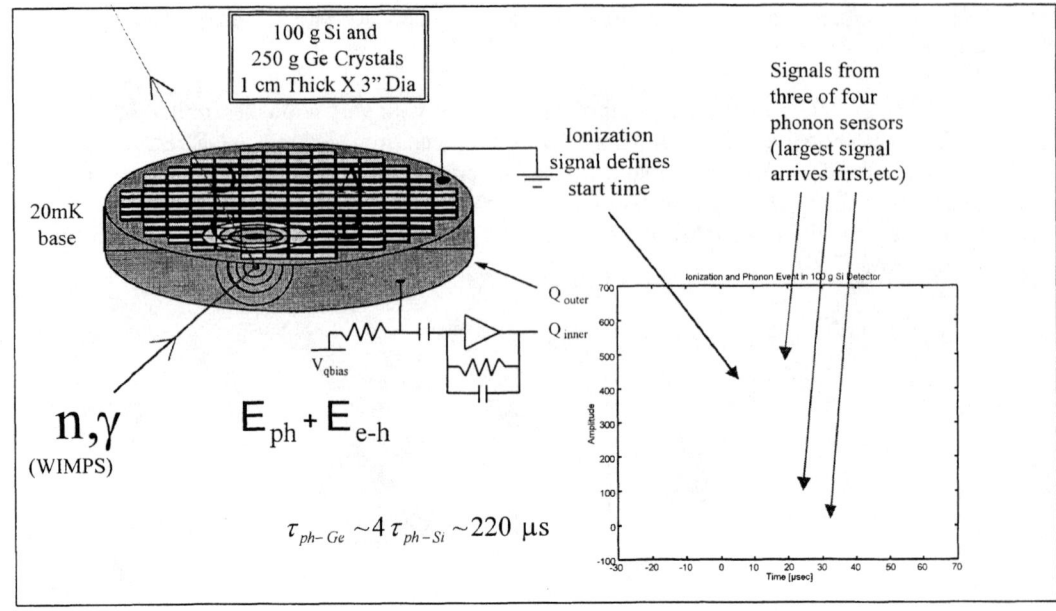

Figure 5

Phonons created by the recoiling nucleus in the crystal make their way to the surface where they excite quasi particle pairs in the aluminum pads. The excited pairs make their way to some tungsten meanders, which are maintained at their transition temperature in a voltage biased mode. The quasi particle pairs give up their energy to the tungsten driving it toward its normal state. As it heats up the resistance increases and the current drops since the tungsten is voltage biased. This causes the temperature to fall again rendering an intrinsically stable detector. Meanwhile, the current pulse produced in the tungsten is sensed by an array of SQUID detectors which amplify it and pass it on for processing. The integral of the phonon pulse measures the energy of the recoiling nucleus. The efficiency of these detectors is very high allowing for very low threshold operation down to a few keV.

Figure 6 shows the separate bands for nuclear recoils and gamma interactions in a 100 g silicon detector. The plots indicate that event by event discrimination is possible down to a few keV. Similar germanium detectors have recently been produced that behave in a similar fashion.

Close examination of the data reveals that there are some events which leak down from the gamma band toward the nuclear recoil band. The problem is more severe in germanium. It has been determined that these events are the result of low energy electrons which do not penetrate very far into the surface of the detectors. Ionization collection is poor for these events. Since they look very similar to nuclear recoils they will cause an undesirable background.

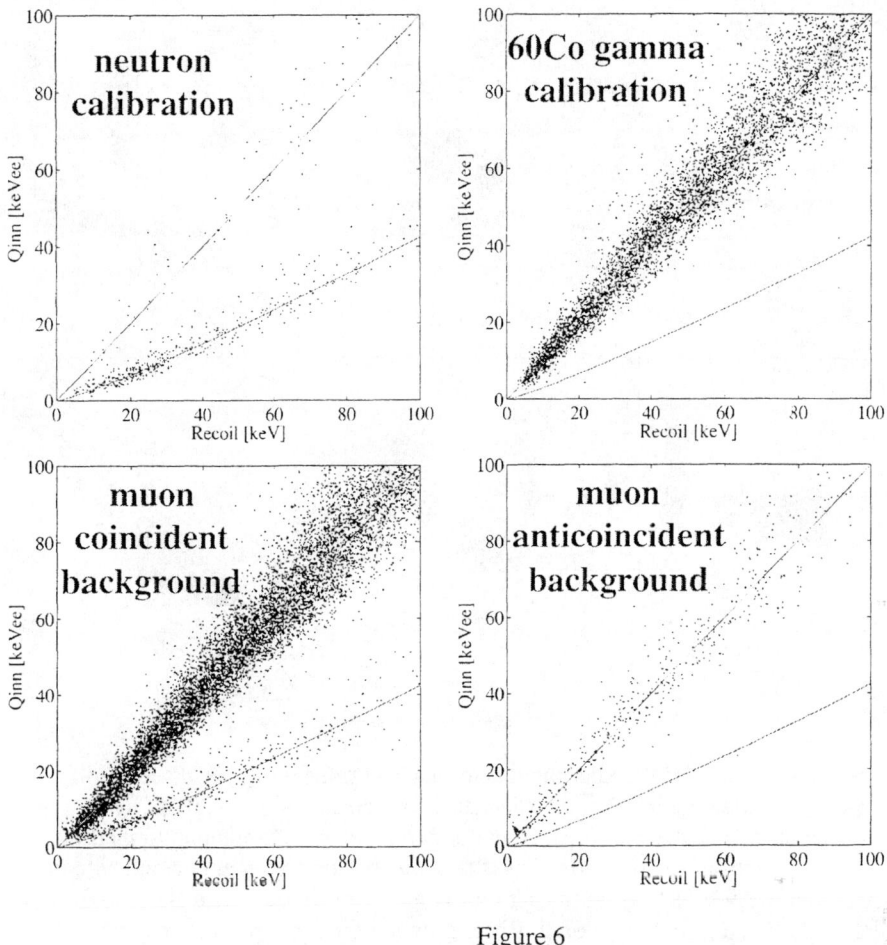

Figure 6

Fortunately, there are two ways to eliminate these background events. The first is to add an amorphous silicon layer to the surface of the detector. The second technique is to make a rise time cut on the phonon pulses from the events. By experimenting with electrons it has been found that the phonon pulse rise times for the low energy beta events are faster than the rise times for nuclear recoil events which allows us to effectively remove this background. This is shown in figure 7 where two dimensional plots shows that betas, gammas, and neutrons all fall into separate areas of the plot.

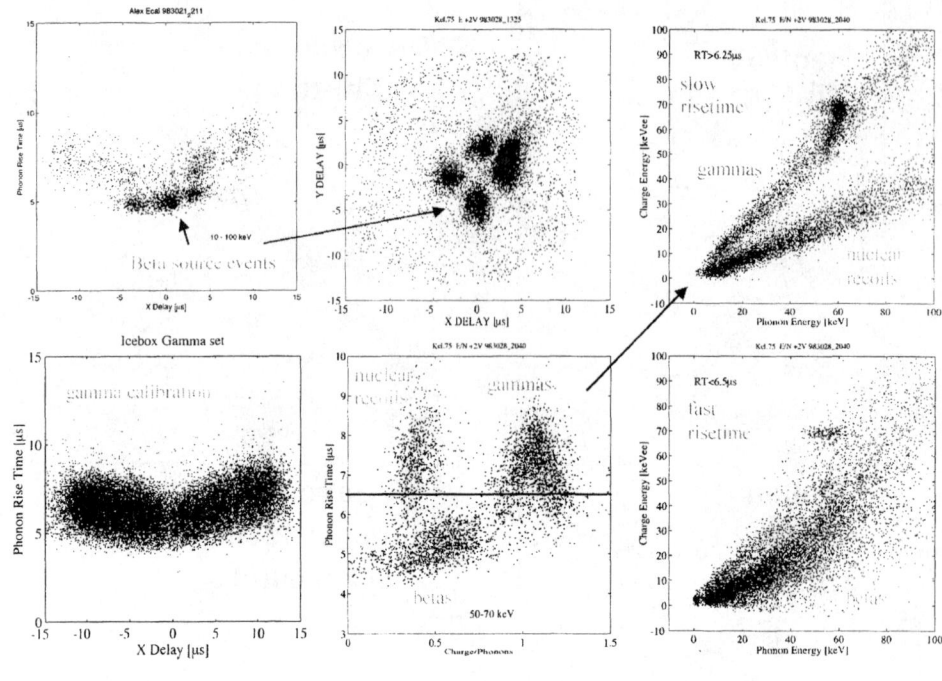

Figure 7

The plan for the CDMS experiment is to place 42 transition edge detectors in a cryostat, which is called the Icebox. The detectors will be cooled to to 20 mk by a dilution refrigerator. The detectors will be half silicon and half germanium to facilitate separation of the neutron background. This is possible because germanium and silicon have very similar responses to neutrons per unit volume, yet silicon is much less sensitive to WIMPs. Additional information on this background will come from events which scatter in more than one detector. Monte Carlo calculations indicate that almost 50% of the neutron events will scatter in two detectors depending on the particular geometry. This will allow us to check whether our neutron simulations are correct.

The Icebox is surrounded by a thick shield made of polyethelene and lead. This passive shield is surrounded by scintillation counters to veto events which occur when a muon enters the detector region creating the possibility of producing a neutron through interactions with the shield material or the cryostat. The experiment is then assembled deep in the Soudan Iron Mine located in northern Minnesota in the USA. The overburden of earth provided by the mine will attenuate the cosmic ray flux incident on the detector by a factor of $10^4$.

Figure 8 shows the expected sensitivities for the CDMS experiment and several other dark matter experiments. Of particular note here is the limit curve from the DAMA group, which shows a both limit and a heart shaped region where there is some evidence for a signal. The evidence comes from observing a very small annual modulation in their signal, which could be produced by the earth's velocity adding and subtracting to the velocity vector of the solar system as it makes its way around the center of the galaxy. This change in velocity with time of year would result in a small change in the flux of WIMPs incident on the detector.

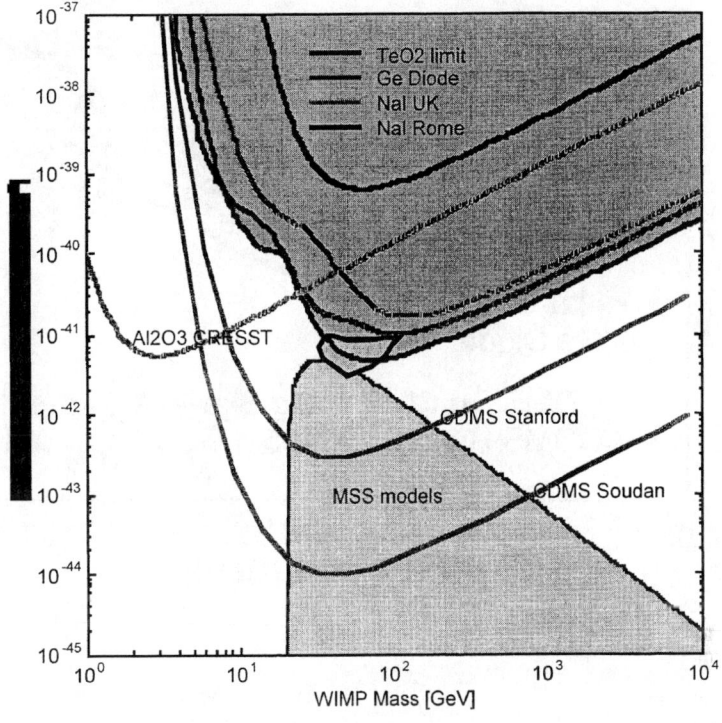

Figure 8

Preliminary results have been obtained in this region by the CDMS group. A small number of an earlier version of the CDMS detectors has been run in the Stanford Underground Facility at Stanford University. The preliminary limits from that running are shown in figure 9 along with the limits from several other experiments. So far there is no indication of a confirmation of the DAMA evidence.

In conclusion, the search for dark matter is being fully addressed at the present time. There are a number of promising techniques including the cryogenic detector efforts discussed in this talk. Both the NaI groups and the cryogenic detector groups are beginning to enter the theoretically interesting region of parameter space.

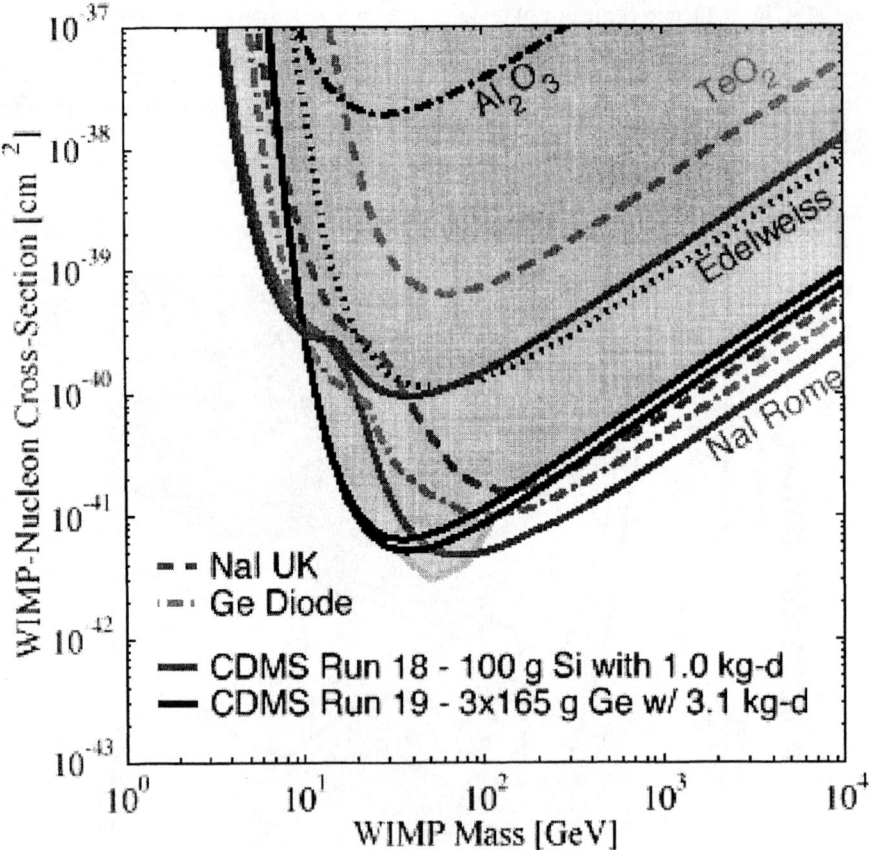

Figure 9

## References

1.) G. Jungman, M. Kamionkowski and K. Griest, Phys. Rep. **267,** 195 (1996)

2.) B. Sadoulet, Direct Searches for Dark Matter, Neutrino 98 Conference

3.) DS Akerib et al., astro-ph/9712343 (30 Dec 1997)

4.) R. Bernabei, et al., University of Rome preprint ROM2F/98/34 (27 August 1998)

5.) R. Bernabei, et al., Instituto Nazionale di Fisica Nucleare Preprint INFN/AE-98/23 (12 Nov 1998)

6.) L. Baudis, et al., hep-ex/989811045 (27 Nov 1998)

7.) M. Bravin et al., hep-ex9904005 (6 APR 1999)

8.) S. Perlmutter et al., Nature, 391:51, 1998

9.) P. M. Garnavich et al. ApJ, 493:L53, 1998

# Review of HERA Experiments

G. Herrera Corral[1]

*Institut für Experimentelle Physik*
*Universität Dortmund*
*Otto Hahn-Str. 4*
*44221 Dortmund, Germany*

**Abstract.** The Hadron-Electron Ring Accelerator facility (HERA) consists of two storage rings. In one of them, protons circulate counterclockwise with an energy of 900 GeV, in the other ring electrons or positrons circulate clockwise with an energy of 27.5 GeV. The two rings are brought together in two points to produce collisions. On each of these collision points detectors are placed to study the interaction of electron and protons.
HERA offers the possibility of carrying out experiments with polarised beams. The degree of polarization of electrons and/or positrons of up to 60 % is very stable. Since 1995 the collaboration HERMES is running an experiment designed to provide information on the spin of the nucleon. A fourth experiment makes use of the proton beam halo to produce B mesons. HERA-B intend to investigate the violation of CP symmetry in the decay of B mesons.
Here we try to illustrate concepts of particle physics and the experimental methods used at HERA and do not intend to give an exhaustive summary of the physics program.

## INTRODUCTION

Fig. 1 shows the accelerators at DESY. The electron accelerator in HERA consists of 82 normal conducting radio-frequency cavities supplemented by 16 superconducting cavities, which ensure that electrons can reach an energy of 30 GeV. Every cavity not only accelerates but also limit the number of particles in the beam, that is why the number of accelerating sections should be kept to a minimum. In a straight section of HERA's electron ring there are eight helium cryostats, each with 2 four-cell niobium cavities working at 500 MHz. These generate an accelerating field of 6 MV per meter. Normal conducting resonators reach only about 1 MV/m. Increasing the energy from 26 GeV to 30 GeV allows polarisation of electrons in a shorter time. The polarising process can be completed in only 18 minutes.

In the 6.3 Km proton ring, there are 422 bending dipole magnets. These are superconducting magnets because a strength of 4.7 tesla is needed and normal iron

---

[1] On leave from Physics Department, CINVESTAV, Mexico.

magnets are limited to 2 tesla. The bending dipoles are joined by 224 superconducting quadrupoles which focus the beam.

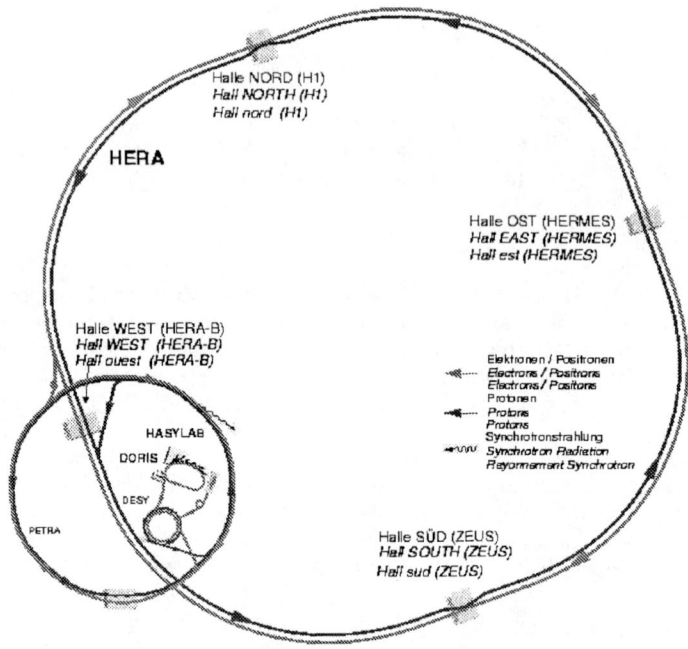

**FIGURE 1.** DESY facilities.

The magnets were designed to have field strength of 4.7 tesla but it was found during tests that they would run up to 6 tesla. This makes possible to push the proton beam energy, orginally designed for a particle energy of 820 GeV to 1000 GeV.

It takes about 20 minutes to fill up the protons. They circulate for about 10 hours before the next filling. Electron injection takes only 15 minutes but has to be repeated every three hours.

In superconducting magnets the shape of the field does not depend on the iron yoke but on the superconducting coils. Therefore the windings should be very accurate. In order to achieve a field with a small error ($< 0.01\%$), the individual coil cables should be placed with a 0.02 mm precision.

Electric currents of 5000 amps producing magnetic fields of 5 tesla in a dipole leads the two halves to repel each other with a force of 134 tonnes per meter of coil length. So the coils are held together by aluminum support collars anchored in the iron yoke of the magnet. The design and manufacture of the support collars should take into account the thermal shrinkage resulting from extreme cooling.

With 200 bunches and 50,000 orbits per second, some 10 million particle bunches cross through each other every second. Only about one in every 100 such crossings can be expected to produce a collision, *i.e.* one has 100,000 particle collisions per second.

The most of these events are of no interest to physicist since they come together with unwanted reactions. A maximum of 5 collisions per second are worth registering for subsequent analysis. Fig. 2 shows the integrated luminosity up to now.

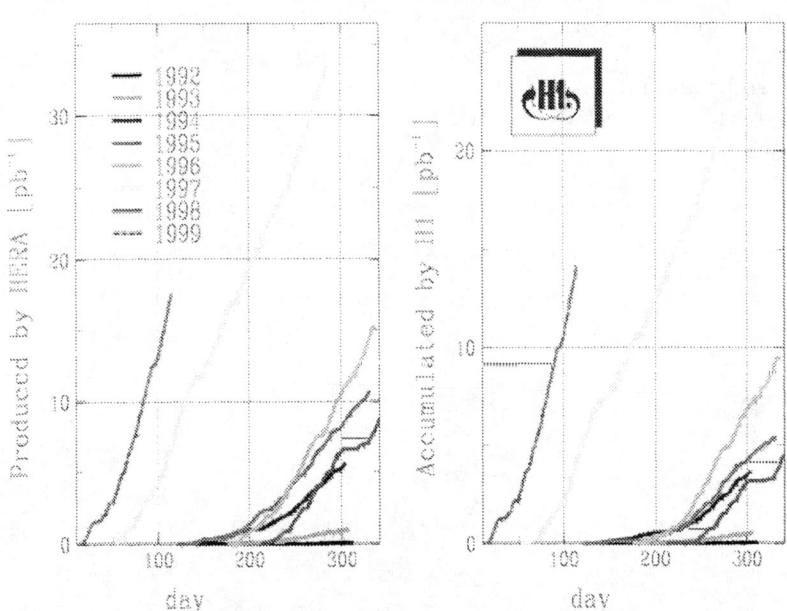

**FIGURE 2.** Performance of the HERA storage ring facility. The curves show the luminosity added up over the operational year in question.

HERA is now operating at design values. Further increase in luminosity requieres an upgrade of the machine. The luminosity is given by

$$L = \frac{1}{4\pi e} \frac{N_p}{\epsilon_p} I_e \frac{1}{\beta_{xp}^* \beta_{yp}^*}$$

where $\epsilon_p$ represents the proton beam brigthness. It is limited in DESY III Synchrotron. $\beta_{yp}^*$ and $\beta_{xp}^*$ are the proton beta functions which are limited by aperture of the beta quadrupoles. The total e-beam current $I_e$ is limited by rf power.

The HERA upgrade program intends to reduce the beta functions at the interaction point. This is underway at HERA and an improvement by a factor of almost five compared to designed values is expected.

There are also plans to introduce spin rotators at H1 and ZEUS in order to polarize protons. This opens a wide range of physics topics with polarized beams.

# I H1 AND ZEUS, THE STRUCTURE OF THE PROTON

Two big experiments: H1 and ZEUS study the collisions of electrons and proton in two opposite points of the HERA machine. Figs. 3 and 4 show longitudinal cuts of H1 and ZEUS respectively.

When electrons and protons collide two reactions can ocurr that are of physical interest. The cross section for a Neutral Current event (NC), $ep \to eX$ (see Fig. 5) is given by

$$\frac{d^2\sigma_{NC}(e^\pm)}{dxQ^2} = \frac{4\pi\alpha^2}{xQ^4}(y^2 x F_1(x,Q^2) + (1-y)F_2(x,Q^2) \mp (y - \frac{y^2}{2})xF_3(x,Q^2))$$

where,

$$F_2(x,Q^2) = \sum_q A_q(Q^2)(xq(x) + x\bar{q}(x))$$

and

$$xF_3(x,Q^2) = \sum_q B_q(Q^2)(xq(x) - x\bar{q}(x)).$$

Here,

$$A_q(Q^2) = e_l^2 e_q^2 + 2 \mid e_l \mid\mid e_q \mid v_l v_q \left(\frac{1}{4sin^2\theta_W cos^2\theta_W}\right)\left(\frac{Q^2}{Q^2 + M_Z^2}\right)$$

$$+(v_l^2 + a_l^2)(v_q^2 + a_q^2)\left(\frac{1}{4sin^2\theta_W cos^2\theta_W}\right)\left(\frac{Q^4}{(Q^2 + M_Z^2)^2}\right)$$

$$B_q(Q^2) = 2 \mid e_l \mid\mid e_q \mid a_l a_q \left(\frac{1}{4sin^2\theta_W cos^2\theta_W}\right)\left(\frac{Q^2}{Q^2 + M_Z^2}\right) +$$

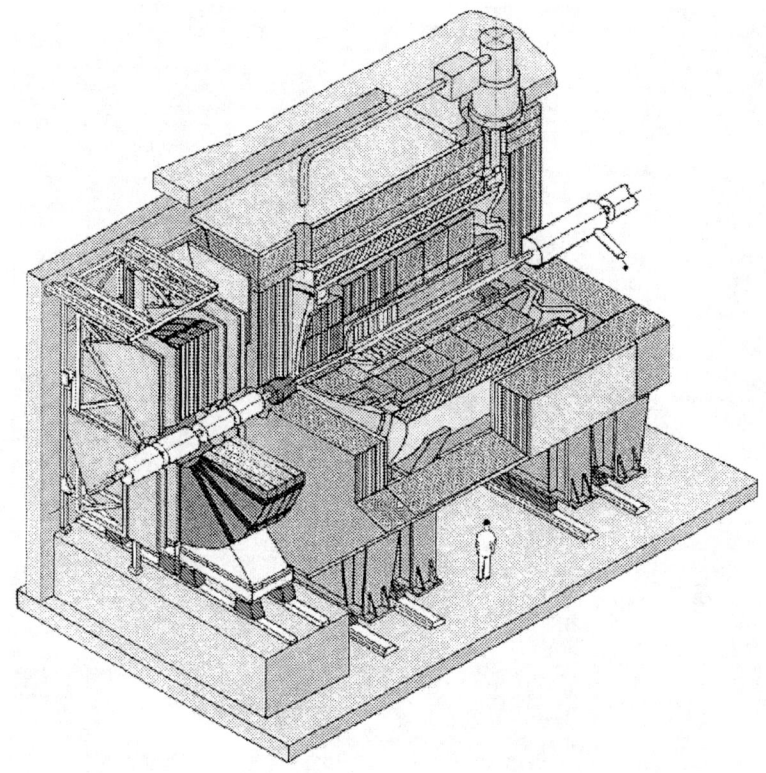

**FIGURE 3.** The H1 detector.

$$4v_l a_l v_q a_q \left(\frac{1}{4sin^2\theta_W cos^2\theta_W}\right)\left(\frac{Q^4}{(Q^2+M_Z^2)^2}\right)$$

For Charged Current events CC, $ep \to \nu X$ (see Fig. 5) the cross section is:

$$\frac{d^2\sigma_{CC}(e^\pm)}{dxQ^2} = \frac{\pi\alpha^2}{8sin^4\theta_W}\frac{1}{(Q^2+M_W^2)^2}\left((1+(1-y)^2)W_2^\pm \mp (1-(1-y)^2)W_3^\pm\right)$$

with

$$W_2^+ = x\sum_i(d_i(x)+\bar{u}_i(x))$$

**FIGURE 4.** The ZEUS detector.

$$W_2^- = x \sum_i (u_i(x) + \bar{d}_i(x))$$

$$W_3^+ = x \sum_i (d_i(x) - \bar{u}_i(x))$$

$$W_3^- = x \sum_i (u_i(x) - \bar{d}_i(x)).$$

The experimentally measured cross sections for both neutral and charged currents as a function of $Q^2$ is shown in Fig. 6. One can see that for $Q^2 \sim M_Z^2$ weak and electromagnetic interactions become comparable as expected from electroweak unification.

### The parton distribution in the proton

The square of the four momentum transfer $q^2$, *i.e.* the mass squared of the virtual boson, determines the resolving power $\Delta b$ of the interaction.

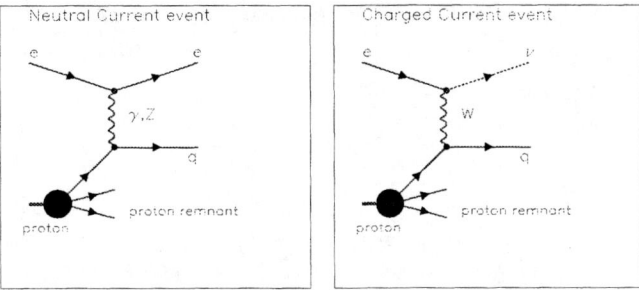

**FIGURE 5.** Neutral and charged current events.

$$\Delta b \sim \frac{\hbar c}{\sqrt{Q^2}} = \frac{0.197}{\sqrt{Q^2}} GeV fm$$

where we introduce $Q^2 = -q^2$. Fig. 7 shows an event observed recently by the H1 detector with a $Q^2 = 94000$ GeV. In this event the boson was able to see the scattered parton as deep as $\Delta b \sim 0.6 \times 10^{-18} mts = 0.0006 fm$, i.e. a scale one thousand times smaller than the size of a proton.

How does the proton looks like to a boson penetrating deep into its interior. Following [1] let us approach the parton distribution functions starting with a rough model and introducing step by step new aspects of the real nature of proton.

In a rough model of the proton structure one assumes that it consists of valence quarks only i.e.

$$p = uud$$

in this picture, the number of $u$ and $d$ quarks is 2 and 1 respectively. Therefore

$$\int_0^1 u(x)dx = 2$$

$$\int_0^1 d(x)dx = 1$$

and the sum of the momentum fraction carried by each quark in the proton should be 1,

$$x_{u_1} + x_{u_2} + x_d = 1$$

and therefore,

$$\int_0^1 xu(x)dx + \int_0^1 xd(x)dx = 1$$

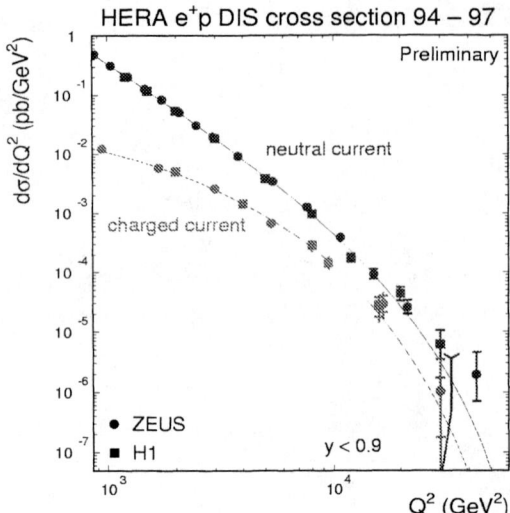

**FIGURE 6.** Single differential cross section for positron proton scattering of the H1 and ZEUS experiment.

(momentum sum rule). A uniform distribution in the plane defined by $x_{u_1} + x_{u_2} + x_d = 1$, with the normalisation imposed by the sum rule leads to:

$$u(x) = 4(1-x) \qquad d(x) = 2(1-x)$$

with

$$\langle x_u \rangle = 2/3$$
$$\langle x_d \rangle = 1/3$$

In a somewhat more realistic model quarks carry half of the proton momentum and gluons the other half.

$$p = uud + gluons$$

in this case, the momentum sum rule becomes

$$\int_0^1 x(u(x) + d(x))dx = 0.5$$

and

$$\int_0^1 xg(x)dx = 0.5 \qquad (A)$$

assuming $\langle x_u \rangle = 1/3, \langle x_d \rangle = 1/6$ and making the ansatz $u(x) = a(1-x)^n$ one gets

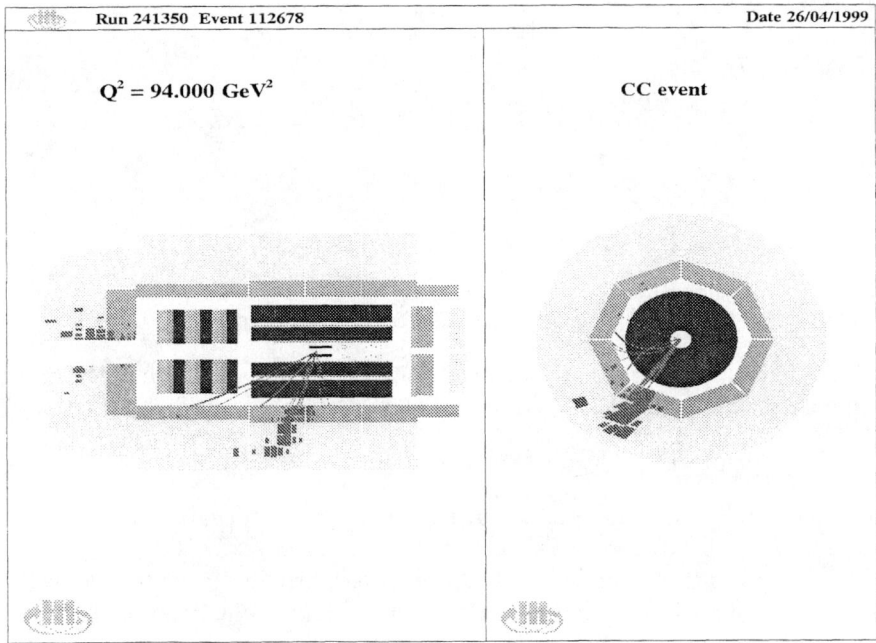

**FIGURE 7.** H1 event with the highest momentum transfer ever seen.

$$u(x) = 10(1-x)^4 \qquad d(x) = 5(1-x)^4$$

assuming that the gluon spectrum is in part similar to $u$ and $d$ quarks and part comes from gluon bremsstrahlung by quarks (this depends on $Q^2$ and $\alpha_s$)

$$g(x) = a(1-x)^4 + bx^{-1}(1-x)^4$$

then, from (A): $a/6 + b = 2.5$. A possible choice $a = 3, b = 2$ gives

$$g(x) = 3(1-x)^4 + 2x^{-1}(1-x)^4$$

We may go on and take into account the existence of a "sea" by considering that gluons can split into quark-antiquark pairs. This means that besides the valence quarks ($q_V$) there will be sea quarks $q_s$.

$$p = uud + gluons + sea$$

We introduce now the following constraints:

$$u_s(x) = \bar{u}_s(x) = \bar{u}(x) \qquad d_s(x) = \bar{d}_s(x) = \bar{d}(x)$$

$$u(x) = u_V(x) + u_s(x) \qquad d(x) = d_V(x) + d_s(x)$$

$$\int_0^1 (u(x) - \bar{u}(x))dx = \int_0^1 u_V(x)dx = 2$$

$$\int_0^1 (d(x) - \bar{d}(x))dx = \int_0^1 d_V(x)dx = 1$$

$$\int_0^1 (s(x) - \bar{s}(x))dx = 0$$

$$\int_0^1 x(u(x) + \bar{u}(x) + d(x) + \bar{d}(x) + s(x) + \bar{s}(x))dx = 0.5$$

$$\int_0^1 xg(x)dx = 0.5$$

since $g \to q_s\bar{q}_s$ one may assume that a sea quark carries half of the gluon momentum, so we can replace $x$ by $2x$ in the gluon distribution to get the sea quark distribution.

$$q_s(x) = K[3(1 - 2x)^4 + x^{-1}(1 - 2x)^4] \qquad x < 0.5$$

where the constant $K$ can be determined by fitting experimental data, *e.g.* Martin, Stirling and Roberts (1993) assumed that the total momentum carried by sea quarks at $Q^2 = 4 GeV^2$ is about 0.18 provided all the sea quark distributions are identical. The valence quark distribution should now be corrected for the momentum carried by the sea, with the ansatz $u_V(x) = a(1-x)^n$ one gets

$$u_V(x) = 16(1-x)^7 \qquad d_V(x) = 8(1-x)^7$$

$$q_s(x) = (1 - 2x)^4 + (3x)^{-1}(1 - 2x)^4 \qquad x < 0.5$$

$$g(x) = 3(1-x)^4 + 2x^{-1}(1-x)^4$$

This parametrization of the parton densities in the proton agrees very well with the more sophisticated parametrizations in use nowdays [2,3] (see Fig.8).

## $F_2$ and the gluon structure function

The structure function $F_2$ is defined as

$$F_2(x, Q^2) = x \sum_q e_q^2 q(x, Q^2).$$

Fig. 9 shows the experimental measurement of $F_2$.
An important aspect of $F_2$ is that it contains information on the gluon contents of the proton.
Since

$$\frac{dq_i(x, Q^2)}{dlnQ^2} = \frac{\alpha_s(Q^2)}{2\pi} \int_x^1 \frac{dy}{y} [P_{qq}(x/y)q_i(y, Q^2) + P_{qg}(x/y)g(y, Q^2)]$$

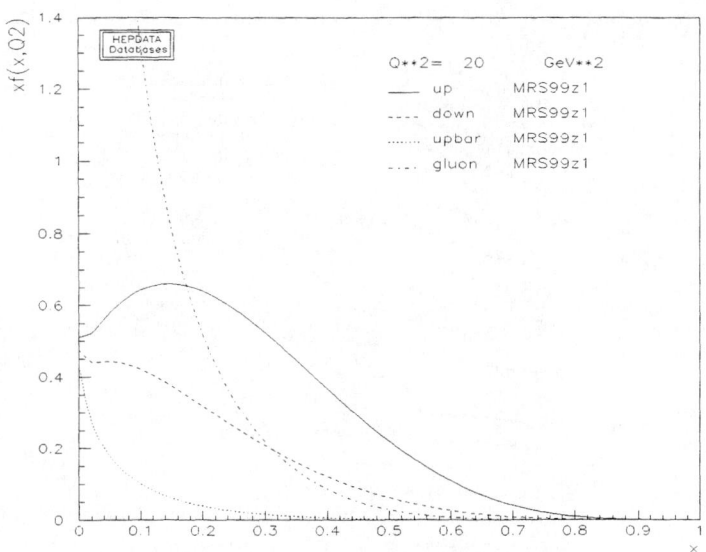

**FIGURE 8.** Parton distributions in the MRS parametrization

where (see Fig. 10),

$$\alpha_s P_{qq}(x/y) \sim \text{probability that a quark with } x \text{ comes from an initial state quark with } y \text{ which has radiated a gluon}$$
$$\alpha_s P_{qg}(x/y) \sim \text{probability that a quark with } x \text{ comes from a } q\bar{q} \text{ created by a gluon with } y$$

then,

$$\frac{dF_2}{d\ln Q^2} = \sum_q e_q^2 \alpha_s(Q^2)/2\pi \int_x^1 \frac{dy}{y} [P_{qq}(x/y)q_i(y,Q^2) + P_{qg}(x/y)g(y,Q^2)]$$

at very low $x$ the quark contribution can be neglected. Using the LO result for $P_{qg}(u) = 0.5[(1-u)^2 + u^2]$ with the replacement of $y = x/(1-z)$ and defining $G(x,Q^2) = xg(x,Q^2)$ one gets

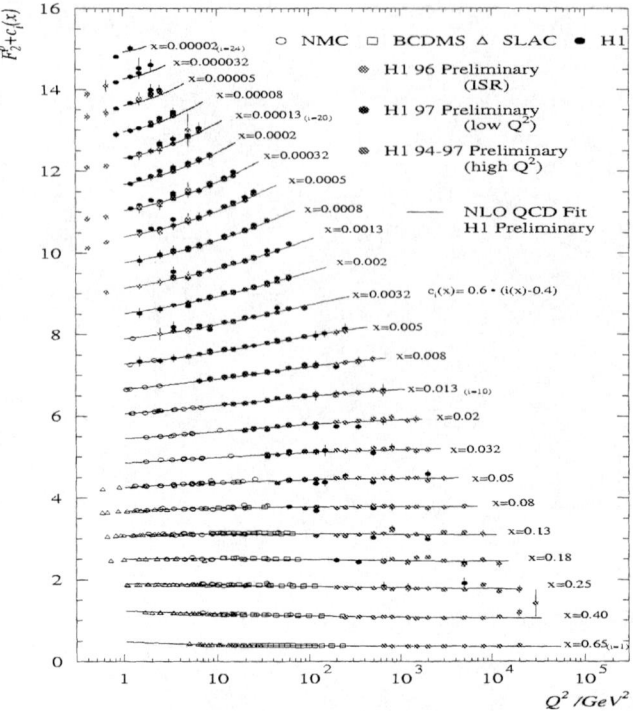

**FIGURE 9.** The H1 measurement of the structure function. The momentum region includes sea quarks at momenta $x = 10^{-5}$ up to the valence quark region. The virtuality $Q^2$ spans five orders of magnitude up to $Q^2 = 30000 GeV^2$. At these large $Q^2$ values the scattering process is sensitive to structures of size $10^{-3}$ fm.

$$\frac{dF_2}{dlnQ^2} = \sum_q e_q^2 \alpha_s(Q^2)/2\pi \int_0^{1-x} dz P_{qg}(z) G(x/(1-z), Q^2)$$

A Taylor expansion of $G(y, Q^2)$ around $z = 1/2$ is made ( $P_{qg}$ is symmetric around $z = 1/2$ ). If $G(y, Q^2) = y^d(1-y)^a$, the result is

$$G(2x, Q^2) = \frac{27\pi}{10\alpha_s(Q^2)} \frac{dF_2(x, Q^2)}{dlnQ^2}$$

Fig. 11 shows the gluon distribution [5] at different values of $Q^2$

## II  HERMES, THE SPIN OF THE PROTON

The HERMES (HERa MEasurement of nucleon Spin) experiment has been designed to make precision measurements of spin physics. It uses a polarized electron

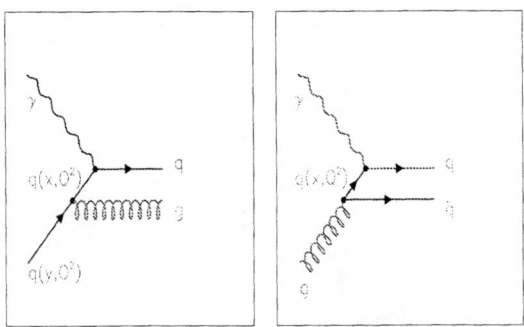

**FIGURE 10.** Diagrams representing $P_{qq}(x/y)$ (left) and $P_{qg}(x/y)$ (right)

storage ring and a polarized gas target. The proton beam from HERA passes through the apparatus without interaction.

In order to enhance the thickness the target gas is injected into a T-shaped aluminum tube. It is 40 cm long and has an elliptical cross section $3 \times 1$ cm. In order to reduce the speed of the gas in the cell the tube is cooled down to 20 K for $^3He$, and 100 K for $H$ and deuterium. With this, the density of the gas increases and the number of collisions too. Fig.12 shows the beam profile.

The $^3He$ is polarized by optical pumping, which consists of a circularly polarized laser beam which in a photon absortion and emission process transfers its polarization to the $^3He$ atoms, (see Fig. 13)

The hydrogen is polarized with a Stern Gerlach spin separation in a non homogen field of a sixpole magnet. Polarizations of up to 90% can be achieved.

The Sokolov-Ternov effect produces transverse electron polarization due to an asymmetric spin flip amplitude in synchrotron radiation. As shown in Fig. 14, the beam polarization grows asymptotically with a time constant of about 30 minutes (at HERA beam energy 27 GeV).

The longitudinal beam polarization is then obtained using two spin rotators mounted in front and behind of the HERMES spectrometer.

In 1987 the EMC experiment announced the surprising result that only a small fraction of the nucleon's spin is carried by the quark spins. This result stimulated a number of experiments at CERN, SLAC and at DESY. HERMES pretends to give an answer to many questions around the "proton spin crises" [6].

In the HERMES experiment an electron is deflected when it exchanges a photon with one of the quarks in the proton. Electrons (or positrons) with spin aligned along the beam direction interact only with quarks that have the opposite spin alignment. By reversing the polarization of the electrons one sees interactions with quarks which spin points on the other direction. The difference reveals the asym-

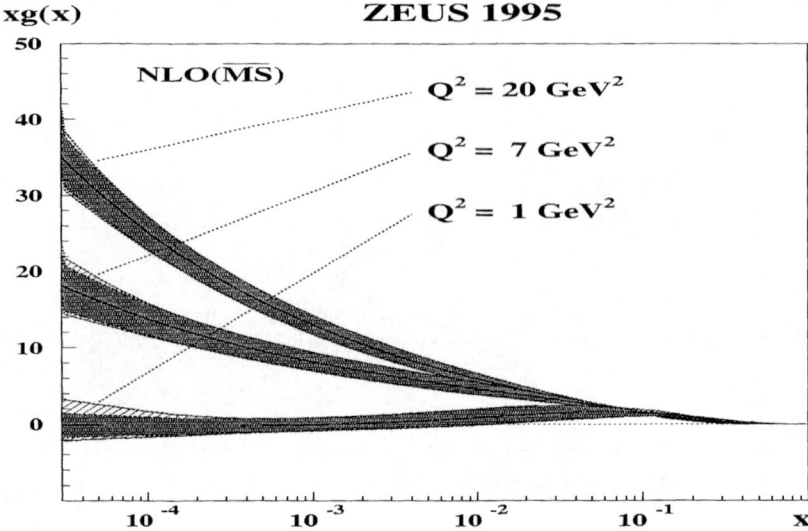

**FIGURE 11.** The gluon momentum distribution as a function of $x$ for different values of $Q^2$ from ZEUS QCD fit

metry of the quark spin in the proton. Fig. 15 shows the experimental results recently published by HERMES [7] on quark spin asymmetries.

If a negative kaon is observed in the final state (see Fig.16) it means that with some probability the photon struck [8] one of the quarks that built the kaon. Observing the kaon and the deflected electron one is able to determine the spin asymmetry of the quarks forming the kaon.

The HERMES experiment has started to produce very interesting information about the spin of the proton. It confirmed that about 30 % of the proton' s spin comes from the spin of the quarks. The missing 70 % can come from the gluon spin and/or from the orbital angular momentum due to the quark and gluon motion inside the proton.

According to present theoretical models the crisis would disappear if gluons contribute to proton's spin with one or two quantum units of spin and say an orbital angular momentum of about -1.

HERMES is trying to develop a way of measuring the gluon contribution. Experiments at RHIC (Relativistic Heavy ion Collider) at Brookhaven will certainly be able to measure and are starting now to take data. Soon we will see if the spin crises continues or just disappers on the light of these totally unexpected contribution of gluons.

## III  HERA-B

The HERA-B [9] experiment uses the 920 GeV proton beam of HERA and a system of wires as target to produce B mesons. The goal is to measure CP violation in "The golden decay" of B mesons $B^0 \to J/\Psi K_s$. This decay is shown in Fig. 17. The idea is to use the interference of $B^0 \to J/\Psi K_s^0$ and $B^0 \to \bar{B}^0 \to J/\Psi K_s$. By looking at the decay products of the $\bar{b}$ quark one can tag the $b$ quarks and see if it has oscillated before decaying.

This system is particularly interesting because there is only one dominating diagram and very small hadronic amplitudes. The theoretical uncertainties are small and it offers good trigger conditions too.

The principle of measurement is based on the fact that the number of $B^0$ is given as
$$n(t) \propto e^{-t}(1 + \sin 2\beta \sin xt)$$
and the number of $\bar{B}^0$
$$\bar{n}(t) \propto e^{-t}(1 - \sin 2\beta \sin xt)$$
where $x$ represents the mixing parameter $x = \frac{\Delta M}{\Gamma} \approx 0.67$, and then the asymmetry as a function of time,
$$a(t) = \frac{n(t) - \bar{n}(t)}{n(t) + \bar{n}(t)} = \sin 2\beta \sin xt$$
may be integrated to give
$$A(t) = \frac{\int n(t)dt - \int \bar{n}(t)dt}{\int n(t)dt + \int \bar{n}(t)dt} = \sin 2\beta \frac{x}{1 + x^2}$$
which is proportional to $\sin 2\beta$ and $\frac{x}{1+x^2} \approx 0.46$ is known to be $\approx 0.46$.

This means that one just have to count the number of decays $B^0 \to J/\Psi K_s^0$ and $\bar{B}^0 \to J/\Psi K_s$ tagging the initial flavour ($b$ or $\bar{b}$ with the second meson (see Fig. 17).

One expects about $3 \times 10^5$, $B^0 \to J/\Psi X$ decays for every year of running at HERA-B. From that something like 17 000 would be reconstucted. About 1500 of them would be "golden" and the rest offers a bunch of interesting physics like: other CP violating decays, QCD tests etc.

The target wires made out of aluminum (cooper and carbon were also tested) are actually $50 \mu m$ thick and $500 \mu m$ wide bands. They are placed a few millimeter from the proton beam to interact with its halo, leaving the proton beam unperturbed.

The HERA-B spectrometer is shown in Fig. 18. It consists of Silicon Vertex Detectors (SVD) to reconstruct the vertex and measure impact parameter of B meson decays, Inner and outer Tracker (ITR and OTR) for pattern recognition and momentum measurements, Transition Radiation Detectors (TRD) and Electromagnetic Calorimeter to identify electrons, Muon System (MUON) for hadron separation and a Ring Imaging Cherenkov Detector (RICH) for hadron identification.

There are many research topics that HERA-B will address in addition to CP violation. Among them

- $B^0$ mixing which provides a measurement of $V_{td}$
- $B_s^0$ mixing which provides a measurement of $V_{ts}$
- Semileptonic $B$ decays which allow to measure $V_{cb}$ and $V_{ub}$
- Rare $B$ decays which test standard model and opens a window for new physics
- Charm physics like measuring $D^0$ mixing
- Exotic states like for example $B_c$ decays and the determination of mass and lifetimes

etc

The HERA-B will reach its design performance by the end of the year. The exact schedule depends on the HERA shutdowns. The detector has been continuously upgraded and is now becoming a full scale experiment.

## ACKNOWLEDGMENTS

I want to thank Prof. D. Wegener for his kind hospitality during my sabbatical stay at the University of Dortmund. At the Lehrstuhl fuer Experimentelle Physik V, I was provided with the best conditions and environment to work. I also want to thank him for the very inspiring discussions.

I would like to thank the organizers of the VIII ICFA School for inviting me to deliver this talk. In particular to Schban Kartal and Nizamettin Erduran for their hospitality and warm welcome.

This work was supported by the Alexander von Humboldt Foundation.

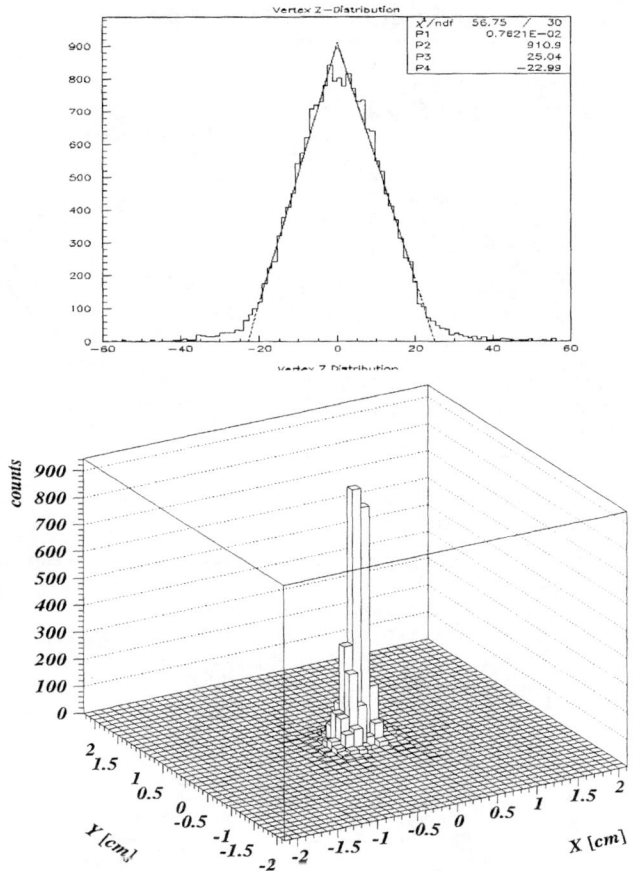

**FIGURE 12.** Vertex reconstruction (up) along and (down) transverse to the beam.

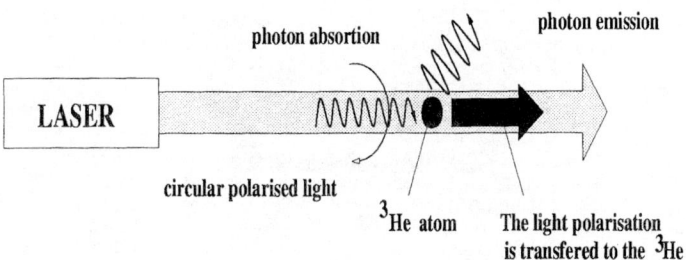

**FIGURE 13.** Optical pumping

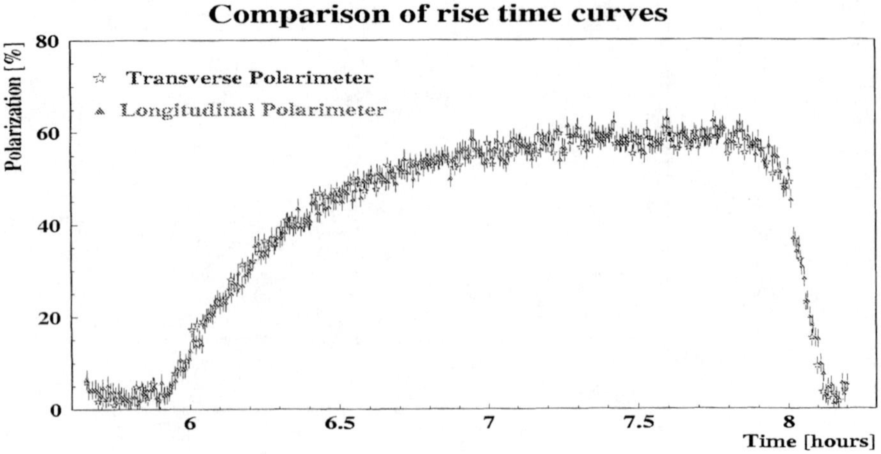

## Comparison of rise time curves

**FIGURE 14.** Electron polarization as a function of time. The longitudinal polarization is produced by the first rotor in the HERMES straight section by a series of magnets. The beam polarization is then rotated back to transverse direction downstream of the experiment.

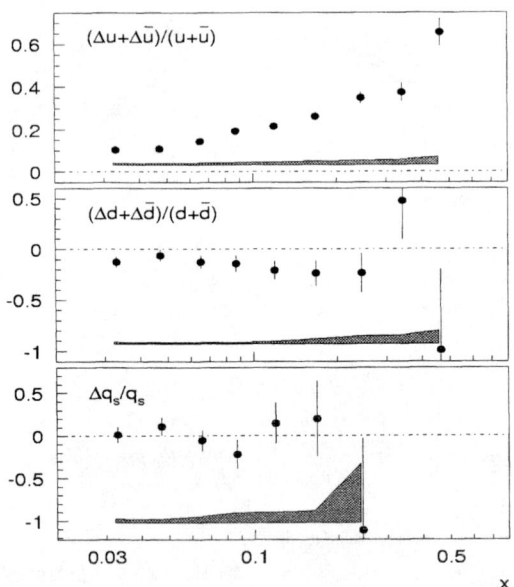

**FIGURE 15.** Flavor decomposition of the quark spin asymmetry as a function of $x$ from HERMES

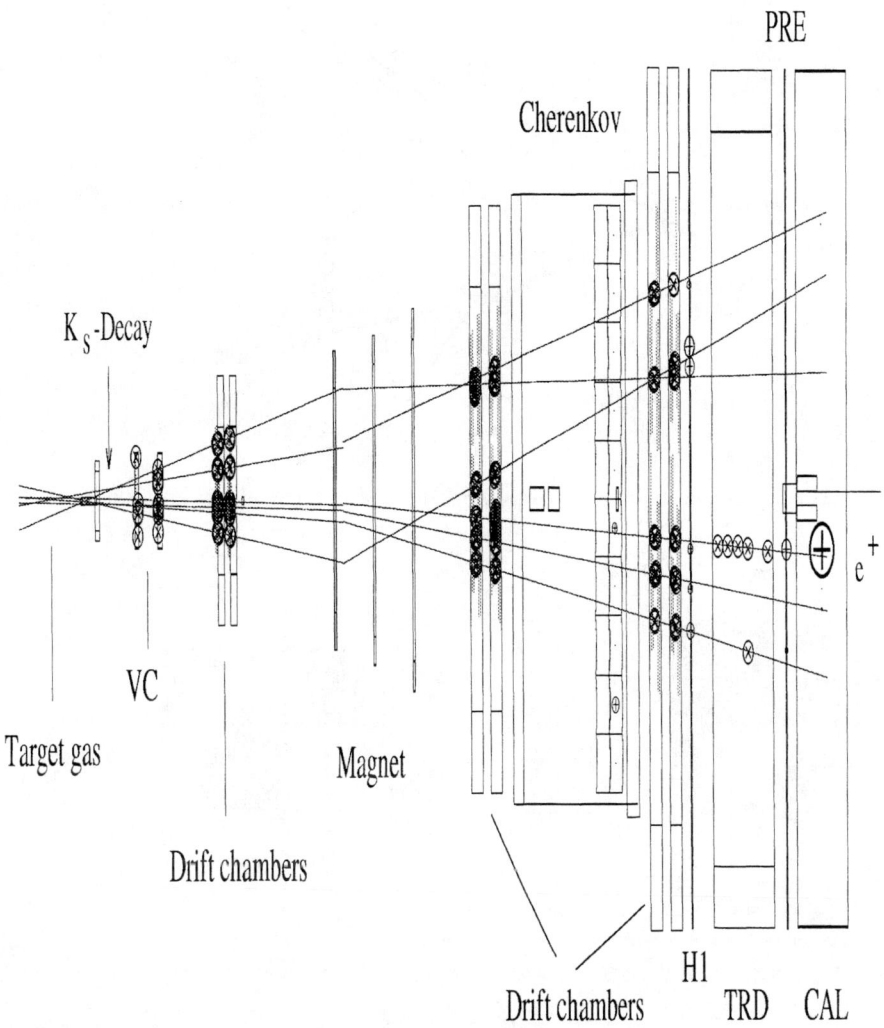

**FIGURE 16.** Deep inelastic event with a $K_s$ in the final state as seen by HERMES. The secondary vertex is clearly visible.

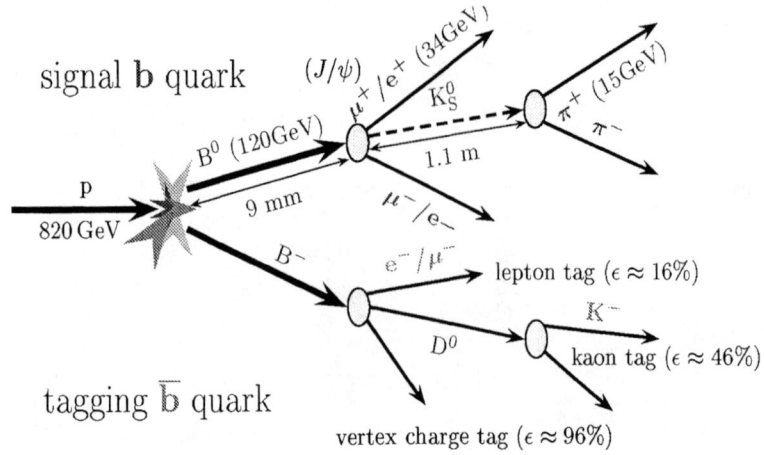

**FIGURE 17.** The golden decay, $B^0 \to J/\Psi K_s^0$ and typical values of efficiencies for the tagging $\bar{b}$ quark.

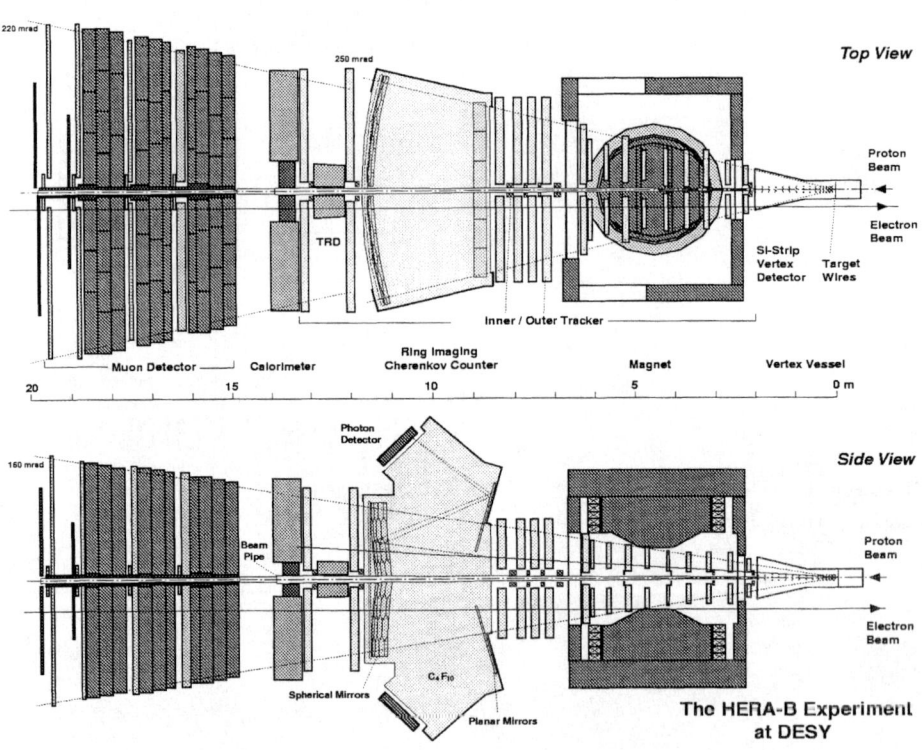

**FIGURE 18.** HERA-B detector.

# REFERENCES

1. G. Wolf, *HERA Physics*, DESY 94-022, ISSN 0418-9833.
2. Martin, A.D., Stirling, W.J. and Roberts *Phys. Rev.* **D47** (1993)867, also http://durpdg.dur.ac.uk/HEPDATA/PDF3.html
3. Glück, M., Reya, E. and Vogt, A. *Phys. Lett.* **B306** (1993)391
4. H. Abramowicz and A. Caldwell, *HERA Collider Physics*, DESY 98-102, ISSN 0418-9833.
5. ZEUS Collaboration, *Eur. Phys. J.*, **C7** (1999)609.
6. HERMES Collaboration, *HERMES Technical Design Report*, DESY, Hamburg, July 1993.
7. HERMES Collaboration, (K. Ackerstaff, et al.) *Phys. Lett.*, **B464** (1999)123.
8. Klaus Rith and Andreas Schäfer, *Scientific American*, July 1999.
9. W. Hofmann, *Nucl. Instr. and Meth.*, **A333** (1993)153
   P. Krizan et al. *Nucl. Instr. and Meth.*, **A351** (1994)111
   W. Schmidt-Parzefall, *Nucl. Instr. and Meth.*, **A368** (1995)124
   J. Spengler, *Nucl. Instr. and Meth.*, **A384** (1996)106
   T. Lohse, *Nucl. Instr. and Meth.*, **A408** (1998)154

# LABORATORY SESSIONS

# Lifetime of Cosmic Ray Muons

J. Anderson, M. Atac and S. Cihangir
Fermilab
N. Erduran and S. Kartal
Istanbul University
S. Erturk (Nigde University)

## Introduction

The experiment is designed to teach the techniques to measure the muon lifetime to an accuracy of 1-2%. The decay rate of cosmic ray muons is studied by detecting those muons that stop in a liquid scintillation counter and measuring the time between the signal from the stopping muon and the signal from the decay electron emitted in the muon decay. Because of lepton number conservation there are two other particles emitted in the decay:

$$\mu^{\pm} \rightarrow e^{\pm} \nu_{\mu} \nu_{e}$$

The muons are produced in the decay of pions produced by primary cosmic rays (protons and heavier nuclei) high in the atmosphere. From accelerator experiments, the charged pion lifetime is about .026 microseconds so that the pions decay high in the atmosphere and the fact that we see muons at sea level is dramatic proof of time dilation for relativistic particles. For simplicity, we will ignore the decay rate difference between positive and negative muons and the effect of negative muon capture by the scintillator material nuclei.

## Description of the Experiment

Most cosmic ray muons have enough energy to pass through the barrel of liquid scintillator without stopping, but those having a kinetic energy of less than approximately 100 MeV stop. The stopping of a muon and its subsequent decay into an electron (and neutrinos) is signaled by a double pulse from a photomultiplier tube viewing the scintillator. Light produced by the muon (the first pulse) and by the electron (the second pulse) bounce around inside the barrel to reach the phototube. The pulses are amplified and processed by a discriminator. The TDC is started by the first pulse and stopped by the second. The time interval between the stopping pulse and the decay pulse is measured by the TDC if the decay pulse occurs within 10 microsec of the stopping pulse. A gate defines this time interval, and another gate assures that the TDC would not receive another start signal within that 10 microseconds. This time interval measurement is then transferred to the on-line computer by the serial interface. A block diagram of the apparatus is given in Fig. 1.

The program running in the computer then bins the events into bins of 0.1 microsec width and these data are displayed as a scattered plot. The plot is the raw decay distribution which must be analyzed to give the muon lifetime.

The plot consists of two components, the exponential decay and a uniform background due to random coincidences of to the second muon going through the scintillator. One can argue from the cosmic ray rate that this background can be ignored. The next step in the analysis is then to plot the data on an analysis program (like Excel) to determine the decay lifetime of muons. The data can be copied to a floppy disk file called MUON.CSV by hitting P on the keyboard.

## Data Analysis and Decay Lifetime

The decay rate $\Gamma$ is the probability per time that a particular muon will decay. For N muons, $N\Gamma dt$ would be number of decays in a time interval of dt:

$$dN = -\Gamma N dt$$

which leads to

$$N(t) = N_o \exp(-\Gamma t)$$

or with life time $\lambda = 1/\Gamma$,

$$N(t) = N_o \exp(-t/\lambda)$$

Here, $N_o$ is the number of muons at t=0 and N(t) is the number of muons at time t. In this experiment we do not have coexisting muons that we observe to decay and count. Therefore, the meaning of N(t) and No are slightly different.

In this experiment No is not the number of muons present at t=0, instead it is the number of muons captured and decayed within the detector. N(t) is then the number of muons that have decayed in time interval t when a single muon entered the detector and decayed.

In Fig. 2 a sample data is plotted on a semi-log scale. From the above equation, we obtain that,

$$\log(N) = \log(N_o) - t \log(e) /\lambda$$

and therefore, the plot would have a slope equal to the coefficient of t at the above equation:

$$\text{slope} = -\log(e) /\lambda$$

which then yields the lifetime $\lambda$.

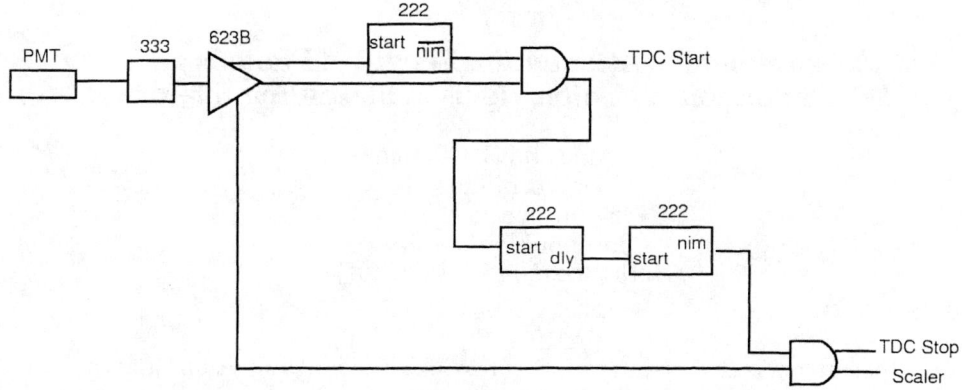

Fig. 1 The block diagram of the muon decay experiment electronics.

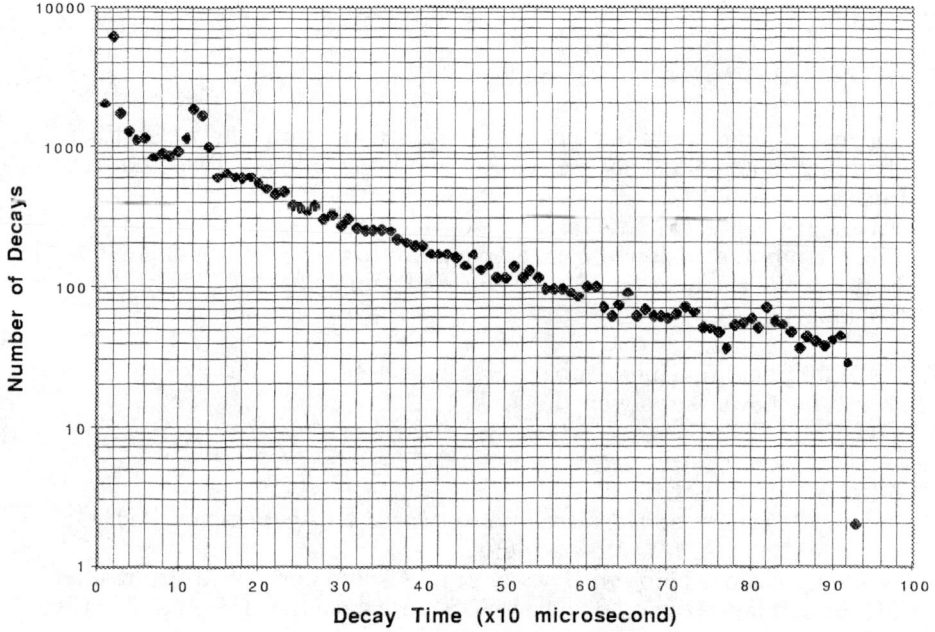

Fig.2 Sample data.

# Measurement of Attenuation Length of Photons and Determination of Photon Yields in Plastic Scintillators

M. Atac and S. Cihangir
Fermilab
N. Erduran and S. Kartal
Istanbul University
S. Erturk (Nigde University)

## Introduction

This experiment is to measure photon yields and attenuation lengths for photons in a polystyrene based fiber, doped with Butyl PBD and POPOP scintillating compounds. The scintillator emits photons efficiently at a characteristic wavelength of 418 nm (blue light). An RCA 31000 photomultiplier tube with a bialkali photocathode with peak quantum efficiency around 420 nm is used to detect the scintillation photons. The photomultiplier has a first focusing dynode arranged to make it uniquely capable of detecting single photons. For this reason, it is called a "Quantacon," (Quantum Counter).

## Materials

o Butyl PBD + POPOP scintillating fiber ribbons,

- 8 layers of staggered fiber ribbons composed of 2 mm singly clad fibers Cladding material is poly-methyl-metacrylate (PMMA). Trade name Lucite.
- Fibers are about 150 cm long .
- Fiber is painted with white reflective paint to prevent cross talk.
- Both ends of the fiber bundle are polished using a special diamond cutting tool (at Fermilab).

o RCA 31000 photomultiplier tube (PMI) with base.
o High voltage power supply for the PMT.
o Electronics,

- NIM amplifier: LeCroy 234,
- LeCroy 621 pulse height discriminator (used as a gate generator),
- LeCroy 2249 CAMAC 12 channel ADC,
- NIM and CAMAC bins with power supplies and Jorway 73A controller,
- Macintosh laptop computer with SCSI interface to CAMAC and software,
- In-line pulse inverter, attenuators, and cables,

o Bismuth-207 beta source (provides ~ 1 MeV electrons) with lead collimator.

# The experimental arrangement:

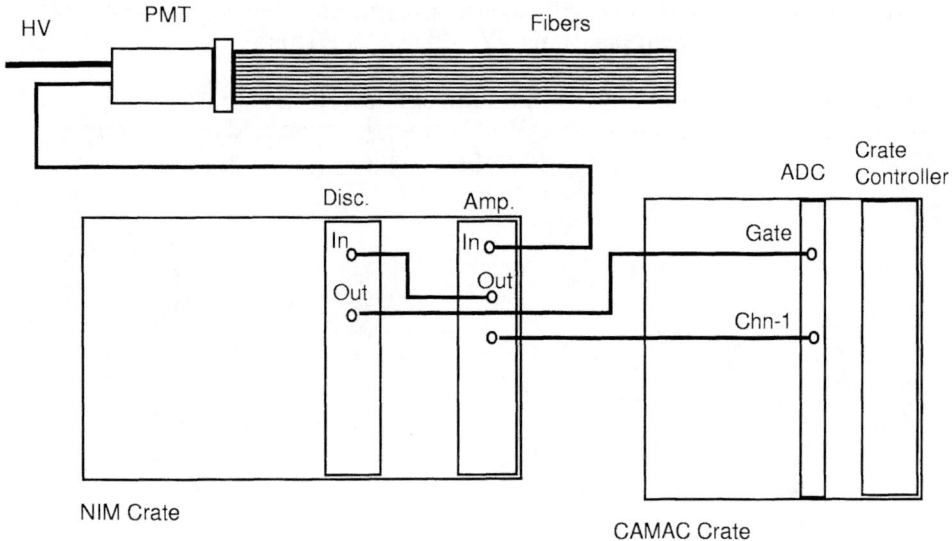

Fig.1 Diagram of Setup

## Procedure:

o Become familiar with the experimental setup. Examine the two inputs to the ADC, the gate trigger, and the PMT signal on an oscilloscope. Review the Kmax appendix to learn about the data acquisition program.

o Using the Kmax program on the Macintosh, determine the pedestal and the first photoelectron peak of the PMT, without using the beta source. What is causing the single electron peak? Is it possible to distinguish the second photoelectron peak? What can you tell from the peak sum?

o Take pulse height spectra with the beta source at various distances from the PMT: 5,10,15, 20, 30, 50, 70,100, 125, and 132 cm. Record the weighted average of each spectrum. Be sure to set the cursors properly to trim off the pedestal, or try using an attenuator on the PMT signal before the ADC. What do you use and why?

o Display and plot each of the results in semi-log form. Knowing that there are two causes for the attenuation, what do you expect to see? (See Figure 2.)

o Determine the attenuation lengths for the fiber bundle between 5 and 30 cm, and from the 30 to 132 cm positions. Why do we ask this?

o With the knowledge of the pedestal and first photoelectron peak values, determine the photoelectron yield in the fiber at the 5 cm and 30 cm points. Assume a 1 MeV electron travels 5 mm in the fibers.

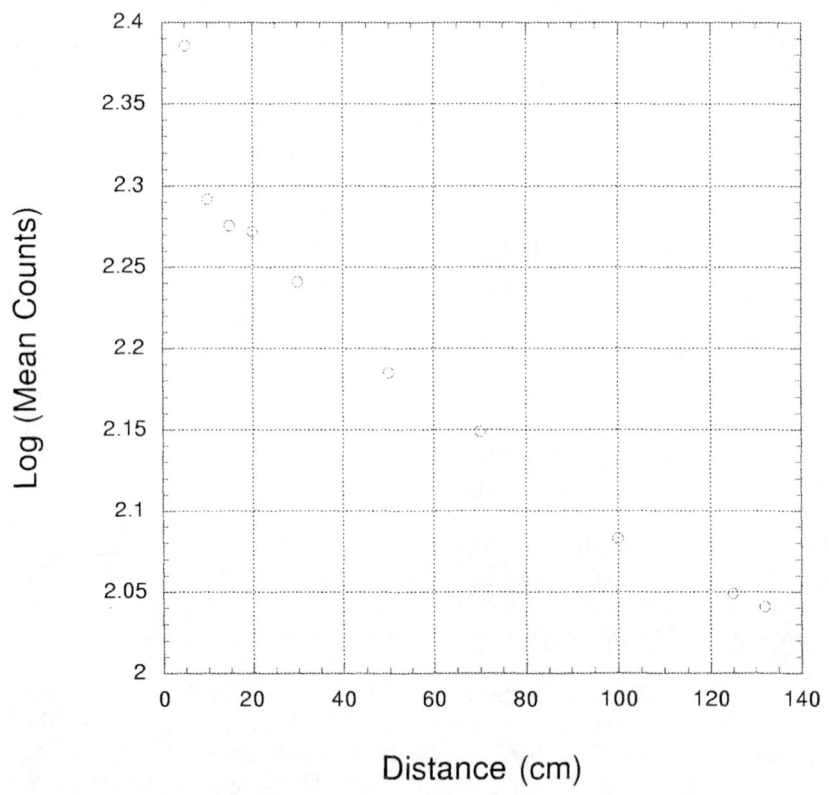

Fig. 2 Semi-log plot of ADC counts vs. distance (of the source). Note the appearance of two trends governing the attenuation.

The intensity of scintillation photons is expected to behave as a function of distance from the source in a form like below. (See Figure 3)

$I = I_o [ B \exp(-x/l_1) + (1-B) \exp(-x/l_2)]$

B represents the relative fraction of the attenuation from the two types of attenuation: absorption by the core material or self-absorption by the fluor, or imperfections in the cladding-core interface.

$l_1, l_2$ are the two attenuation lengths, $I_o$ is the unattenuated photon intensity, equivalent, then, to 0 cm.

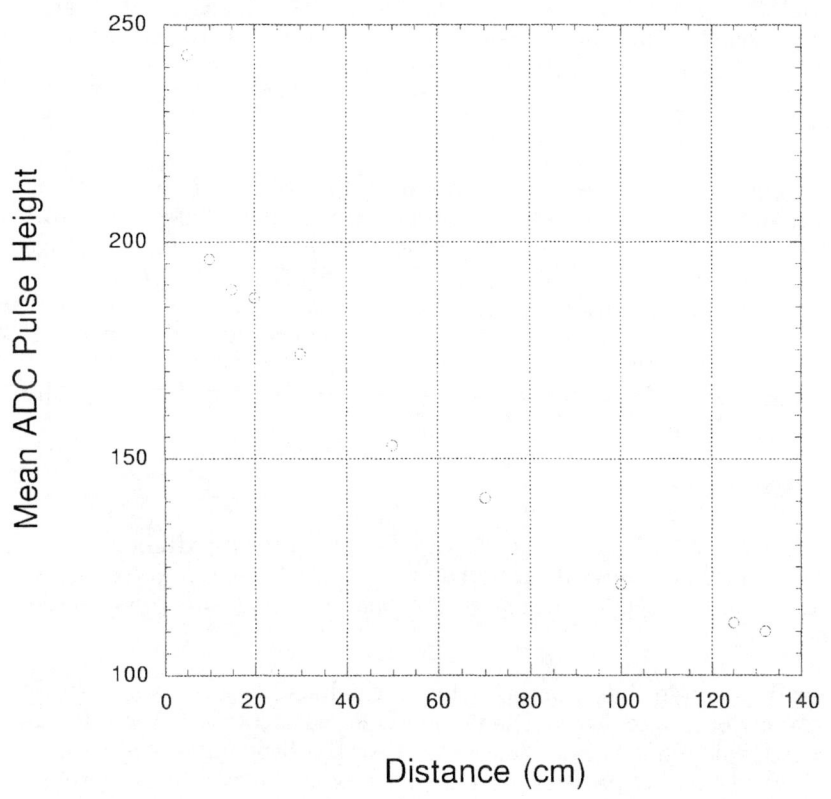

Fig. 3 Data plotted with ADC count. Note two different regions.

# Appendix: Kmax Basics

## Setup

Connect SCSI Cable from laptop to crate controller with in-line terminator. Make sure cable is inserted properly into laptop. Turn on the CAMAC crate and then pen the laptop and press the start button on the rear left. If the laptop does not boot properly, try the procedure again.

## Running Kmax:

After the desktop has been loaded, double click on the instrument program called "AD2249_SRQ." This launches the Kmax program and the proper instrument file. Once running, there will be three windows present: a Control Panel labeled "AD2249_SRQ", a DATA window, and a Report window. The instrument is turned on by clicking the mouse on the left most button on the control panel. (The attached sheet explains some of the functions of other buttons on the bar.) As the instrument starts, it will ask for the channel number of the ADC from which the data is to be read.

## Controls:

The following controls are available in the control panel. The pair of buttons labeled Run and Stop start and stop the data acquisition sequence. The Run Timer button is used to set the length of time for which data is to be acquired. At the end of the time, Kmax will stop gathering data. To force it to be free running, enter a time of 0 minutes. The run timer takes priority over the event counter control. The event counter control allows a user entered number of events to be accumulated, after which Kmax stops. The event counter will count only data between the cursors in the data window. Any data acquisition sequence can be stopped with the Stop button.

## Other Information

To save a plot to disk, select the DATA window (the title bar will be highlighted), and choose Save As from the File menu. These can be opened by double-clicking on the icon, or choosing the Open -> Histogram option in the File menu list.

The printer uses single sheets of paper fed in the back. Insert a sheet and press the yellow button to feed in the paper. Then with the DATA window in front, choose Print from the File menu list. After the printing is completed, stick in a second sheet of paper, and answer okay to the computer's prompt.

Statistics on the histogram can be obtained by holding down the SHIFT key and clicking on the blank area of the menu bar on the instrument's Control Panel. Another SHIFT-click will bring more information. The vertical scale can be adjusted by choosing Vertical Scale from the Format menu. The vertical scale can also be automatically scaled by choosing Turn Auto Scale On from the format menu. The cursors denote the active region of the histogram, from which statistics are calculated. They can be set from the Set Cursors option from the Format menu, or by using the mouse on the active DATA window. Using the Expand command from the Format menu will fit the selected portion of the histogram to the DATA window.

The instrument must be stopped before Kmax can be exited. This can be done by stopping the data acquisition by stopping the data gathering, and then clicking the Kmax halt button on the leftmost side of the control panel menu bar. When exiting the program, do NOT save changes to the instrument. Saving histograms must be done before exiting the program.

**TO START UP THE MAC**

- CONNECT SCSI BUS CABLE FROM CAMAC TO LAPTOP;

- CONNECT REQ TO G IN WITH LEMO CABLE;

- CONNECT THE BIPOLAR OUTPUT OF THE AMPLIFIER TO THE DISCRIMINATOR INPUT TO FORM THE TRIGGER GATE;

- CONNECT TRIGGER FROM THE DISCRIMINATOR TO THE GATE ON 2249A ADC;

- CONNECT SIGNAL FROM THE AMPLIFIER'S UNIPOLAR OUTPUT TO ONE OF THE ADC INPUTS;

- TURN ON CAMAC CRATE;

**TO START THE MACINTOSH POWERBOOK COMPUTER**

- USE ALIAS FOR AD2249_SRQ IN MACINTOSH HD/APPLICATIONS Kmax; DOUBLE CLICK ON IT;

- CLICK ON OK TO PRINTER ERROR;

- TO START THE PROGRAM, CLICK ON THE HISTOGRAM AND THEN THE ARROW ON THE CONTROL BAR;

- PUT IN CHANNEL NUMBER TO BE USED, THEN HIT RETURN;

- CLICK ON READ;

- YOU CAN RUN ON TOTAL COUNTS OR TIME;

- TO ANALYZE DATA, CLICK ON DATA WINDOW, MAKE SURE THAT DATA WINDOW HAS BARS;

- PUT MOUSE IN CONTROL BAR AREA, HOLD SHIFT DOWN AND CLICK MOUSE;

- REPEAT THE SHIFT-CONTROL BAR AND CLICK TO GET FURTHER DATA ANALYSIS INFORMATION ON PLOT;

- APPLE AND B WILL CLEAR HISTORGRAM.

# MUON LIFETIME MEASUREMENT

Serkant Ali Çetin[a], Tuba Çonka-Nurdan[a], Arif Mailov[a,b]

a) Department of Physics, Boğaziçi University, Istanbul, Turkey
b) Institute of Physics, Academy of Sciences, Baku, Azerbaijan

## 1 What is Muon

The Muon is the second lightest lepton in the Standard Model with an approximate mass of 106 MeV. It has a spin 1/2 and a charge of ±1. Its position in the classification of fermions can be seen in Table 1.1. The mean lifetime of the muon is around 2.2 $\mu$s. Muon has a high penetrating power and its energy loss through matter is less than that of electron's.

Table 1: The Fermions

|  | $1^{st}$ generation | $2^{nd}$ generation | $3^{rd}$ generation |
|---|---|---|---|
| L E P T O N S | $Q=-1$  $S=1/2$  $e$  $M \approx 0.511 MeV/c^2$  electron | $Q=-1$  $S=1/2$  $\mu$  $M \approx 106 MeV/c^2$  muon | $Q=-1$  $S=1/2$  $\tau$  $M \approx 1.78 GeV/c^2$  tau |
|  | $Q=0$  $S=1/2$  $\nu_e$  $M \leq 15\,eV$  e neutrino | $Q=0$  $S=1/2$  $\nu_\mu$  $M \leq 0.17\,MeV$  $\mu$ neutrino | $Q=0$  $S=1/2$  $\nu_\tau$  $M \leq 24\,MeV$  $\tau$ neutrino |
| Q U A R K S | $Q=2/3$  $S=1/2$  $u$  $M \approx 0.3 GeV/c^2$  up | $Q=2/3$  $S=1/2$  $c$  $M \approx 1.5 GeV/c^2$  charm | $Q=2/3$  $S=1/2$  $t$  $M \approx 174 GeV/c^2$  top |
|  | $Q=-1/3$  $S=1/2$  $d$  $M \approx 0.3 GeV/c^2$  down | $Q=-1/3$  $S=1/2$  $s$  $M \approx 0.5 GeV/c^2$  strange | $Q=-1/3$  $S=1/2$  $b$  $M \approx 4.7 GeV/c^2$  bottom |

## 2 How Are Cosmic Muons Created

Muons were first identified in cosmic ray experiments by Anderson and Neddermeyer in 1936. The cosmic radiation, which consists of high energy particles that are mostly protons, enters the earth's atmosphere and interacts with the atmospheric nuclei resulting in secondary particles (Figure 2.1). Collisions of cosmic rays with atoms in the upper atmosphere produce mostly neutral and charged pions. Each neutral pion decays into a pair of gamma rays in less than $10^{-6}$ seconds. The charged pions each decay, within less than $30\ ns$, into a muon and a muon neutrino. So, the cosmic muons produced by the decay of pions, which are produced high up in the atmosphere (typically on 20,000 meters of amplitude). The pion decay, characterized by weak interaction is;

$$\pi \to \mu + \nu_\mu$$

The muons then decay in about 2 $\mu s$ into an electron, a muon neutrino and an electron anti-neutrino. Decay of charged pions into muons is as shown in Figure 2.1. It is primarily these muons that are observed by the detector.

Since the muon has a relatively long life time and the reaction mechanism is characterized by weak interaction, a large amount of the muons produced on high altitutes will reach the surface of the earth.

In matter there are two competing processes for muon:

- Decay (Figure 2.2)

$$\mu \to e + \nu_e + \nu_\mu$$

- Capture by the nucleus (Figure 2.3)

$$\mu + p \to \nu_\mu + n$$

Due to Coulomb repulsion, nuclear capture is not very likely for the positive muons. So, for $\mu^+$ decay is the dominant process. For the negative muon, however, capture by the nucleus is the most possible process. This leads to the shorter life time for $\mu^-$.

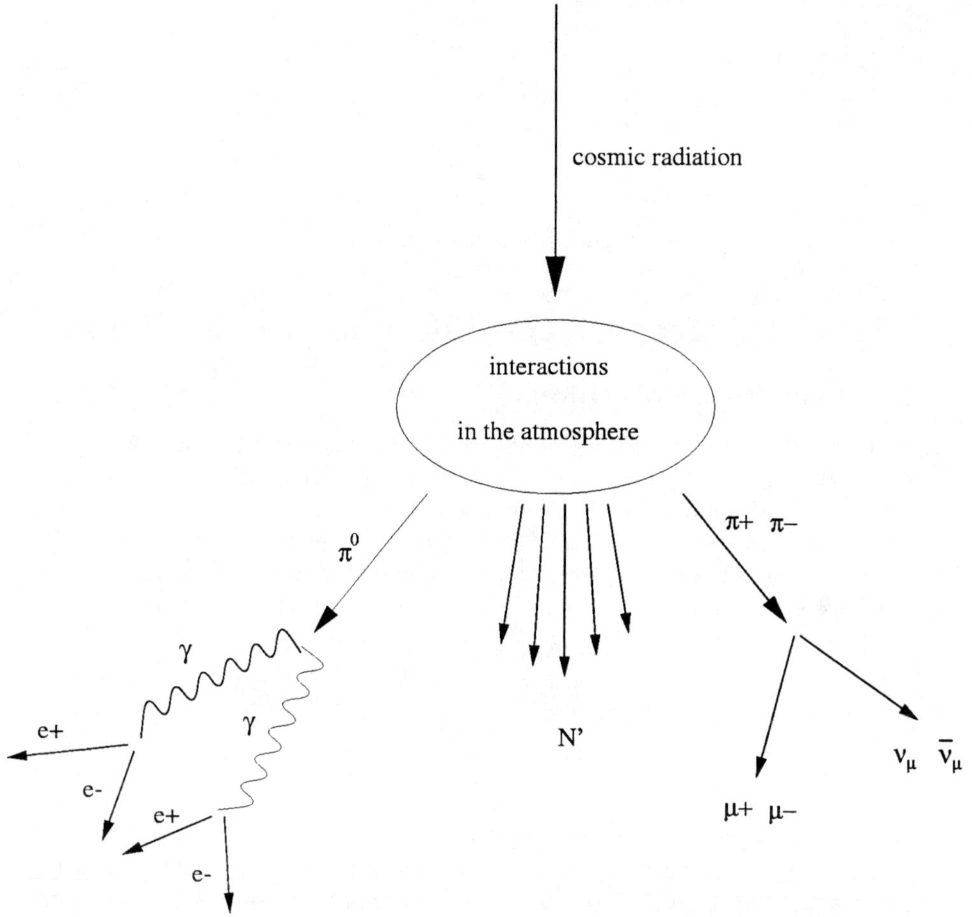

Figure 1: creation

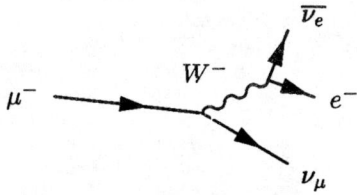

Figure 2: Decay of $\mu^-$

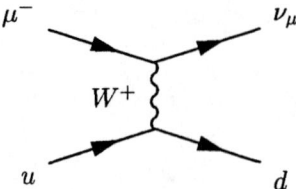

Figure 3: Capture of $\mu^-$

# 3 How To Measure the Lifetime of the Muon

## 3.1 Mean Range of Muons

Muons are assumed to decay spontaneously and therefore the probability of a muon decaying during any time interval $dt$ is the same, and this probability is independent of the history of the muon.

Consider N muons at the time $t = 0$. The average number of muons still intact at a time $t$ is denoted by $N(t)$; the number of muons decaying in the interval $t, t + dt$ is

$$-dN(t) = N(t)\frac{dt}{\tau_0} \qquad (1)$$

Integrating equation (1) we find

$$N(t) = N e^{\frac{-t}{\tau_0}} \qquad (2)$$

Here $\tau_0$ is the mean life of the muon.

To obtain the mean range of a fast muon, one has to distinguish clearly between apparent life and proper life. The lifetime of a muon is equal to $\tau_0$ in the rest system of the muon. Due to relativistic time dilation, a muon with a velocity $V$ appears to have a lifetime

$$\tau_0' = \tau_0 \gamma, \quad \gamma = \frac{1}{\sqrt{1 - \frac{V^2}{c^2}}} \qquad (3)$$

The mean range of a muon with velocity $V$ is therefore

$$R(V) = V \cdot \tau_0' = V \cdot \tau_0 \cdot \gamma \qquad (4)$$

or

$$R(V) = \frac{\tau_0}{m_\mu} \cdot p \qquad (5)$$

$$p = mV, \quad m = \frac{m_\mu}{\sqrt{1 - \frac{V^2}{c^2}}} = m_\mu \cdot \gamma$$

where $m_\mu$ and $p$ are rest mass and momentum of the muon. The range $R$ of a muon increases therefore more rapidly than its velocity.

The following remark may be useful. In the system of reference where the muon is moving, the apparent life of the muon is prolonged due to time dilation and therefore the muon can cover a range $R$, which could not be covered during the time $\tau_0$. In the rest-system of the muon, however, the time available for the muon is only equal to $\tau_0$. The range which was found to be equal to $R$ in the system where the muon was moving appears to be contracted by a factor $\frac{1}{\gamma}$ in the rest-system. Therefore the muon has to live only for the time needed to pass the contracted range, $R/\gamma$.

## 3.2 Muons Traversing an Absorber

We calculate the mean range of a muon passing through an absorber. Consider muons with momentum $p$ entering an absorber. Due to absorption the muons reaching a depth $\eta$ will have a momentum:

$$p(\eta) < p. \tag{6}$$

Denote the average number of muons reaching $\eta$ by $N(\eta)$. The average number decaying in the interval $\eta$, $\eta + d\eta$ is obtained from (1.1) and (1.3)

$$-dN = N(\eta) \frac{m_\mu}{\tau_0} \frac{d\eta}{p(\eta)} \tag{7}$$

Integrating over $\eta$ we find $N(\eta)$:

$$N(\eta) = N(0) \cdot e^{\frac{m_\mu}{\tau_0} \int_0^\eta \frac{d\eta}{p(\eta)}} \tag{8}$$

### 3.2.1 Muons Traversing a Homogeneous Absorber

For fast muons ($p \gg m_\mu c$) we may assume

$$\frac{dp}{d\eta} = constant = \frac{m_\mu \cdot c}{a} \tag{9}$$

and therefore

$$p(\eta) = p_0 - \frac{\eta}{\eta_1}(p_0 - p_1) \tag{10}$$

where $p_0$ is the momentum of the incident muons and $p_1$ is the momentum of the muons which have reached a depth $\eta = \eta_1$.

Using Equation 2.9 and Equation 2.10 we find

$$N(\eta_1) = N(0)(\frac{p_0}{p_1})^{\frac{-m_\mu \eta_1}{\tau_0(p_0-p_1)}}. \tag{11}$$

### 3.2.2 Muons Nearing the End of Their Range

We may use the approximate range-momentum relation which comes from energy losses of charged particles due to inelastic collisions. (Rossi and Greisen (1941))

$$-\frac{dE}{d\theta} = \frac{A}{(\frac{V}{c})^2} \tag{12}$$

with

$$A = 0.153 \cdot \frac{Z}{A} \cdot \begin{cases} 20.2 + 3\log\frac{p}{m_e c} - 2\log z & for\ electrons \\ 20.5 + 4\log\frac{p}{m_e c} - 2\log z & for\ protons,\ muons \end{cases}$$

Neglecting the terms $\log\frac{p}{m_e c}$, $A$ can be regarded as constant, and introducing $E = mc^2(\gamma - 1)$ and $(\frac{V}{c})^2 = 1 - \frac{1}{\gamma^2}$ the equation (1.12) can be integrated

$$R(E) = \frac{E^2}{A(E + mc^2)} \tag{13}$$

The following two approximations are useful

- Nonrelativistic Case:

$$R(E) \sim \frac{E^2}{Amc^2} \sim \frac{p^4}{4Am^3c^2} \tag{14}$$

- Extreme Relativistic Case:

$$R(E) \sim \frac{1}{A}(E - mc^2) \tag{15}$$

Taking into account non-relativistic case we find

$$R(p) \sim \frac{1}{4}(\frac{p}{m_\mu c})^4 a \tag{16}$$

So from (1.8)

$$N(\eta) = N(0) \cdot e^{-\frac{a}{3c\tau_0} \cdot \frac{p(0)^3 - p(\eta)^3}{(m_\mu c)^3}} \tag{17}$$

The fraction of muons with initial momentum $m_\mu c$ which decay only after being brought to rest is therefore:

$$N(p=0)/N(p=m_\mu c) = e^{-\frac{a}{3c\tau_0}}$$

In the case of air we have approximately;

$$\frac{dp}{d\eta} = 2.5 \cdot 10^{-3}$$

$$a = 4 \cdot 10^4, \quad c\tau_0 = 6.3 \cdot 10^4$$

therefore,

$$N(p=0)/N(p=m_\mu c) = 0.81.$$

Thus most slow muons moving in air will be brought to rest before decaying.

### 3.2.3 Muons Traversing the Free Atmosphere

The change in air density with height has to be taken into account. For simplicity we assume the density distribution to be exponential.

$$\theta(x) = \theta_0 e^{-\frac{x}{x_0}} \qquad (18)$$

where $x_0$ is the height of the homogeneous atmosphere. $\theta(x)$ is the absorption equivalent of the atmosphere at the height $x$. $\theta_0$ is the absorption equivalent of the whole atmosphere.

Fast muons arriving at sea-level with a momentum $p_1$ must be supposed to have a momentum

$$p(x) = p_1 + \mu c \frac{x_0}{a}(1 - e^{-\frac{x}{x_0}}) \qquad (19)$$

at the height $x$. Denoting the loss of momentum throughout the whole of the atmosphere by

$$p_A = \frac{m_\mu \cdot c \cdot x_0}{a} \qquad (20)$$

we find

$$N(0) = N(x) \left( \frac{p_A}{p_1} \cdot \frac{p(x)}{p_A + p_1 - p(x)} \right)^{-\frac{x_0}{c\tau_0} \cdot \frac{m_\mu c}{p_1 + p_A}}.$$

## 3.3 Direct Measurement of the Lifetime of the Muon

A classical arrangement (Figure 3.1) to measure the lifetime was used by Maze (1941), Rasetti (1941) and Rossi (1942). The absorber $S$ is placed below a coincidence arrangement I-II. A set of anticoincidence counters $A$ is placed below $S$. Anticoincidences I, II-$A$ are caused mainly by particles coming from above and stopping in $S$. To reduce the effect of showers the top counters are surrounded by lead and a 10 cm. thick lead block is placed between the coincidence counters. Due to this selection the anticoincidences are mainly due to muons stopping in $S$.

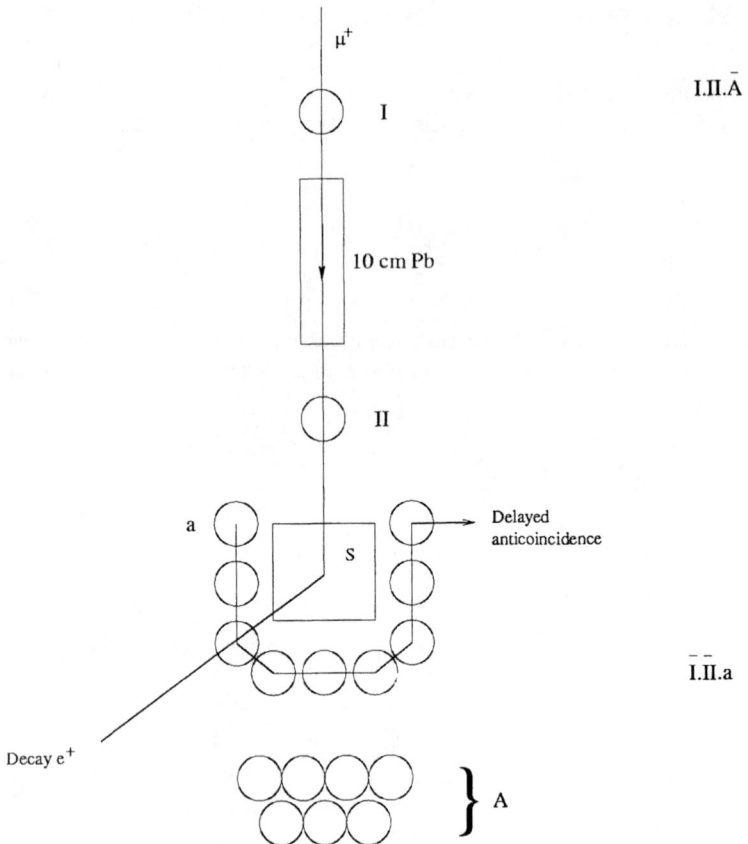

Figure 1: Setup of Rossi

The sides of the absorber $S$ are covered with a counter battery. If any of the counters $a$ are discharged shortly after the coincidence pulse, then the pulse is recorded and the time delay between the anticoincidence and the pulse of the counter $a$ is also recorded. Such delayed pulses are mainly due to muons stopping in $S$ and emitting a decay electron afterwards. The registered time delay can be regarded as the time spent by muon in the absorber before it decayed.

As a result of extensive measurements it was found that the rate of delayed pulses corresponding to delays greater than $t$ is given by:

$$\zeta(>t) \sim e^{-t/\tau_0}, \qquad (21)$$

$$\tau_0 = \begin{cases} 1.5 \pm 0.4 \mu S & \text{Rasetti} \\ 2.2 \pm 0.2 \mu s & \text{Maze} \\ 2.15 \pm 0.15 \mu s & \text{Rossi} \\ 2.33 \pm 0.15 \mu s & \text{Conversi} \end{cases}$$

whereas the latest measurements give an average of 2.19703 ± 0.00004 $\mu$s.

## 4 Experimental Apparatus

The main parts of the experimental set-up are three scintillation counters and a thick aluminium block. The experimental arrangement is shown in Figure 4.7. The scintillators are for detecting the incoming and outgoing particles, the aluminium plate causes some of the incoming cosmic muons to stop. Aluminium was chosen as a stopper because it has a long radiation length, it is non-magnetic and it is also relatively cheap. The scintillator counters (Figure 4.2) consist of a scintillator(Figure 4.1) connected to the photo multiplier tube (pmt) through a light guide. As the particles pass through the scintillators electrons of the doped molecules are excited. Then these excited electrons emit photons during deexcitation. Since the scintillator plates are covered by aluminium foil no light can escape from side walls and in addition to that they are covered by black tape which prevents any escape or intake of light. In this way all the photons are directed to the light guide and then to the photocathode of the pmt where the photoelectrons are formed and then amplified by negative high voltage to produce negative signal output.

Since the aim is to find the lifetime of the muon, we need to book the time interval between each incoming muon and outgoing electron. Then fitting the data using Equation 3.2 will give us the lifetime.

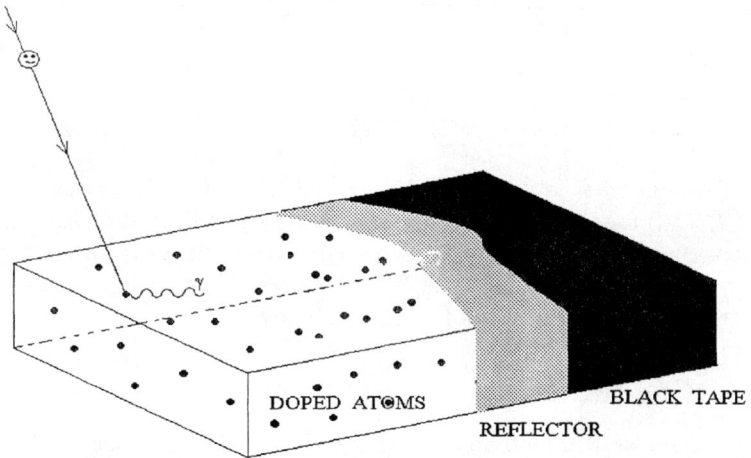

Figure 1: Scintillator

## 4.1 Hardware Instruments

Other than the detector itself, we use a NIM (Nuclear Instrument Modules) and a CAMAC (Computer Automated Measurement And Control) crate with appropriate cards and modules. All algorithm and tunings of the experiment are done within the NIM crate and CAMAC is used for the read-out. The NIM modules to be used can be summarized as follows:

- Discriminator:
  Accepts negative input signals. If the input has a larger magnitude than its adjustable threshold, it gives out a -0.8 V square output (NIM level) with a desired width.

- FI/FO (Fan-in Fan-out):
  In general is used to sum several channels and invert the result if necessary. It doesn't use NIM level, it has an adjustable DC off-set.

- TAC (Time to Amplitude Converter):
  Works like a clock to calculate the time between two successive events (start/stop). It has an adjustable interval range for which an output of 0-10 V (positive) is generated with a fixed width of 2 $\mu$s. It accepts start/stop signals in NIM level.

- Coincidence:
  Uses NIM level both for output and input. Gives out a signal with an adjustable width if the input signals coincide (overlap in their widths) for a period.

- Dual Timer:
  Allows to change the width of a NIM standard input and also give very long delays (upto seconds).

- Counter:
  Counts the number of NIM level signals for a given time interval.

The CAMAC crate is connected to a PC through GPIB (General Purpose Interface Board) for the read-out. An ADC (Analog to Digital Converter) card which has a gate and signal input is placed in the crate. It can integrate a maximum value of 12.5 V.ns which has to be considered while adjusting the width of the gate signal that has to be in NIM level.

## 4.2 Getting Started

In this section we will be familiarizing ourselves with the experimental setup. It is important to go through all the steps mentioned below before you start the experiment, especially if it is your first time with such instruments.

### 4.2.1 Analog signal

There are three layers of scintillating counters in our experiment. In this part, the top counter will be used to see the signal on the analog oscilloscope. It is crucial to see the level of signal and level of noise, in order to determine the threshold to be set for the discriminator. Make sure that you have the below connections made, and observe the behaviour of the PMT output.

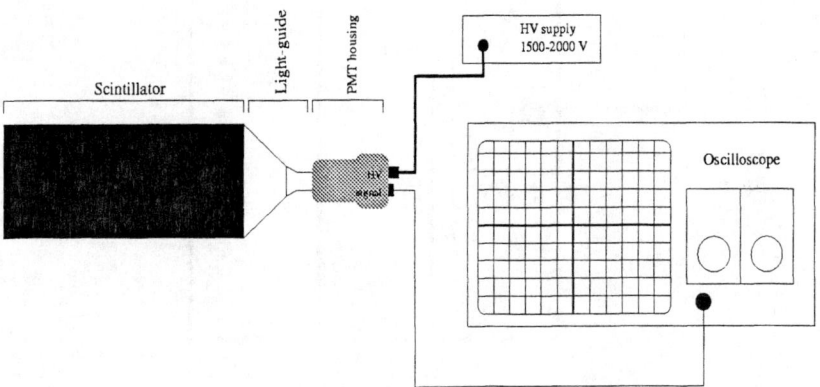

Figure 2: set1

### 4.2.2 Plato Curve Measurement

Every scintillating counter has a dependence on the high-voltage (HV) applied to the PMT. To explore the behaviour of the number of counts as a function of the applied HV, you will take data with the below settings. The threshold of the discriminator must be set to the value that you have determined in the previous part. Adjust the time interval of the counter to 10 seconds and take data at every 25 Volts between 1700-2500 Volts, wait for about 10 seconds after changing the voltage for stabilazation. Record and then plot your results on the sheet at the back. You should observe a stable region (plato) on your graph.

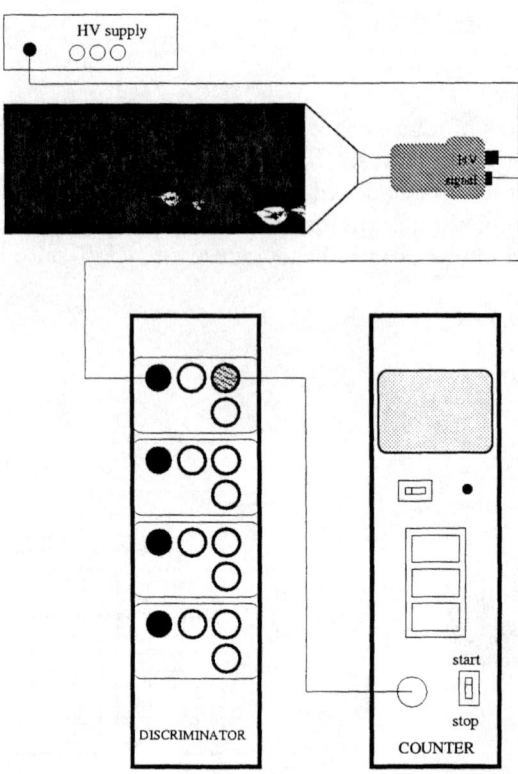

Figure 3: set2

| HV (volts) | RATE (Hz) | HV (volts) | RATE (Hz) | HV (volts) | RATE (Hz) | HV (volts) | RATE (Hz) |
|---|---|---|---|---|---|---|---|
| | | | | | | | |
| | | | | | | | |
| | | | | | | | |
| | | | | | | | |
| | | | | | | | |
| | | | | | | | |
| | | | | | | | |
| | | | | | | | |
| | | | | | | | |

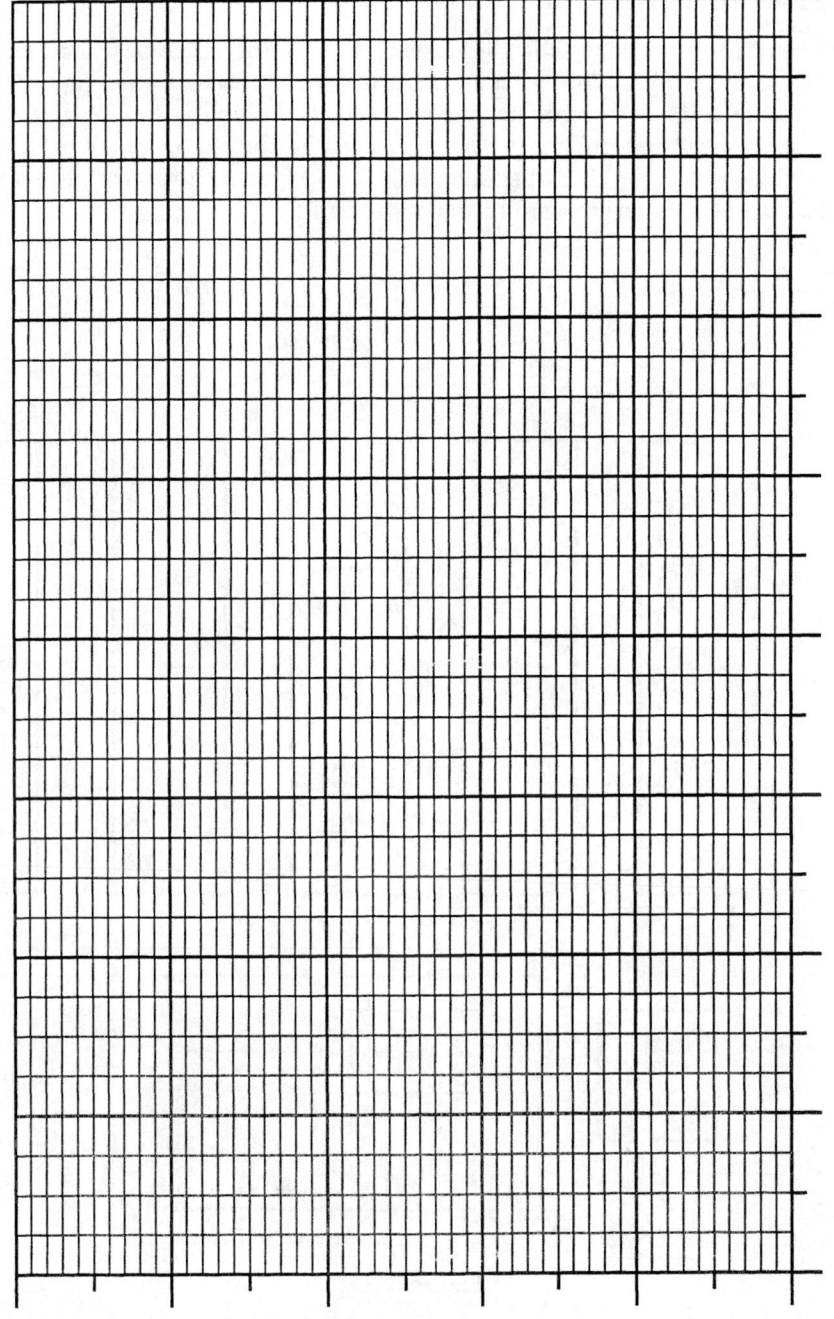

Figure 4: Plato

### 4.2.3 TAC Calibration

This part is performed to show the linear dependence of the TAC output to its input. In order to check this, we must generate two successive signals (start/stop) with a known time difference and record the output amplitude of TAC. If this is done for several time differences the results can be plotted to see the behaviour. Carefully connect the cicuit below and set the TAC range to 10 $\mu$s. Follow your assistants instructions to perform the data read-out.

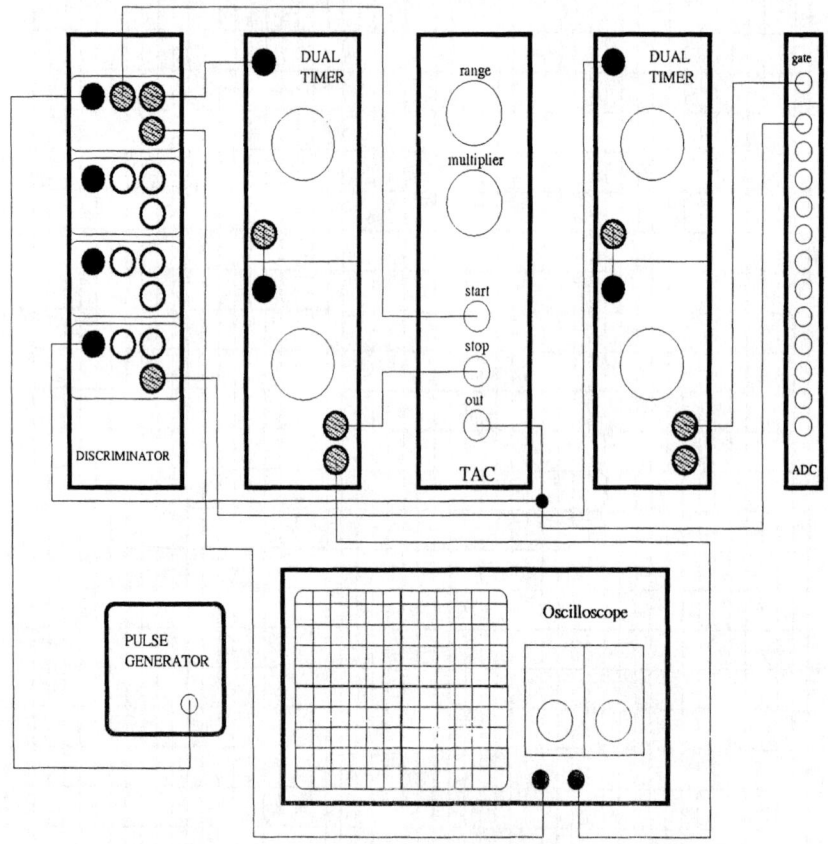

Figure 5: set3

| Time difference ($\mu$s) | | | | | | | | | |
|---|---|---|---|---|---|---|---|---|---|
| TAC output (-Volts) | | | | | | | | | |

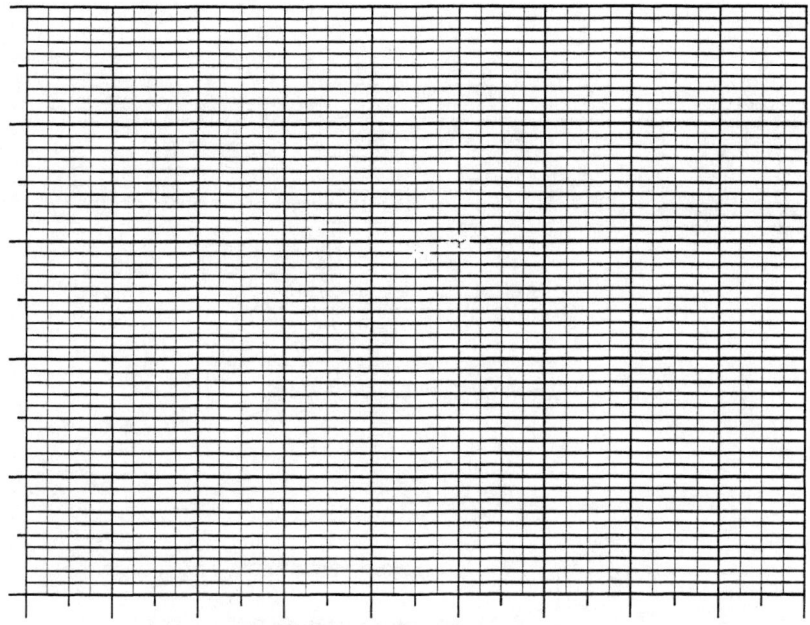

Figure 6: TAC

## 4.3 Experimental Set-up

Our set-up consists of three layers of scintillator counters each with an area of $\approx 0.4\ m^2$ placed on top of each other with a spacing of $\approx 15$ cm. In between the middle and bottom layers, an aliminium block of 7 cm thick is placed. This aliminium medium is used to stop the incoming muons and cause them to decay (Figure 2.2). The incoming muon will be the start for the TAC and the decay positron will be the stop, hence the algorithm (**Figure 4.7**) is set as follows: The start signal is generated when there is a signal from both top (A) and middle (B) counters, but nothing from the bottom (C), this enables us to identify a muon that stops in the aliminium. The stop signal is generated when there is a signal from C but nothing from A and B which shows us that this event belongs to the decay positron that proceeds to the forward cone of the direction of the incoming muon.

All detectors, modules and necessary connections for our experiment are summarized in Figure 4.8. Try to make the connections and understand the purpose of each. Before you start data taking show your setup to the instructor and make sure that you adjust the following operational values which we had optimized for the best performance:

| Detector | A | B | C |
|---|---|---|---|
| Threshold | -40mV | -28mV | -22mV |
| Width | 10 ns | 10 ns | 10 ns |
| High Voltage | -1900V | -1900V | -1750V |

| | |
|---|---|
| Gate threshold | -18.8mV |
| Gate width | 14 ns |
| Gate delay | 400 ns |

| | |
|---|---|
| TAC Range | 100 ns |
| TAC Multiplier | 100 |

Table 2: Operating values

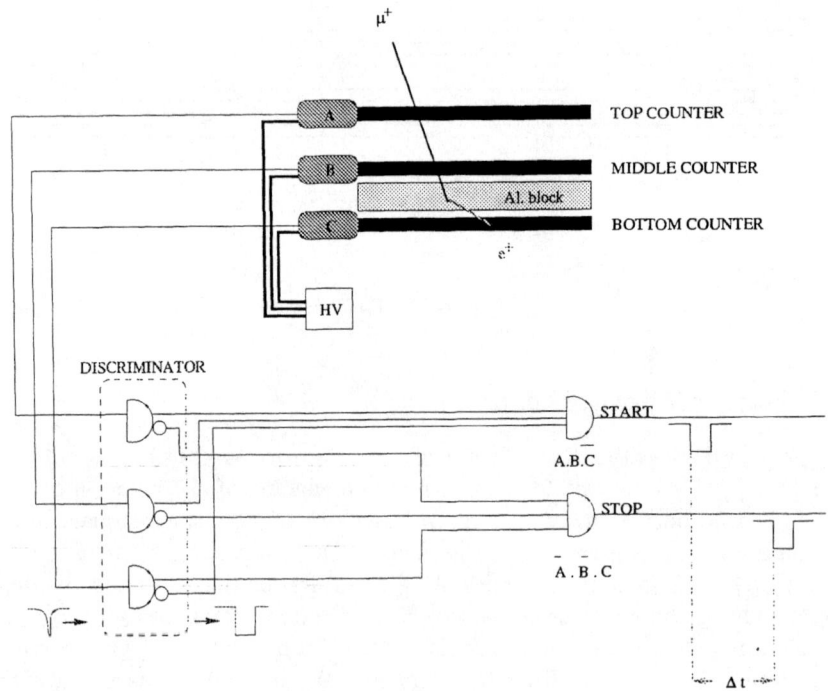

Figure 7: Algorithm

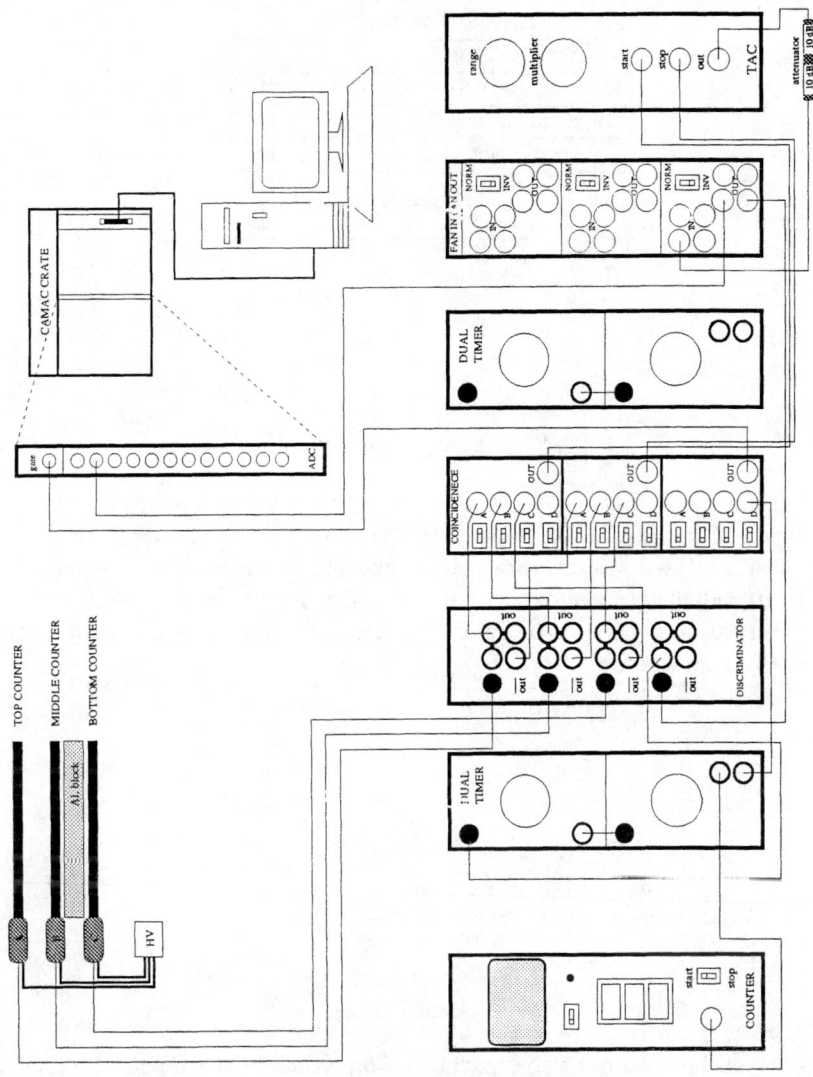

Figure 8: setup

## 5 Data Taking and Analysis

Before starting the real run, it is best to check some event rates with the detector. The first thing is determining the event rates for each layer (A, B, and C) individually, after that the START, STOP and trigger rates can also be checked. You should use the counter on the crate and make the necessary connection to get the rates:

Table 3: Event Rates

| | |
|---|---|
| Top detector (A) | Hz |
| Middle detector (B) | Hz |
| Bottom detector (C) | Hz |
| Start | Hz |
| Stop | Hz |
| Trigger | Hz |

The rate of trigger, as you have just measured, does not supply us a good statistics within the few hours that you are expected to perform this experiment. At least 10.000 trigger events are necessary for a reliable analyses. For this reason, you can analyse an old set of data given to you by using the following function to make the fit:

$$N(t) = P_1 . e^{-t/P_2} + P_3 . e^{-t/P_4} + Constant \tag{22}$$

Note that you have to use two separate exponentials in the function which, in fact, is a special case for our set-up. The exponentials are used to differentiate the $\mu^+$ decays from the $\mu^-$ captures. The rate of $\mu^-$ capture by the nucleus is significantly higher in aliminium than in any other stopping medium, this enforces us to introduce the second exponential. An example of the fit to a set of data ($\approx$14000 events) is shown in Figure 5.1. Note the logarithmic scale and the error bars on each point. It is important to estimate the level of random trigger that introduces the constant level to our function, and put that constant by hand. You can now start data taking using the read-out program named "cosmic_run".

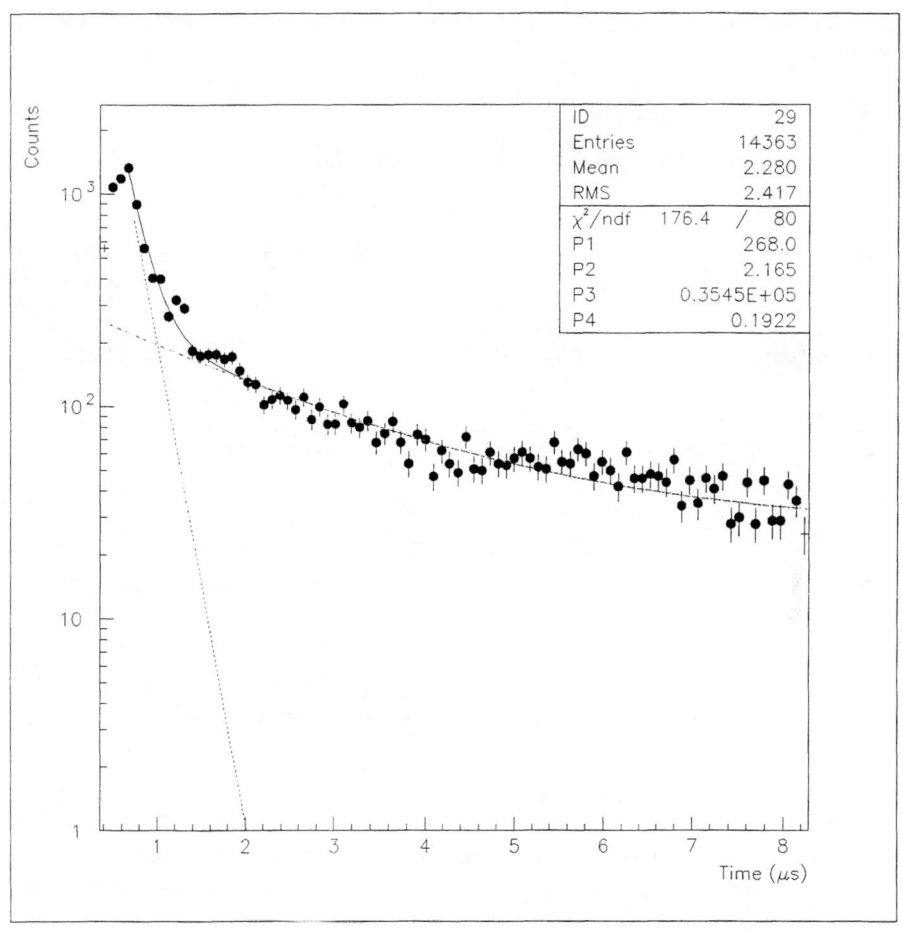

Figure 1: Data Analysis

# 6  Simulation with GEANT

Simulation of the decay of the cosmic muon is examined by using *GEANT* simulation package of CERN. GEANT is developed for designing and optimizing detectors in high energy physics.

The GEANT program simulates the passage of elementary particles through the matter and the particles trajectories and detectors can be visualized either interactively or in a batch mode. We used a Motif-based graphical user-interface (GUI) version of GEANT++.

All user subroutines with data describing the experimental setup were written in FORTRAN 77 programming language. The user routines are classified as follows:

- In UGINIT subroutine the user opens files and initialize GEANT

- In UGEOM the user define needed materials, compounds and volumes

- GUKINE routine has used to generate kinematics of the initial track

- GUSTEP is called at the end of each tracking step and provides the user by hits in detectors

- GDXYZ is used to visualized the tracks during the transport process

At interaction of the incoming particle with matter the each physics process is generated by Lund event generator will accepted by GEANT for further analysis.

The decay of cosmic muons is simulated by using GEANT in which first of all volumes have to be determined. The outermost medium is defined to be *universe*. The Master Reference System is attached to it. All detectors and stopper are positioned inside the *universe*. The *concrete* of the building is also included in consideration. As in the case of real muonscope, there are three scintillator plates and a stopper. After the definition of the volumes, kinematic properties of the particles are defined in the related subroutines. The incoming particles are positively charged muons having the average momentum of the order of 3 GeV when arriving at the *concrete*. The *concrete* is taken with thickness $\approx$ 1.5 m will filter out the incoming muons and therefore the only 0.75 − 0.9 GeV momentum interval is important for study the decay of muons in *Al* absorber. Then the conditions have to be given for selecting events which will create START and STOP signals for measuring the lifetime of cosmic muons. The logic goes as follows: cosmic muons pass through the *universe*, *concrete*, top plate, middle plate and stopper. In the stopper they decay into positrons and these positrons are supposed to be detected in bottom plate. In GEANT dotted blue lines represent gammas, solid red line stands for charged particles, except muons, black blank/dotted line is for neutral hadrons and neutrinos, dashed green line for muons. The result of GEANT simulation for 5 muon events is shown in Figure 5.1.

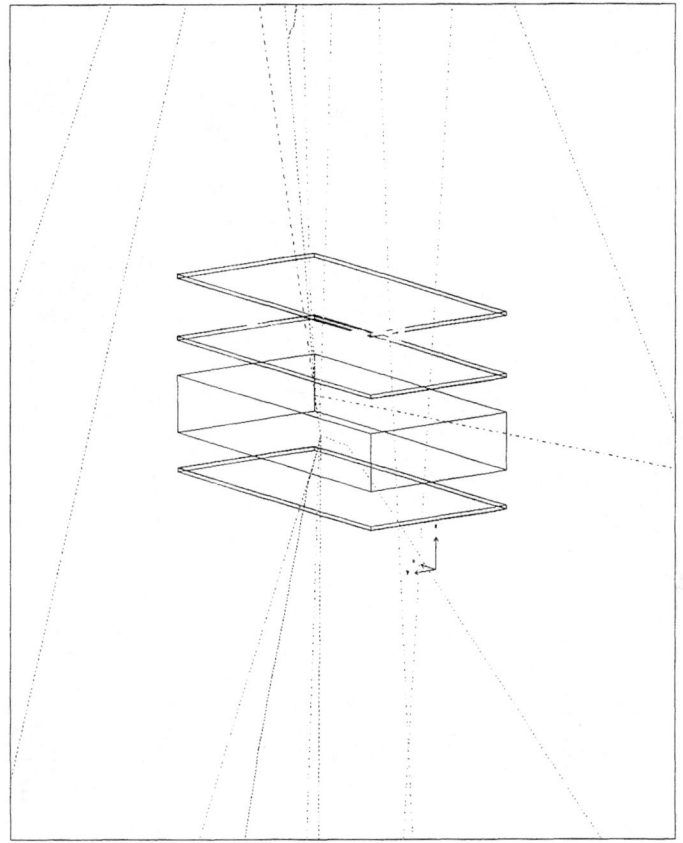

Figure 1: GEANT Simulation

In the figure incoming green lines are $\mu^+$'s. The red line represents the decayed positron. Our studies show that out of 150 muon events the only one of them fulfills our condition, namely the incoming muon decays into positron and neutrinos in the stopper and reaches bottom plate.

The distribution in lifetime of positive muons is presented in Figure 5.2. The result of fitting the simulation data with single exponent gives mean lifetime $2.03 \pm 0.1 \mu s$.

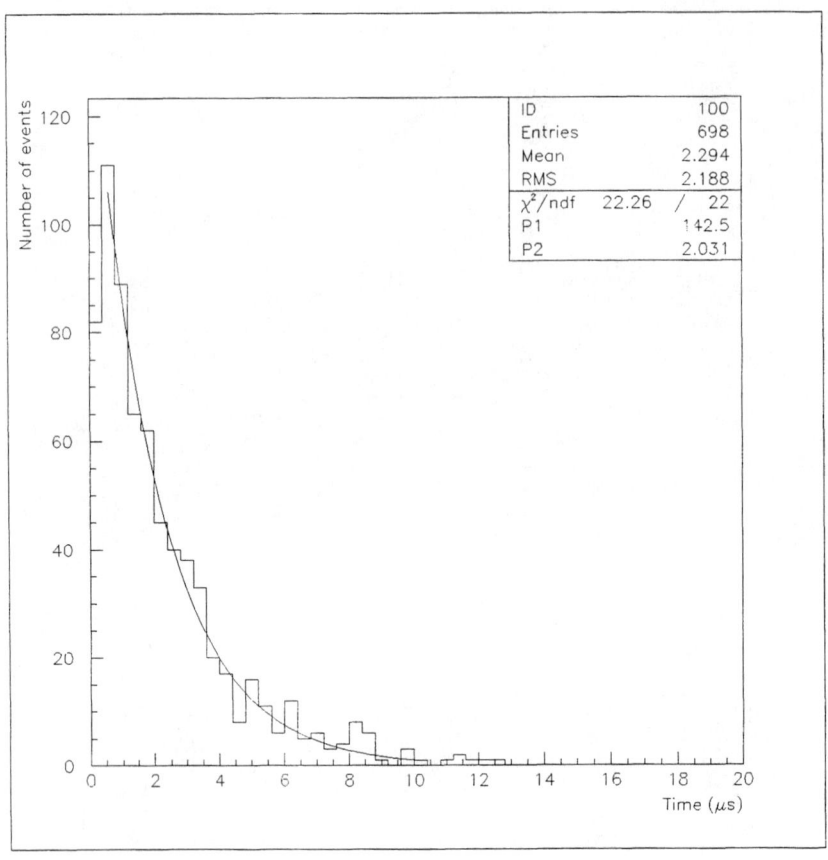

Figure 2: Distribution in lifetime of $\mu^+$

# References

[1] "Cosmic Rays", L. Janossy, (1950).

[2] "Interpretation of Cosmic Ray Phenomena",B. Rossi, Rev.Mod.Phys. 20, 537, (1948).

[3] "Measurement of $\mu^-$ Capture Rates Using Cosmic Rays", W. Slater, Preprint Physics 180F.

[4] "Laboratory Study of the Cosmic Ray Muon Lifetime", T. Ward et. al., Am. J. Phys. 53(6), 542, (June 1985).

[5] R. E. Hall et. al., Am. J. Phys., 38, 1196, (1976).
A. Owens et. al., Am. J. Phys., 46, 859, (1978).

[6] "Lifetime of the Muon", Caltech Senior Physics Laboratory, Experiment 15, (1996).

[7] "Violation of Mirror Symmetry in the Production and Decay of Cosmic Ray Muons and Measurement of the Muon Magnetic Moment", MIT, Dept. of Physics, (1996).
"The Speed and Decay of Cosmic Ray Muons", Junior Physics Laboratory, Experiment 20, MIT, Dept. of Physics, (1996).

[8] "Model 566 Time-to-Amplitude Converter", EG&G ORTEC, Operating and Service Manual.

[9] "PAW Physics Analyses Workstation", CERN Program Library, CERN-Q121, (1995).

[10] "GEANT Detector Description and Simulation Tool", CERN Program Library, CERN-W5013, (1993).

[11] "Review of Particle Physics", The European Physical Journal C, Vol.3, No.1-4, (1998).

[12] "Particle Physics", B.R. Martin, G. Shaw, (1992).

[13] "Techniques for Nuclear and Particle Physics Experiments: A How-to Approach", (1994).

# Principle of Operation of Micro Pattern Detectors

A. Cattai, R. Malina[1]

*CERN, 1211 Geneva 23, CH*

**Abstract.** Microstrip Gas Chambers (MSGC's), Small Gap and Groove detectors with an optional GEM foil are a recent development of gaseous micro-pattern detectors providing a cost efficient solution for a tracking detector. They provide good spatial resolution ($\sim 35$ $\mu m$) and can operate in high radiation environments (up to $\sim 10^4$ particles/$mm^2$). The laboratory course on micro pattern detectors tried to give an understanding of the layout and basic functioning of these detectors.

## I INTRODUCTION

MSGC's were invented by A. Oed in 1988 [1]. They are an evolution of the Nobel-prize winning MWPC (MultiWire Proportional Chamber) of G. Charpak [2]. The sense wires of the MWPC's are replaced by gold strips that are etched on a glass substrate in a photolithographic process. This allowed a reduction of the distance between electrodes from a former 1 $mm$ to a few 100 $\mu m$. Therefore, much higher particle densities can be dealt with.

Fig. 1 shows the cross section of an MSGC [3], [4]. A gas volume filled with a mixture of a noble gas (e.g. Neon, Argon) and a quencher (e.g. $CO_2$, DME (Di-Methyl-Ether)) is enclosed by an aluminized plane (the drift plane) and a glass substrate on which the gold strips – anodes and cathodes alternatingly – are etched. Putting a negative voltage of $\sim 3$ $kV$ on the drift plane generates an electric field within the gas filled volume between the drift plane and the substrate of $\sim 10$ $kV/cm$. Particles that traverse the chamber leave a track of ionized atoms within the gas. In the electric field the corresponding electrons are drifted towards the substrate. A voltage of about $-500$ $V$ applied to the cathodes generates a strong field gradient around the anodes which are at ground potential, leading to electron multiplication [5]. A simulation of the field configuration is shown in Fig. 2.

The ions that are left over from the avalanche process drift away from the anodes, thereby inducing the electrical signal in the anodes which is read out and amplified.

---

[1] roman.malina@cern.ch

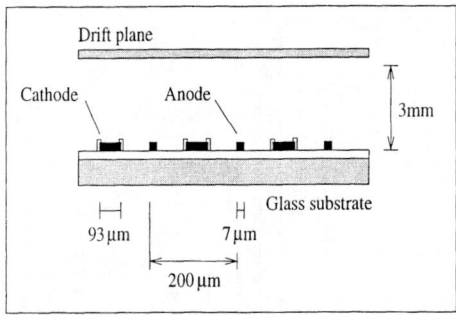

**FIGURE 1.** Schematic cross section of an MSGC. A gas volume is enclosed by a drift plane and the substrate bearing the micro pattern structure.

Thus the passing of a traversing particle can be detected. By reading out all the anodes at one time, the point of traversal can – by means of charge interpolation – be determined to a precision of down to 30 $\mu m$ thus making the MSGC's well suitable for application in a tracking environment.

## II  THE EXPERIMENT

The multiplication of electrons, i.e. the gain of an MSGC, depends on several factors: composition of the gas mixture, temperature, pressure, geometry of the micro pattern, drift and cathode voltage. In this laboratory course we tried to get an understanding on how the latter three affect the performance of a gaseous micro pattern device. For two different $V_{drift}$ settings a $V_{cathode}$ scan was performed, followed by a $V_{drift}$ scan at constant $V_{cathode}$.

Three groups of students worked on three different micro-pattern detectors. The standard MSGC has already been explained in the introduction; the different layout of a Groove and Small Gap substrate are shown in Fig. 3. The Small Gap chamber has broader cathodes than the MSGC, thus reducing the gap between anodes and cathodes without altering the readout pitch. This leads to a steeper field gradient close to the anodes and therefore a higher gain for a given cathode voltage. The Groove detector has geometrical dimensions similar to the Small Gap chamber (and should therefore have similar gain), but anodes and cathodes are not on the same plane. Instead, a kapton foil is used as a spacer between the electrodes.

In our laboratory course the Groove detector was also equipped with a GEM foil (gas electron multiplier, [6]). This is a kapton foil suspended in the gas volume of the detector and metallized on both sides. A potential difference of $\sim 400\ V$ is applied between the surfaces. Small holes of $\sim 70\ \mu m$ spaced $\sim 120\ \mu m$ from each other over the entire surface channel the field lines through them, generating

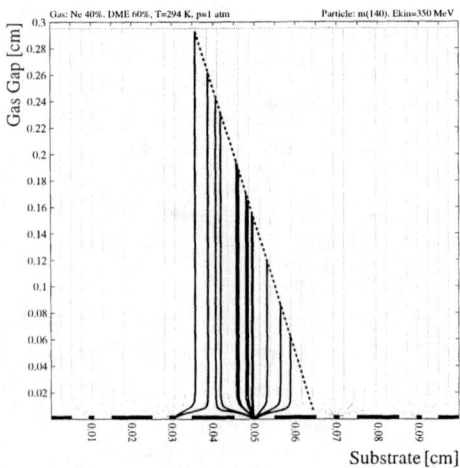

**FIGURE 2.** The electric field in an MSGC. The electrons liberated in the gas along the track of a traversing particle drift along the field lines to the anodes. There, the signal is generated in an avalanche process (not shown).

locations of high field gradient in which electron multiplication takes place. Thus, the combination of Groove + GEM is a two stage amplification device, the gain of the detector is the product of the amplification factors of the GEM foil and the Groove pattern.

## III  RESULTS

If an electron needs, at a given voltage $V_0$, a distance $d_0$ to gain enough energy to knock out another electron from a gas molecule, this necessary distance is halved when the voltage is doubled, i.e. the mean free path can loosely be written as $d = l/V = const \times 1/E$. Assuming that in every collision one electron is liberated, we will have $2^N$ electrons generated in the N-th collision, and therefore with $N = \Delta x/d$ we count $2^{\Delta x/d} \propto e^E$ electrons for a fixed length $\Delta x$ traveled in the gas. Thus, integrating over the total length of the avalanche from the position of the primary electron to the surface of the relevant anode, we see that the gain of the chamber depends exponentially on the strength of the electric field around the anode which is a function of cathode and drift voltage $V_{cath}$ and $V_{drift}$,

$$\text{gain} \propto e^{k_1(V_{cath}+k_2 V_{drift})}. \tag{1}$$

In the experiment this behavior was verified for all three detectors; the results are shown in Fig. 4 and Fig. 5.

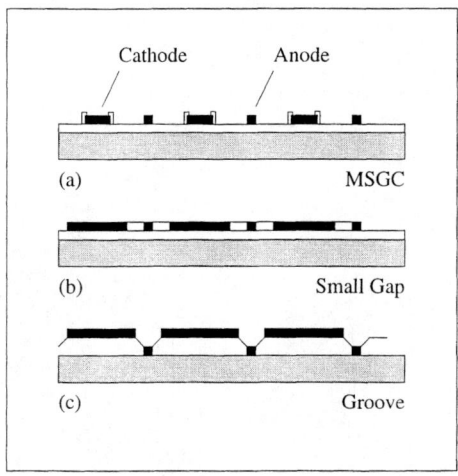

**FIGURE 3.** Comparison of the different micro patterns used in the laboratory session. (*a*) shows the layout of the substrate of an MSGC, (*b*) and (*c*) corresponds to a Small Gap and a Groove detector, respectively.

In the case of the Groove+GEM detector the 'cathode' voltage was supplied via a voltage divider, i.e. $V_{cath} = 300\ V$ corresponds to 300 $V$ on the cathode, 600 $V$ on the lower surface of the GEM and 900 $V$ on the upper GEM plane. Thus, micro-pattern and GEM foil amplification factors where raised synchronously. We note that indeed, due to the smaller space between the electrodes the Small Gap chamber has a higher gain than the MSGC at the same $V_{cath}$. We also see the substantial contribution to the gain of the GEM foil when comparing Small Gap and Groove+GEM detector.

# REFERENCES

1. A. Oed. Photon-Sensitive Detector With Microstrip Anode for Electron Multiplication With Gases. *Nucl. Instr. Meth.*, A263:351, 1988.
2. G. Charpak et al. The Use of Multiwire Proportional Counters to Select and Localize Charged Particles. *Nucl. Instr. Meth.*, 62:262, 1968.
3. *The Tracker Project – Technical Design Report.* CERN LHCC/98-6, 1998.
4. *CMS – Technical Proposal.* CERN LHCC/94-43, 1994.
5. W. Blum and L. Rolandi. *Particle Detection with Drift Chambers.* Springer Verlag, 1994.
6. J. Benelloch et al. Development of the gas electron multiplier (GEM). *IEEE Trans. Nucl. Sci.*, 45:234, 1998.

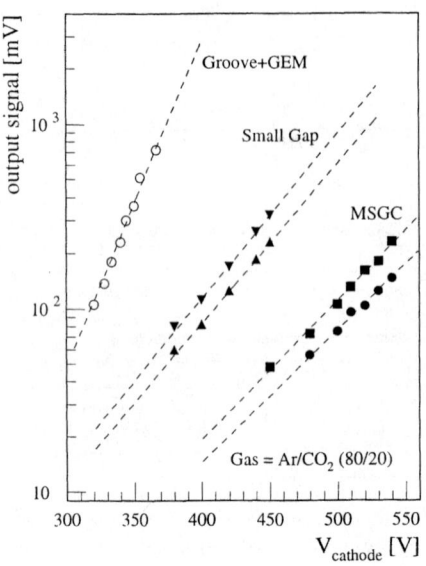

**FIGURE 4.** The output signal in $mV$ (proportional to the gain) of an MSGC, a Small Gap Chamber and a Groove detector with GEM foil are shown. The exponential behavior of the gain as a function of the cathode voltage is observed. For the first two detectors the curves were taken for two different $V_{drift}$ settings of 1 $kV$ and 2 $kV$.

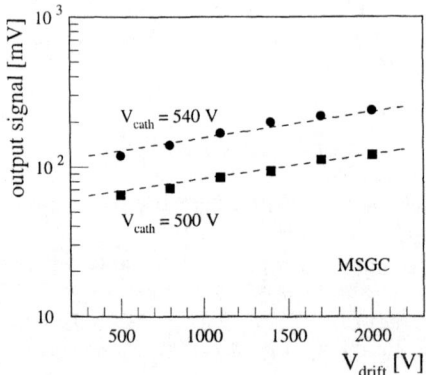

**FIGURE 5.** The output signal (i.e. gain) of the MSGC as a function of $V_{drift}$ for two different settings of $V_{cath}$. Again, the exponential dependence of the gain also on $V_{drift}$ is seen.

# Use of De-Randomizing Buffers in a Data Acquisition System [1]

## Marvin Johnson

*Fermilab, P.O. Box 500, Batavia, IL, 60510, USA*

## Marleigh Sheaff

*Departamento de Fisica, CINVESTAV-IPN, Apdo. Postal 14-740, 07000, Mexico, D. F., MEXICO*

**Abstract.** This is the course manual for a laboratory course that was designed to demonstrate a basic technique employed in data acquisition systems for the purpose of reducing deadtime and thereby increasing throughput. A custom CAMAC module was designed and built at Fermilab for use in these measurements. The use of queueing theory to predict the performance of the system is discussed, including some equations which can be used for that purpose.

## INTRODUCTION

Modern high energy physics experiments typically require sophisticated, complex hardware triggers at first level to select events for subsequent processing, since the events of interest represent only a very small fraction of the total cross section. Although the rejection factors provided by these triggers can be large, $\sim 10^4$ or more in some cases, the data throughput to higher level triggers still must have large bandwidth to keep the deadtime at an acceptable level. Since this has to be achieved within rather stringent financial constraints, the design of the data acquisition path plays an important role when planning a new (or upgraded) experiment. The purpose of this laboratory experiment is to measure the significant reduction in deadtime (or, equivalently, increase in bandwidth) that can be achieved by adding de-randomizing buffers to each channel of front end electronics. A custom CAMAC board has been designed and built at Fermilab for these measurements.

---

[1] Laboratory Course presented at the *1999 ICFA School on Instrumentation in Elementary Particle Physics*, University of Istanbul, Istanbul, Turkey, 28 June - 9 July, 1999.

An example of the use of this technique is the SVXIII chip chosen for the CDF Upgrade, which contains four buffers, i.e., four sample-and-hold devices, per channel. The D0 experiment will use an earlier model, the SVXII, which contains only a single sample-and-hold stage. The difference in cost is approximately $1M. The use of 4 buffers instead of just one is expected to increase the data throughput by a factor of approximately 5, i.e., 50,000 events per second out of level 1 for CDF versus 10,000 out of level 1 for D0 at the expected luminosity of $2 \times 10^{32}$. Since the physics goals of D0 are somewhat different from those of CDF, the use of the more expensive chip was felt to be a less effective use of the funds available for the D0 upgrade than other detector upgrades more specific to those goals.

In a high energy physics experiment, both trigger rate and event length (and thus, output rate) are almost always randomly distributed. It is possible to predict the behavior of a data acquisition system for such an experiment using queueing theory. The results of the measurements made using the CAMAC board designed for this course can be compared to the predictions made for the system implemented on the board, which is a single queue containing a variable number of buffers. The match between the throughput predicted and that actually achieved may not be exact because of other factors, for example noise pickup in the system, that alter the input and/or output rate distributions so that they are no longer Poisson-like.

# BACKGROUND

A queueing process is one in which events accumulate at some input stream, are serviced by a facility or process, and then exit in an output stream [1]. A simple example is a line at the grocery store. The customers in line form the queue, and the check-out cashier acts as the service facility. As customers pay for their goods, they successfully exit the queue and the line advances. Measures of effectiveness are important even to the customers in the grocery store. How long must they wait before receiving service? As one can imagine, the problem becomes more complicated when several cashiers are involved or when a person with an extremely large number of groceries is in line. The customer has to decide which line appears to provide the fastest service, and the store manager has to decide how many check-out registers need to be open to satisfy the customers and yet keep costs down.

Even the simplest data acquisition system represents a queueing process. Information is gathered at various detectors and transmitted to a central processing point, where it is assembled and then recorded onto a storage medium. An understanding of the queues formed at each stage of a data acquisition system is critical for design optimization. Rarely will the interarrival time of data be intrinsically well-matched to the computer readout time. Since physics processes tend to occur randomly, the exact arrival time of each event can not be known in advance. The inter-arrival times will follow a probability distribution instead. Event length differs from event to event, and thus processing (read out) time is also randomly distributed. A queueing model provides a means of studying a proposed data acquistion system. The goal is to minimize the idle time of the processors for a given system configuration and thus to maximize the throughput it can achieve taking into account the costs involved. Often general trends and steady-state behaviors can be predicted, and, in some instances, even transient solutions can be obtained explicitly.

Six characteristics are required to adequately describe a queue. Both the distribution of arrival times of events into the queue and the distribution of processing times for events must be specified. The number of parallel service facilities affects queue dynamics as well. (A store with three open check-out lines can service more customers per unit time than a store with only one.) In practical queues, there is usually a restriction on the system capacity. After a certain number of events have gathered in the queue, all subsequent events are ignored or rejected until the next event exits the queue. In some cases, the queue discipline can be important. Typically, the queue is first in, first out (FIFO) but there are many situations where certain events take priority upon arrival and some where service takes place on a last in, first out (LIFO) basis (i.e. a stack).

Using the standard shorthand notation, the data acquisition system implemented on the custom CAMAC board is an example of an M/M/1/(1-255)/FIFO queue. The first M refers to the exponential distribution of the interarrival times of the cosmic ray events (equivalently, a Poisson distribution in the arrival rate [2]). The second M denotes the approximately exponential distribution of the read out times, 1 specifies the number of parallel servers in the queue, and (1-255) is the user-specified maximum number of spaces (buffers) in the queue. When the number of buffers is set to 3, for example, no more data can be collected as soon as there are three events in the queue. All incoming events are rejected until the event currently being processed has been read out and the queue advances. The queue discipline is traditional first in, first out (FIFO). This queue is a very common model, which has well-known analytic solutions.

Let the event arrival rate and readout rate be Poisson distributed with mean rates $\lambda$ and $\mu$, respectively. Then, the inter-arrival times and processing times follow exponential distributions with means $1/\lambda$ and $1/\mu$. The ratio $\rho = \lambda/\mu$ is known as the utilization factor. Let K be the maximum number of events the queue can contain before further events are rejected, i.e., the total number of buffers. Using this notation, we can determine $p_n$, the probability of n events being in the queue at time t. (For details of the mathematics, see [3].) We do this by first writing down the set of difference equations in both n and t for the steady-state system. Then, taking the limit of small intervals in time, we solve equations differential in t to produce a set of relations of the form:

$$p_n = (1-\rho)\rho^n/(1-\rho^{K+1}) \tag{1}$$

for $\rho \neq 1$ and ($n \leq K$). Note that

$$\sum_{n=0}^{K} p_n = 1 \tag{2}$$

and that the fractional deadtime is just the probability that the system is in the state where all K buffers are full, i.e., the probability that there are K events in the queue. Substituting K for n in the above equation, we see that this is:

$$p_K = (1-\rho)\rho^K/(1-\rho^{K+1}) \tag{3}$$

## EXPERIMENTAL SETUP

### The Equipment

The overall experimental setup is shown schematically in Figure 1. A list of the equipment used for the experiment follows.

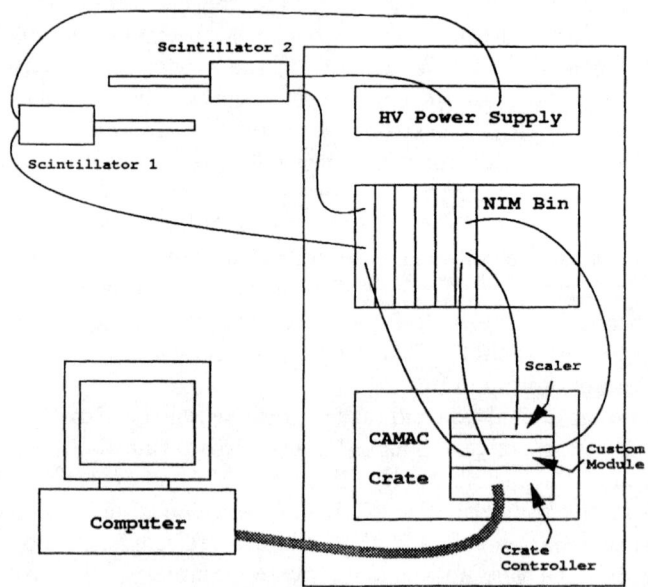

**FIGURE 1.** Experimental Setup

Scintillation Counters:

- Plastic scintillator
- Magnetic shields
- High voltage zener divider
- Photomultiplier tubes
- High voltage power supply

The scintillation counters contain organic plastic compounds doped with materials that scintillate. The passage of a charged particle excites the dopant molecules, which then de-excite by fluorescence. The photomultiplier registers the flourescence photons on a vacuum sealed photocathode optically bonded (with optical epoxy or optical grease) to the plastic scintillator and amplifies the resultant photoelectrons through a series of high voltage dynodes.

NIM Electronics:

- NIM bin with power supply
- NIM cooling fan
- Linear fan in/fan out
- Discriminator
- Majority logic unit
- NIM/TTL level translator
- Pulser
- Delay and interconnect cables

NIM is an international system of logic levels and module standards. Each of the following components can be plugged into a standard NIM bin. A linear fan in/fan out accepts an input signal and reproduces it at multiple output ports. A discriminator sends out a NIM true pulse when the input signal is above a certain voltage threshold (set by using a small screwdriver to adjust a potentiometer through a port on the face of the module). The majority logic unit produces a NIM pulse when a user-selected number of the four possible inputs are NIM true. The user sets the requirement by inserting a pin into one of four small ports marked 1-4 on the front face of the module. TTL is another system of logic levels. The level translator converts between NIM and TTL, both ways. The pulser unit sends out pulses of user defined frequency, amplitude, and width.

CAMAC System:

- CAMAC crate with power supply
- DSP crate controller
- Custom CAMAC module
- CAMAC 12-channel 24-bit blind scaler module

CAMAC is an acronym. It represents an agreed-upon standard for computer-hardware interfaces. The crate is controlled by a crate controller which issues commands on the backplane (the DATAWAY) of the CAMAC crate. Each module plugs into an 86-pin card-edge connector in the crate's backplane (called a station), and communicates with the controller. CAMAC is primarily a command protocol package. It provides a system of commands that can be issued, an addressing scheme, and timing diagrams that must be adhered to, but there is no restriction on what an addressed module can do in response to a command. The specific action performed when a command is received and decoded by a custom module is totally at the discretion of the designer of the module. The custom CAMAC module listed above was designed at Fermilab for this laboratory course. It contains an integration circuit (which produces a voltage proportional to the integrated charge of the input signal), an analog-to-digital converter which converts that voltage into a number (integer), a clock and countdown counter, and scaler displays. The unit is controlled by a Field Programmable Gate Array (FPGA). The FPGA is a very large array of small transistor gates that can be programmed externally to perform a defined set of logical state machine operations. The FPGA on the custom CAMAC board has been programmed to provide the functionality of the M/M/1/(1-255)/FIFO queue described above. Each channel of the blind scaler contains a counter which increments by one every time its input transitions from NIM false to NIM true. The unit can be gated off by an external veto.

Computer:

- Compaq Portable III 386 computer
- DSP Technologies backplane card
- 40-pin interface cable

## Details of Experimental Arrangement

Figure 2 shows details of the NIM electronics used to form the trigger and the data for the experiment. The delay cable shown is approximately 300 ns. Other cables are of appropriate lengths to correctly "time in" the trigger.

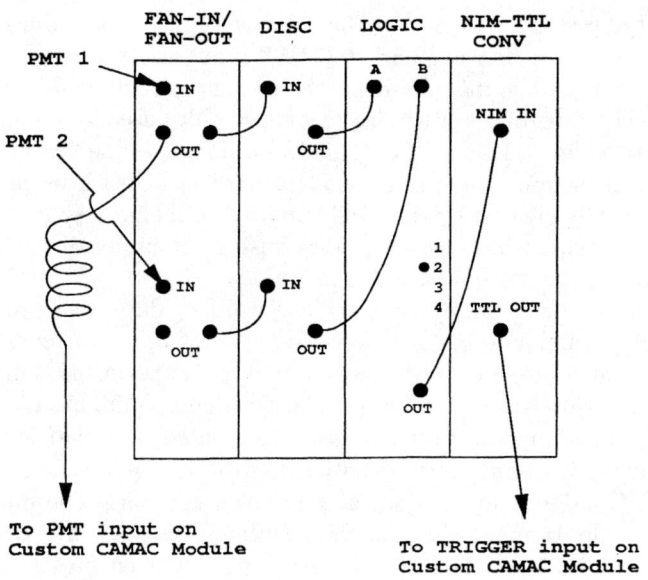

**FIGURE 2.** Detail of NIM electronics and wiring between modules

The events collected in this experiment are cosmic rays. The event trigger is made by forming the coincidence of the signals from two scintillation counters mounted one above the other. A piece of steel of the same size as the counters inserted between the counters ensures that the particles that trigger will be of high enough energy to be minimum ionizing in the upper counter. Anode signals from the two photomultiplier tubes (PMT's) attached to the counters are sent to the NIM fan in/fan out module. Each pulse is then sent through a discriminator. When the pulses are above the set threshold, the discriminators send a NIM true pulse to the coincidence (majority logic) unit. The coincidence unit (with the pin placed at 2 to indicate that both input signals are required to be true in order for the output to be true) performs a logical AND of the two inputs. The output is a NIM true and it is used to form the gate (i.e., event trigger) for the CAMAC board. The integrator on the board is of the switched capacitor design. Since it uses TTL switches, the level translator is needed to switch the trigger from NIM to TTL logic. The TTL signal should be connected to the input port labeled "Trigger" on the custom CAMAC board.

The anode pulse from the PMT of the upper counter is routed through the linear fan in/fan out module to the input labeled "PMT" on the custom CAMAC board. This signal must be delayed by approximately 300 ns. to ensure that it arrives in time with the gate for the integrator. (The switches described above are very slow.) The delay is imposed by adding a long cable between the fan in/fan out and the PMT input on the custom board. The PMT pulse is charge-integrated by the op amp circuit within the CAMAC module. The integrated pulse size (charge = pulse area = height x width) is then digitized by an 12-bit ADC. This 12-bit integer is the "data" entered into the queue (providing there is an available buffer) for each event. The queue itself is actually created by the FPGA in its own memory space. The number of buffers in the queue is set by the user using the upper of two dip switches on the face of the custom CAMAC module.

Since each event contains exactly one word of data, the processing time would not be random if the data word were then simply read out. Instead, the random event processing time expected in a real experiment is modeled using the pulses themselves. The charge collected due to minimum ionization loss is actually Landau rather than Poisson distributed, but that is close enough to give an approximately exponential distribution of read out times. The 12-bit ADC value is loaded into a countdown counter. The on-board clock is used for the count down. As the counter is clocked, the train of pulses that results is output through the output port labeled "AMP" (amplitude) on the front face of the custom CAMAC module. This should be input to channel 0 of the CAMAC blind scaler. Since the train of pulses output by the custom CAMAC module is TTL, use the TTL/NIM converter to convert to NIM before inputting to the blind scaler. When the counter reaches zero, the module issues a CAMAC look-at-me to notify the computer that the data in the scaler is ready to be read out. After the computer reads out the scaler and resets it to 0, it writes a data word to the custom module. The module responds by advancing the FIFO by one and loading the next data word (if at least one is waiting for processing) into the countdown counter. The countdown times can be made longer by using the second (lower) dip switch on the face of the custom CAMAC module. The switch setting effectively multiplies the intrinsic countdown time by a factor equal to the value set.

If, during the processing time for one event, another cosmic ray passes through the scintillation counters, it will generate a trigger and the pulse from the upper counter will be digitized. Then, if there is at least one buffer available in the defined queue, the data for the new event will be placed in an available buffer. If all of the buffers in the queue are full, however, the new event must be ignored, because there is no place to store it. There will only be space for a new event in the queue after the computer has read out the leading event in the FIFO and the queue advances. When the queue advances, the next event in the FIFO is loaded into the countdown timer, and the process repeats.

Figure 3 is a block diagram which shows the main features of the custom CAMAC module. The module is connected to the Dataway in the CAMAC crate through an 86-pin card edge connector through which it receives power and receives and sends CAMAC commands. There are several scaler displays on the front panel of the CAMAC module. They are "Accepted" and "Rejected" event counters, a display of the current event's amplitude and the current number of full buffers in the queue. There is also a "Reset" button and a "Run/Stop" switch, which will enable you to operate the board manually.

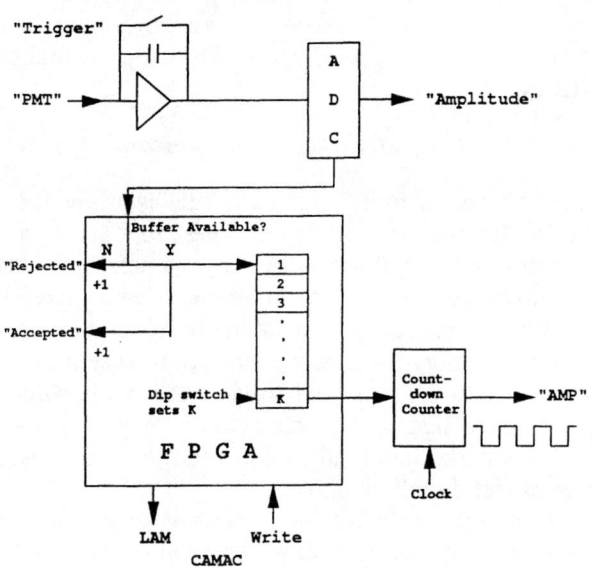

**FIGURE 3.** Block Diagram of Custom CAMAC board

## PROCEDURE

Before beginning to experiment with this setup, be sure to read through this manual, and to familiarize yourself with the hardware. Check to make sure that the NIM electronics is correctly set up to form the trigger. Also, be sure the PMT signal you are digitizing comes from the upper counter so it will be certain to come from a minimum ionizing signal. The relative positions of the two scintillation counters should be checked to make sure you are maximizing the coincidences.

You may have to boot the computer before beginning work; the switch is located on the rear left corner of the machine. The central data acquisition program is located in the c:icfa97 directory. To run the program, simply type "readdaq". Enter a suitable output file name when prompted. At the end of a data taking run, this file will contain a listing of amplitudes for the events that were accepted and written to disk and will also contain some summary information at the end.

The main goal of this experiment is to measure the decrease in deadtime that can be achieved by increasing the number of buffers in which data can be stored while it is waiting to be processed. You can select the depth of the queue to use (in the range 1-255, although you will find that there is not a lot of change once you get past 10 or so). At the end of each experimental "run", defined as the time between a "Reset/Start" and "Stop", record the number of accepted and rejected events. The experimental deadtime can then be calulated as:

$$(Rejected)/(Rejected + Accepted) \qquad (4)$$

Select the number of buffers for each run by using the upper set of dip switches on the face of the module. It is best to start with one buffer and to set the lower dip switches, if necessary, to get a deadtime between 10 and 20% for the one buffer case. Then, perform a series of runs increasing the number of buffers by one each time. Since the number of rejected events decreases rapidly as the number of buffers is increased, be sure to take a long enough run each time to get the statistics you need to keep the errors sufficiently small for a meaningful measurement of each data point. A plot of the deadtime versus the number of buffers should indicate a trend that can be explained in terms of the queue model described above.

A plotting program, called "plot" has been provided that will enable you to plot a histogram of the pulse sizes. These are expected to follow a Landau distribution. You can also use the external NIM pulser unit set to a frequency much greater than the event rate to measure the event processing time distribution. Just put the output of the pulser into the input of the blind scaler instead of the "AMP" output of the custom module. The processing times will be seen to be approximately exponentially distributed. Set the number of buffers to a high enough value that the processor deadtime is negligible for this measurement.

## ADDITIONAL OPTIONS

If there is time remaining, there are some additional aspects of the setup that can be varied. The NIM pulser unit can be used to form the trigger and data for the events instead of the cosmic ray telescope. The pulser represents a deterministic arrival time distribution, since the pulse frequency is constant in time. By varying the size of the pulses (amplitude times width, both also constant) issued by the generator, you can control the digitized pulse size (total charge) which is loaded into the countdown timer. In effect then, the pulse size will control the processing time, which will also be the same for each event. The larger the pulse size, the longer the processing time. Using the pulser in this manner can be a good demonstration of what can happen with a poorly matched set of arrival and service time distributions.

Another option is to lengthen the countdown time and thus the average readout time for events by using the lower set of dip switches. The pile-up of events in the queue will grow larger and the loss of data will become more dramatic as the run progresses. This is an example of a transient effect. Once the queue fills up, data will be lost at a constant rate. Try this with a relatively large number of buffers for several different choices of this number. At the other extreme, if the service time is short and the interarrival time between events is much longer than this, events will never accumulate in the queue. These limiting cases should be observable experimentally.

For given arrival and service time distributions, i.e., a fixed value of $\rho$, increasing the number of available buffers decreases the number of events rejected by reducing the idle time of the processors. This is the meaning of the term term "de- randomizing" buffers. With an insufficient number of buffers, the arrival time of events at the processing facility is random, the processors remain idle after reading out one event while they wait for the next, and the system is inefficient. Adding buffers smooths the random arrival times, with the result that the processors are kept busy nearly full-time, as if they were presented with a stream of events matched to the processing rate.

## ACKNOWLEDGEMENTS

This work was made possible by support from the Consejo Nacional de Ciencia y Tecnologia, Mexico, and the U. S. Department of Energy and National Science Foundation. The authors also wish to thank Rafael Gomez and Timothy Meyer for their contributions to hardware and software for this project as well as to parts of this manual.

## REFERENCES

1. D. Gross and C. M. Harris, *Fundamentals of Queueing Theory*, John Wiley and Sons, New York, 1979, Chapters 1-2.
2. Ibid, pp. 23-29, discusses the equivalence between a Poisson distribution of arrival rate and an exponential distribution of interarrival times.
3. Ibid, pp. 64-71, discusses the steady state solution for a M/M/1/K/FIFO queue.

# TESTS OF A POSITION SENSITIVE PHOTOMULTIPLIER AND MEASUREMENT OF DIFFRACTION PATTERN BY COUNTING SINGLE PHOTONS

S.Korpar[1,2], P.Križan[1,3], A.Gorišek[1] and A.Stanovnik[1,4]

[1]Jožef Stefan Institute, Ljubljana, Slovenia
[2]Faculty of Chemistry and Chemical Engineering, University of Maribor, Slovenia
[3]Faculty of Mathematics and Physics, University of Ljubljana, Slovenia
[4]Faculty of Electrical Engineering, University of Ljubljana, Slovenia

## ABSTRACT

The present paper describes a laboratory course held at the ICFA'99 Instrumentation School in Istanbul, Turkey. This course intends to introduce position sensitive multianode photomultiplier tubes (Hamamatsu R5900 type M16 and L16). Using a light emitting diode, the student will measure position sensitivity by scanning the light spot across the M16 PMT. The second part of the exercise consists of measuring a diffraction pattern produced by light passing through a slit. For this purpose the L16 PMT is used and the students attention is drawn to the pedagogical problem of wave-particle duality.

## I INTRODUCTION

Photomultiplier tubes (PMT's), or photomultipliers (PM's) for short, are sensitive detectors of weak light signals capable of detecting even single photons [1,2]. The photomultiplier consists of an evacuated glass vessel containing a photocathode, from which incident photons may eject an electron, and a system of electrodes (dynodes) in which this photoelectron is multiplied to give a measurable electrical signal at the anode. The photocathode, the dynodes and the anode have leads through the glass to the outside of the vessel, enabling connections of high voltage and allowing the signals to be further processed by suitable electronics. The photomultiplier is thus plugged into a photomultiplier base, which consists of a resistor chain providing appropriate voltages for the dynodes and a load resistor, on which the signal appears. In some cases, potentiometers are provided for adjusting the voltage on the electrodes for focusing the photoelectrons to the first dynode and capacitors or Zener diodes to stabilize the voltage on the last dynodes in case of high rate and high gain operation (Fig. 1).

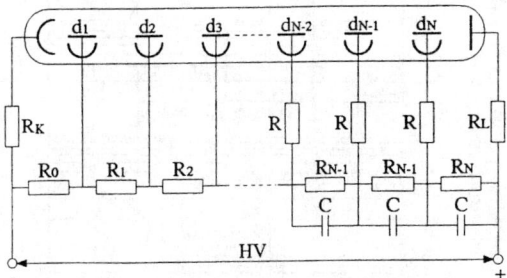

**FIGURE 1.** Voltage divider.

An important parameter of a photomultiplier is the quantum efficiency (QE), defined as the ratio of the number of photoelectrons ejected from the photocathode to the number of photons incident on the photomultiplier. Clearly, this parameter is a function of the energy (or wavelength) of the incident photons and is a product of the probability for the photoelectric effect and the probability for the electron to escape from the photocathode. The most common photocathode materials are semiconductors containing alkali elements. The quantum efficiency $QE(\lambda)$ is connected to the photocathode radiant sensitivity $S(\lambda)$, which is defined as the photocathode current divided by the incident photon power:

$$QE(\lambda) = S(\lambda)\frac{hc}{e_0\lambda}$$

The quantum efficiency is cut off on the low energy (high $\lambda$) side by the vanishing probability for the photoelectron to escape into the vacuum and on the high energy (low $\lambda$) side by photon absorption in the PM glass window.

The photoelectrons ejected from the photocathode are focused to the first dynode, where they eject more electrons. The electron multiplication is given by the secondary emission factor, which depends on the incident electron energy as well as on the dynode material. Usually, there are several dynodes (10 to 12) leading to an overall amplification of about $10^6$ to $10^7$.

In experimental physics, photomultipliers are most often used as detectors of scintillations, which charged particles, neutrons or gamma rays produce when depositing some or all of their energy in special scintillating materials. More recently, since much progress has been achieved, PM's may be used as position sensitive detectors of single photons, especially for the Ring Imaging Cherenkov (RICH) counters in high energy physics experiments [3,4].

The present laboratory course will introduce two such photomultiplier tubes produced by Hamamatsu Photonics K.K.; the R5900-M16 and the R5900-L16 multianode photomultipliers. We will also take advantage of these instruments to gain

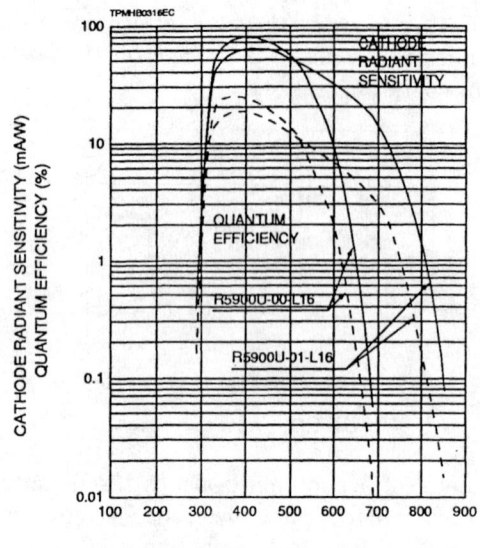

**FIGURE 3.** Typical spectral response of L16 PMT [5].

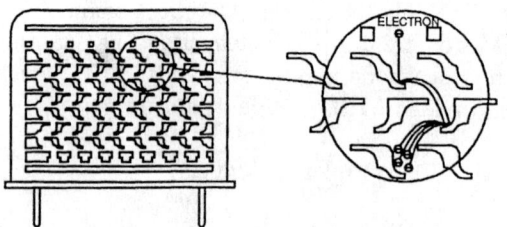

**FIGURE 4.** Metal channel type PMT [6].

For the M16 photomultipliers, measurements have been made of single photo-electron pulse height distributions showing a well resolved single electron peak corresponding to a plateau on the rate-versus-voltage curve [4]. Measurements at high rates (3 MHz/channel during 30 days) show that these photomultipliers operate smoothly even in otherwise hostile environments as are characteristic of the new high energy colliders. According to specifications [5], the pulse rise time is 0.8 ns with a transit time spread of 0.3 ns, so they could also be used for timing purposes.

## III  EXPERIMENTAL SET-UP

The exercise is divided into two parts. The first consists of measuring the high voltage plateau and the position dependence of the M16 count rate for a pencil

some experience with the basic quantum mechanical phenomenon referred to in some text-books as the wave-particle duality.

# II  POSITION SENSITIVE PHOTOMULTIPLIERS

The R5900 series M16 and L16 multianode photomultipliers produced by Hamamatsu Photonics K.K. are shown in Fig. 2. The M16 is divided into 4 x 4 = 16 anode outputs, each covering a pad size of 4 x 4 mm$^2$ with 0.5 mm gaps between pads. The L16 anode, on the other hand, is divided into 16 strips (0.8 x 16 mm$^2$) at 1 mm pitch. The exact dimensions of the photomultipliers and the locations of the electrode pin connectors are given in the data sheets [5].

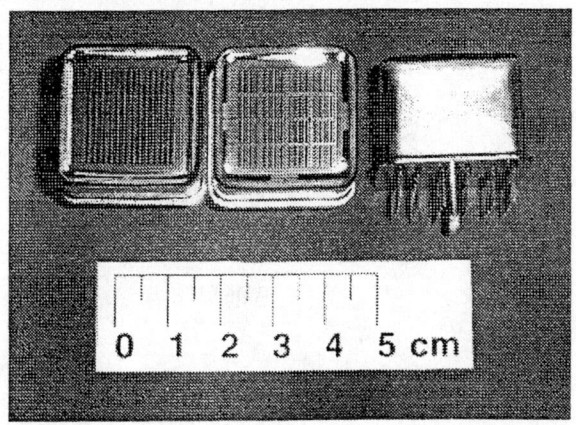

**FIGURE 2.** Hamamatsu multianode photomultipliers (L16, M16, M16 from left to right).

The quantum efficiency and the radiant sensitivity given by the manufacturer for L16 photomultipliers are shown in Fig. 3. It seems that allowance has to be made for an additional efficiency factor due to less than perfect collection and transmission ($\sim$80%) of the photoelectrons by the dynode system [4].

The dynode system in these multianode photomultipliers differs considerably from those in conventional photomultipliers. It consists of foils with specially shaped perforations or channels. On the walls of these channels, secondary emission takes place thus multiplying the number of electrons (Fig. 4). With 10-12 such dynode foils, gains above $10^6$ are reached. The anode dark current is mainly below 200 nA [5]. Attention must be paid not to exceed the maximum allowed voltage of 900 V for L16 and 1000 V for M16 PMT and the maximum allowed current of 0.01 mA [5].

Of special interest e.g. in Cherenkov ring imaging is the position resolution, which is mainly given by the anode pad size. The cross-talk to adjacent channels is small and the response across the photocathode surface seems to be uniform to the level of some 10% [4,5].

beam. The second part of this exercise represents a measurement of a diffraction pattern by counting single photons with the L16 position-sensitive photomultiplier.

## A    M16 - HV plateau and position dependence of the count rate

The experimental set-up for measuring the M16 photomultiplier is shown in Fig. 5. Light from the source (blue LED, $\lambda$=470 nm) is collimated by two pinholes, defining an illuminated spot of about 0.5 mm diameter on the photocathode. The

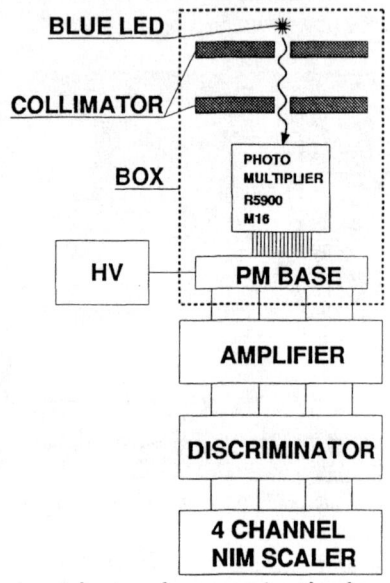

**FIGURE 5.** The experimental set-up for measuring the characteristics of M16 PMT.

photomultiplier is plugged into a PM base and both are enclosed in a light-tight box together with the light source and collimators. High voltage is provided by a HV power supply from which a cable leads to the PM base inside the light-tight box. Cables from four anode pads connect each signal first to an amplifier, then discriminator and finally to a scaler. The plate on which the PMT is fastened, may be displaced in a direction transverse to the light beam by means of a screw thread (1 mm/turn), which could be operated from the outside of the box. The height of the beam is set in order to be centered on one of the four rows with four pads. After observing the set-up the box is closed and the count rate at given threshold is recorded as a function of high voltage (see Fig.6). The voltage is then set on the plateau and count rates of the four pads are measured as a function of the PM position relative to the light spot (Fig.7). From the results of this measurement one may study the position resolution, the cross talk between adjacent pads, the uniformity of pad response and the response variation across a given pad, which reflects the structure of the dynode channels, as also seen in Fig. 2.

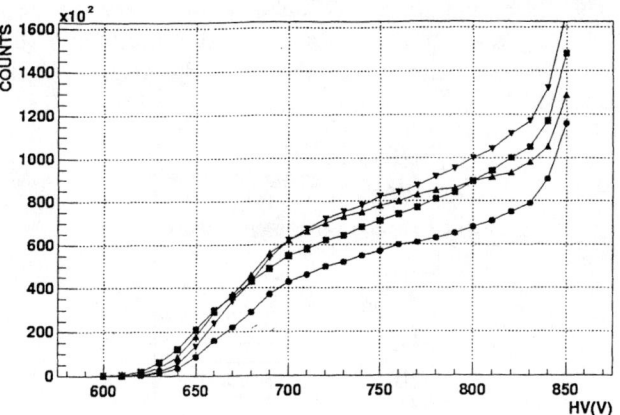

**FIGURE 6.** Plateau curves for 4 channels of the M16 PMT.

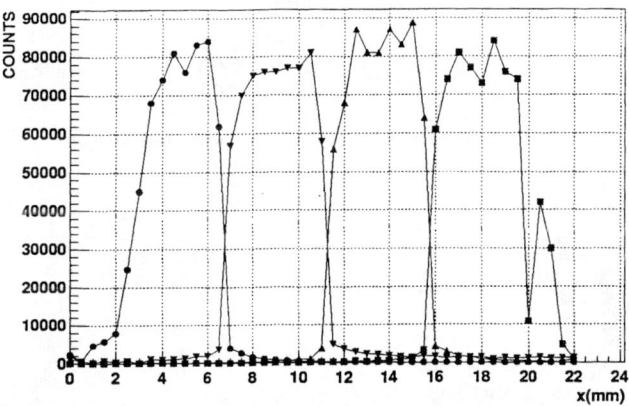

**FIGURE 7.** Count rate on 4 channels of the M16 PMT depending on the light spot position.

# B  L16 - Diffraction pattern

The schematic diagram of this experimental set-up is shown in Fig. 8. The light source is again a blue light emitting diode. This light is passed through a slit of width D, on which diffraction occurs. The diffraction pattern is given by

$$j(\vartheta) = j_0 \frac{sin^2\alpha}{\alpha^2}$$

where $\alpha = \frac{\pi D \sin\vartheta}{\lambda}$ and $\vartheta$ is the diffraction angle with respect to the beam direction. In terms of the distance x from the central maximum and the distance L between the slit and the photomultiplier, this angle is given by $tg\vartheta = x/L$ (see Fig. 9). The first minimum in the diffraction pattern occurs at $sin\vartheta_{min} = \lambda/D$. Assuming that the diffraction angle $\vartheta_{min}$ is small, the x-position of the minimum will be given by $x_{min}/L = \lambda/D$. In the present exercise one measures the position of the minimum and thus determines the slit width $D = \lambda \cdot L/x_{min}$ ($\lambda = 470$ nm, $L \simeq 0.3$ m).

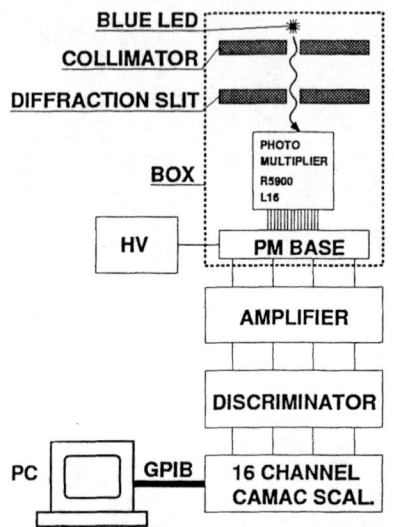

**FIGURE 8.** The experimental set-up for measuring diffraction with the L16 PMT.

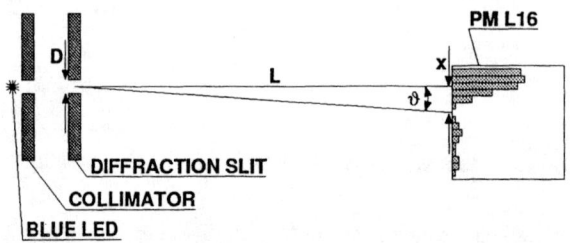

**FIGURE 9.** Geometric parameters for the diffraction measurement.

From the 16 anode strips, the signals are led through amplifiers and discriminators into a 16 channel CAMAC scaler. This scaler is connected via CAMAC and GPIB to a personal computer, which runs a data acquisition programme and displays the diffraction histogram. (Due to a delay in CAMAC crate delivery, the individual channel rates had to be read out by NIM scalers during the ICFA'99 School.) With the 16 channels at 1 mm pitch only a 16 mm portion of the diffraction pattern could be measured simultaneously. In order to cover a broader range of diffraction angles, the photomultiplier may be displaced relative to the light beam by means of a screw thread (1mm/turn) operated from the outside of the light-tight box.

A diffraction pattern is first demonstrated by using a light beam from a laser pointer and slits made from razor blades. The slits are then inserted onto the rails in front of the light emitting diode, the distance $L$ is measured and the box is closed. The high voltage on the PMT is set to approximately 800 V and the current through the LED is adjusted for an acceptable count rate. The diffraction pattern is then measured in at least two different positions of the PMT relative to the light beam and the results are appropriately connected. From the distribution (Fig.10), one determines the position of the first minimum and then calculates the slit width D from the above equation. At this point the student may be reminded of the analogy between this experiment and the measurement of nuclear sizes by the so called diffraction scattering.

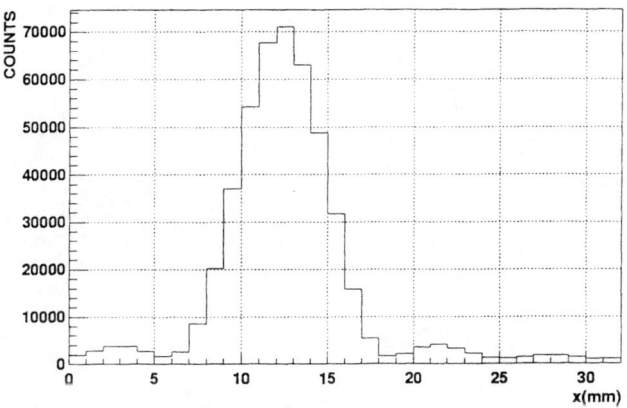

**FIGURE 10.** Measured diffraction distribution.

In this exercise, the pedagogical problem of wave-particle duality is stressed. With sufficiently low counting rate one may in principle simultaneously observe the count increment of individual channels and the appearance of the diffraction histogram (Fig. 10). The individual hit is a manifestation of the particle nature of the photon, while the diffraction distribution speaks of its wave properties.

# ACKNOWLEDGMENT

We are grateful to Hamamatsu Photonics K. K. for donating the four multianode photomultipliers used in the present laboratory course.

# REFERENCES

1. G.F.Knoll, Radiation Detection and Measurement, John Wiley, 1989
2. W.R.Leo, Techniques for Nuclear and Particle Physics Experiments, Springer-Verlag, 1987
3. R. Debbe et al., In-beam tests of a RING Imaging Cherenkov detector with a multianode photomultiplier read-out, Nucl. Inst. and Meth. in Phys. Res. **A362**(1995)253-260
4. P.Križan et al., Tests of a Multianode PMT for the HERA-B RICH, Nucl. Inst. and Meth. in Phys. Res. **A394**(1997)27-34
5. Hamamatsu Photonics K.K., Data Sheet of R5900-L16 and Data Sheet of R5900-M16
6. http://www.hpk.co.jp/hp2e/products/Etd/PDFfiles/PMThd6E.pdf

# Laboratory Course On Silicon Sensors

Richard Brenner*, Shaun Roe¶, Alan Rudge¶

*Uppsala University, Box 535, S-75121 Uppsala, Sweden

¶CERN, CH-1211, Geneva, Switzerland

**Abstract.** The laboratory course consisted of five different mini sessions on various aspects of silicon sensors and related integrated electronics. The course gave a practical approach to the subject in order to give the student some hands-on experience with silicon sensors. Although there was only a short time available for the course, we thought it was more inspiring to have several different experiments rather than one single session going into details in only one subject. The five mini sessions were:

1. Characterisation of silicon diodes for particle detection

2. Study of noise performance of the Viking readout circuit

3. Study of the position resolution of a silicon microstrip sensor

4. Study of charge transport in silicon with a fast amplifier

5. Analysing pulse height spectra from silicon detectors

The data in the following was obtained during the ICFA school by the students.

# 1. CHARACTERISATION OF SILICON DIODES FOR PARTICLE DETECTION

## Introduction

Near intrinsic n-type silicon with a metallised p-doped region, is the most frequently used semiconductor structure for detecting charged tracks in high-energy physics experiments. A polarisation voltage is applied across the diode structure, which depletes the silicon from charge carriers. Charged particles or photons interacting with the silicon will create electron-hole pairs that drift along the electric field lines to the contacts located on the silicon surface.

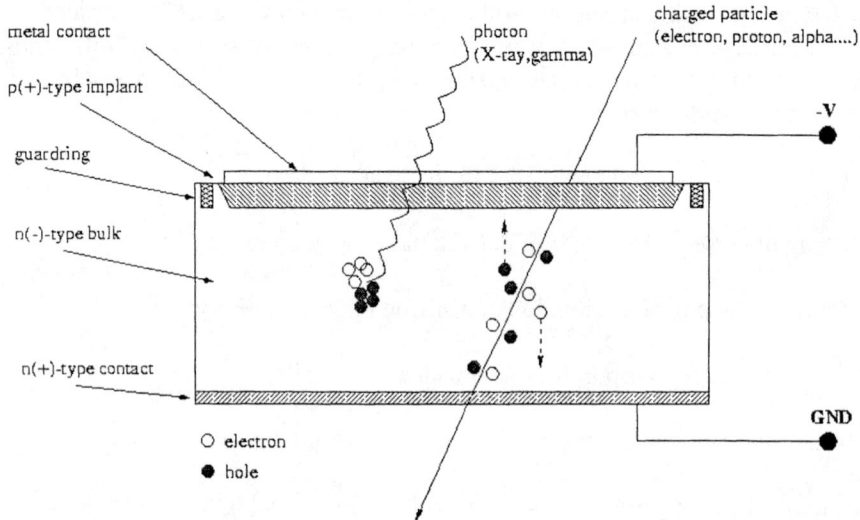

**FIGURE 1.** A schematic picture of a silicon sensor diode

A first step in constructing a particle detector based upon a silicon sensor is to characterise the sensor without readout electronics attached. The static characteristics of a sensor are usually adequate to determine if the sensor can be used for particle detection. The leakage current behaviour as a function of voltage and the voltage needed to fully deplete the sensor are two important parameters. The voltage needed to deplete the sensor can be determined by measuring the capacitance between the diode implant and the backplane of the sensor. In the final particle detector system, both the capacitance and the leakage current will influence the performance of the readout electronics.

The capacitance and leakage current depend on the geometry of the sensor and the quality of the material and manufacturing process. In a well controlled and uniform process sensors with the same geometrical layout, processed on the same substrate, should have the same behaviour. In reality, there may be variation both in the process and in material and therefore there may be sensors which differ largely from what we naïvely would expect. When constructing an experiment consisting of many sensors we have to measure them in the laboratory to find the good sensors that can be built into the experiment.

In this laboratory session the aim is to characterise silicon diodes of different sizes. This session requires some knowledge of the basic principles of diodes and will give experience handling unprotected diodes, operating the microscope and probe manipulators. Because of the short time available we restrict ourselves to simple DC-coupled diodes.

In this experiment we use following equipment:

- probe station
- digital multimeter
- capacitance meter
- peltier element (to show the temperature dependence of leakage current)
- power supply

## Measurement of IV-properties of silicon diodes at room temperature

The main sources of leakage current in silicon sensors are:

1. Diffusion of charge carriers from undepleted regions of the detector to the depleted region

2. Thermal generation of electron-hole pairs in the depleted region

3. Surface currents depending on contamination, surface defects from processing and edge effect from dicing, etc.

Contribution (1) is generally well controlled and small giving a few nA/cm$^2$. The contribution from (2) depends largely on the purity of the material since recombination centres and trapping centres increase the creation of electron-hole pairs. The magnitude is higher than for (1) giving a few µA/cm$^2$. The leakage current originating from thermal generation is of course temperature dependent. By lowering the temperature of the sensor we may reduce the contribution. By decreasing the temperature by 10 °C the leakage current will typically be reduced to a third. In some cases the contribution from (3) may be the dominant source of leakage current. The surface current may be caused by effects on the non-depleted edge region or by a bad processing environment. The leakage current originating from the surface may vary extensively from sensor to sensor. To reduce the effects from surface current a guard ring structure is processed on the silicon. The guard ring can be anything from a single implant around the diode to a complex structure of alternating implants and floating metal rings around the silicon diode.

We will now study the IV-characteristics of two silicon diode sensors with 300 µm thickness but of different sizes listed in table 1. The connection of the setup is shown in the figure 2.

**TABLE I.** Size of the silicon diodes measured

| Diode | Area | Size |
|-------|------|------|
| A | 5 mm x 4 mm | 0.2 cm$^2$ |
| B | 0.24 mm x 0.24 mm | 0.0006 cm$^2$ |

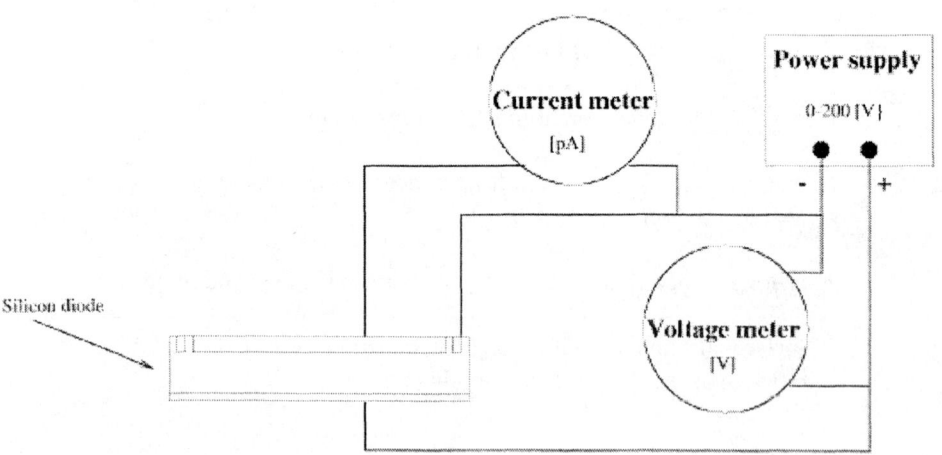

**FIGURE 2.** The setup for IV measurement

## Measurements

Execute the steps below.

-Place the silicon diode on the vacuum chuck with conductive rubber under it

-Connect the diode with the probe needle to the negative pole of the battery pack

-Connect the guard ring of the diode to the negative pole of the battery bypassing the current meter

-Connect the chuck (and thus, the silicon detector backplane) to the positive pole of the battery.

-Cover the probe station to prevent light from generating current in the diodes, note down voltage and current at 0V

-Ramp up the voltage in 10 V steps and note down the voltage and the corresponding current in table 2.

**TABLE 2.** IV-measurements

| Voltage [V] | Current for A [pA] | Current for B [pA] |
|---|---|---|
| 0 | -12 | -8 |
| 10 | 27 | 2 |
| 20 | 56 | 4 |
| 30 | 65 | 6 |
| 40 | 86 | 8 |
| 50 | 114 | 10 |
| 60 | 132 | 14 |
| 70 | 215 | 18 |
| 80 | 360 | 22 |

## Observations

At low voltages a small negative current is observed because the diode is forward biased. If bulk current dominated the leakage current of the silicon diode the measurement should show dependence on the depleted volume. In an ideal case we should expect more than 300 times higher leakage current for the fully depleted silicon diode of type A than for type B. The temperature and humidity conditions during the ICFA school were however not optimal for these measurements. The temperature was over 30 °C and the humidity was close to saturation throughout the school. The currents were clearly dominated by surface effects. Silicon diode A shows a strong increase in leakage current after 40V. This may due to the weather conditions or indication of breakdown in the structure.

## Measurement of CV-properties of silicon diodes

At low reverse bias voltage the capacitance will fall such that $1/C^2$ is proportional to V. When the sensor has reached full depletion the capacitance will not change anymore. This is clearly visible when plotting $1/C^2$ vs. V. The point where the curve shows a kink and does not reduce further gives the point for full depletion.

For large single diodes the capacitance to the back plane is the dominant contribution to the total capacitance. For small and segmented sensors such as pixel and microstrip sensors the inter pixel/strip capacitance will dominate over the back plane capacitance when the sensor is fully depleted. The inter pixel/strip capacitance do not change in the same way as the backplane capacitance when applying reverse bias to the sensor. The capacitance of the silicon diode is measured between the backplane and the p-implant. Figure 3 shows the setup for capacitance measurement.

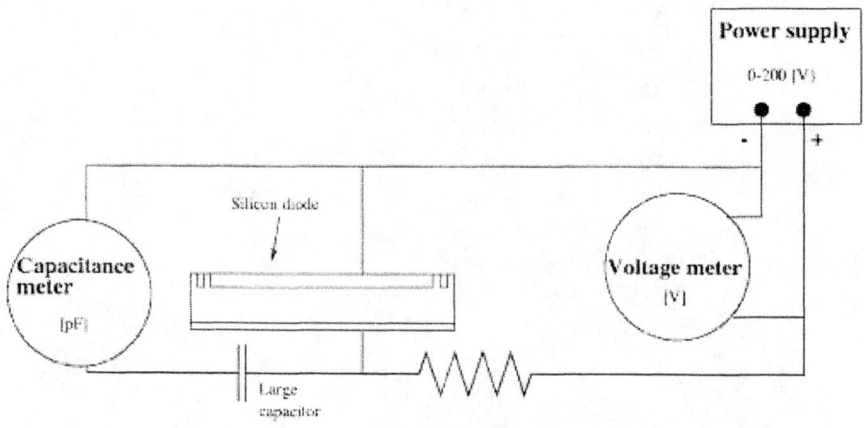

**FIGURE 3.** The setup for capacitance measurement

## Measurements

Execute the steps below:

-Find the capacitance of the setup by leaving the circuit open

-Follow the procedure outlined for IV-measurements, except now the capacitance meter is connected in parallel with the diodes instead of the current meter in series, and note down the measurements in table 3.

**TABLE 3.** CV- measurements

*Capacitance of the setup :    38              pF*
*Measured capacitance of diodes*

| Voltage [V] | Measured A [pF] | Corrected A [pF] | Measured B [pF] | Corrected B [pF] |
|---|---|---|---|---|
| 0 | 89 | 51 | 44 | 5 |
| 10 | 66 | 28 | 42 | 4 |
| 20 | 56 | 18 | 41 | 3 |
| 30 | 54 | 16 | 40 | 2 |
| 40 | 51 | 13 | 40 | 2 |
| 50 | 49 | 11 | 39 | 1 |
| 60 | 48 | 10 | 40 | 2 |
| 70 | 48 | 10 | 39 | 1 |

Corrected capacitance = Measured capacitance - capacitance of the setup

## Observations

The capacitance decreases with higher depletion voltage. The larger silicon diode A shows expected behavior of capacitance against depletion voltage. Figure 4 shows the graph of the corrected measurement points in table 3 for diode A. The silicon diode shows full depletion at 60V. The small diode, B, is too small to be dominated by capacitance to the backplane.

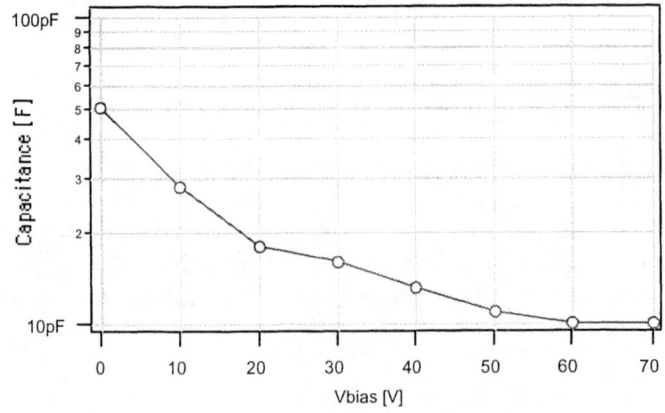

**FIGURE 4.** The logarithm of the corrected capacitance in pF for diode A plotted against depletion voltage in V.

## 2. STUDY OF THE VIKING ARCHITECTURE AND NOISE PERFORMANCE

### The Viking architecture

A number of read out chip methodologies exist. The principle ones are, - the MX series, very successfully used in the DELPHI vertex detector using the double correlated sampling principle. Figure 1 shows the MX principle of douple correlated sampling.

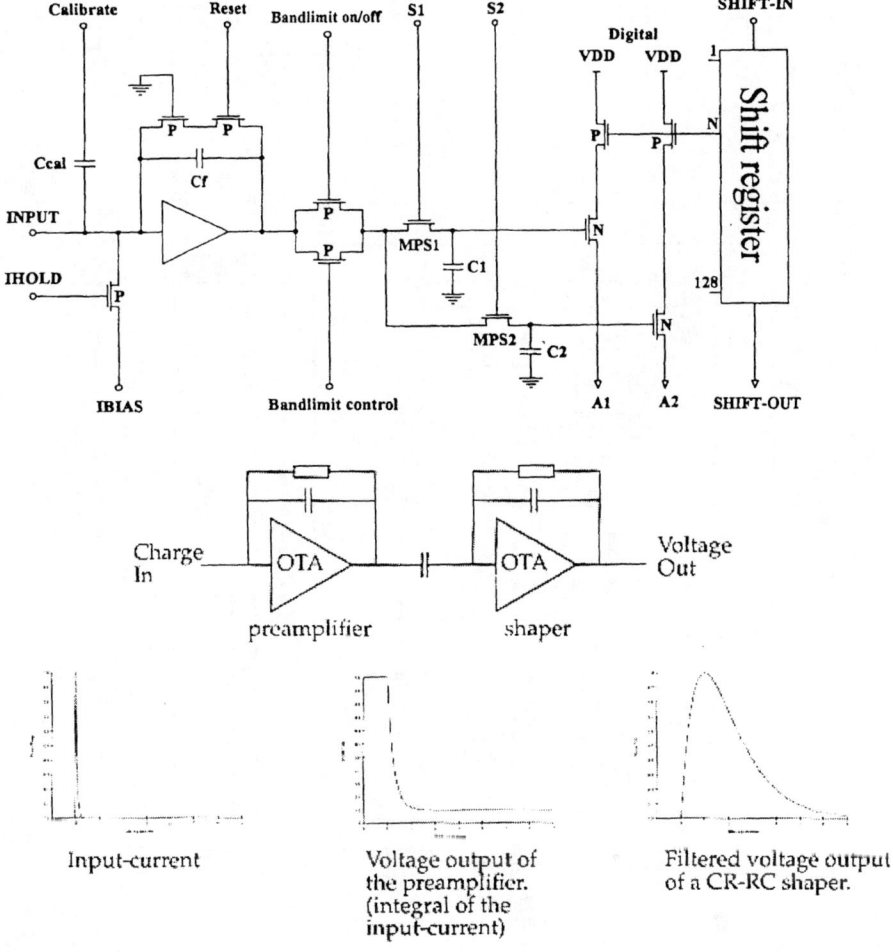

**FIGURE 1.** The MX principle

And the Amplex / Viking "time continuous" shaping is shown in figure 2.

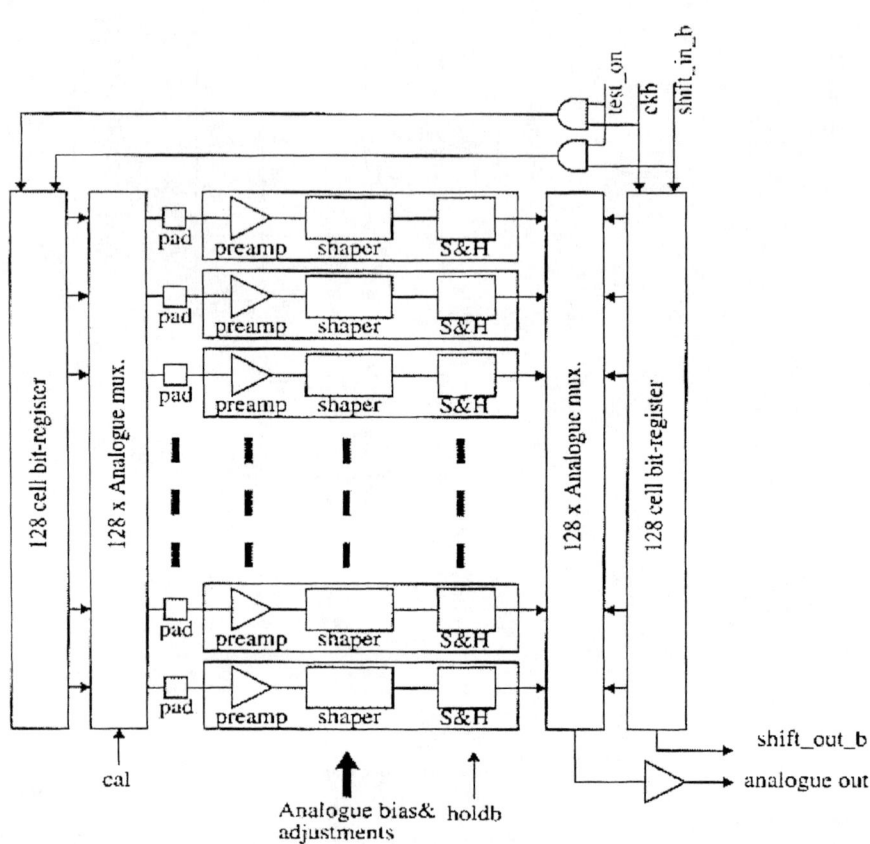

**FIGURE 2.** The AMPLEX/Viking architecture

A voltage corresponding to the input charge is stored on the sample and hold capacitor and read out sequentially via the output multiplexer. It is important to understand the shift in / shift out concept to "daisy chain" many (up to 20) chips. Figure 3 shows a timing diagram for the readout sequence of a Viking chip. Corresponding screen shots from a digital oscilloscope is shown in figure 4 to 6.

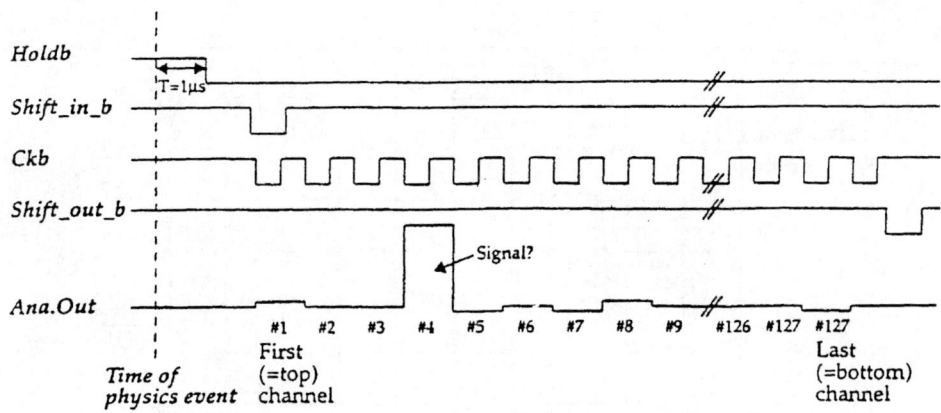

**FIGURE 3.** The timing diagram for the readout sequence of a Viking chip

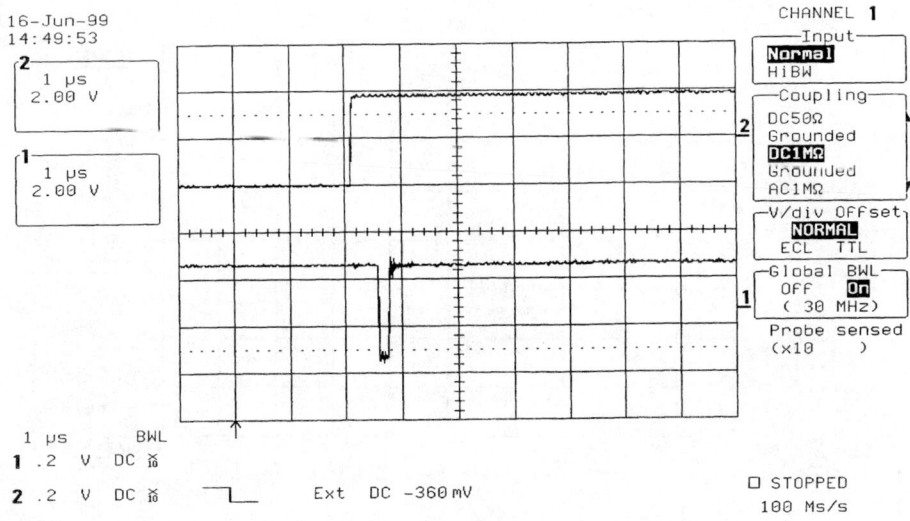

**FIGURE 4.** Screen shot showing hold and shift in signals

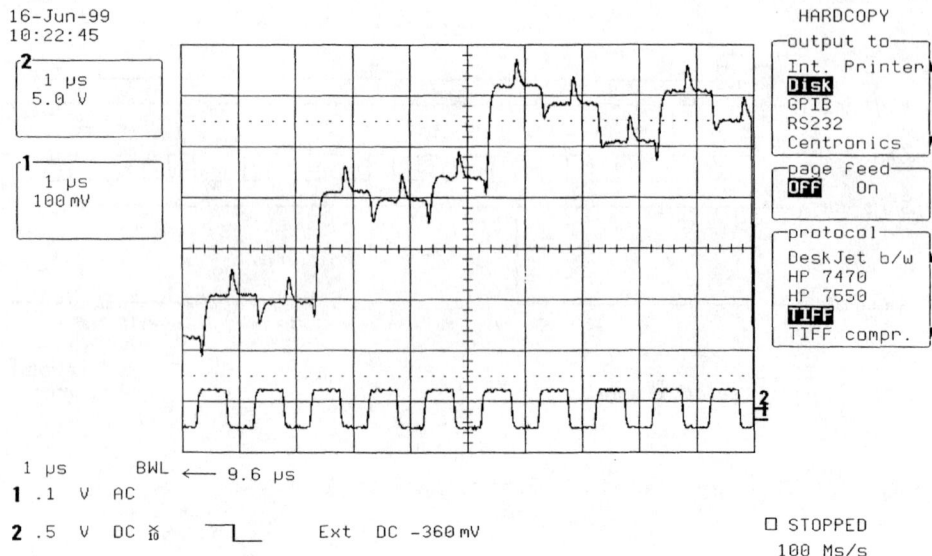

**FIGURE 5.** Screen shot showing clocking and analogue output pulse

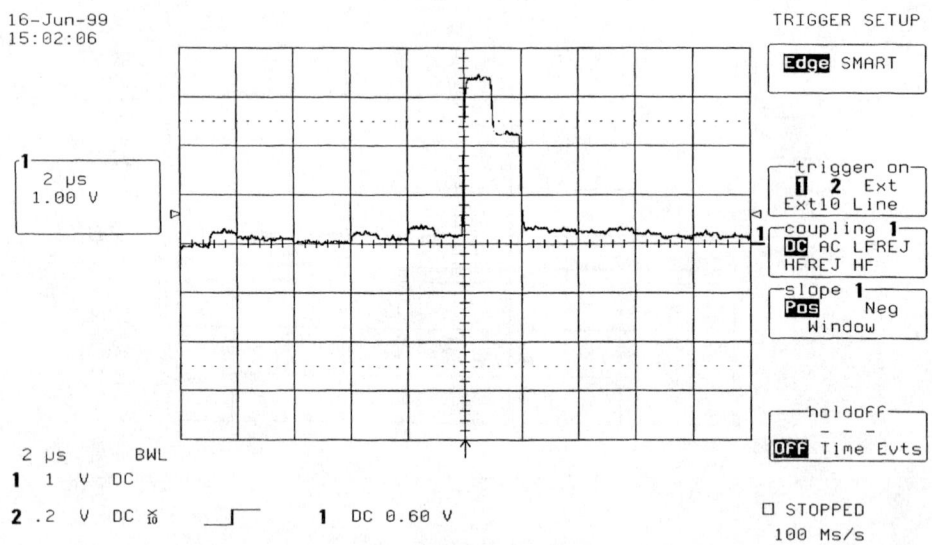

**FIGURE 6.** Screenshot showing output waveform showing hits in two adjacent channels

An important feature of the Viking chip is single channel operation. In figure 7 we see a single channel output from a Americium source.

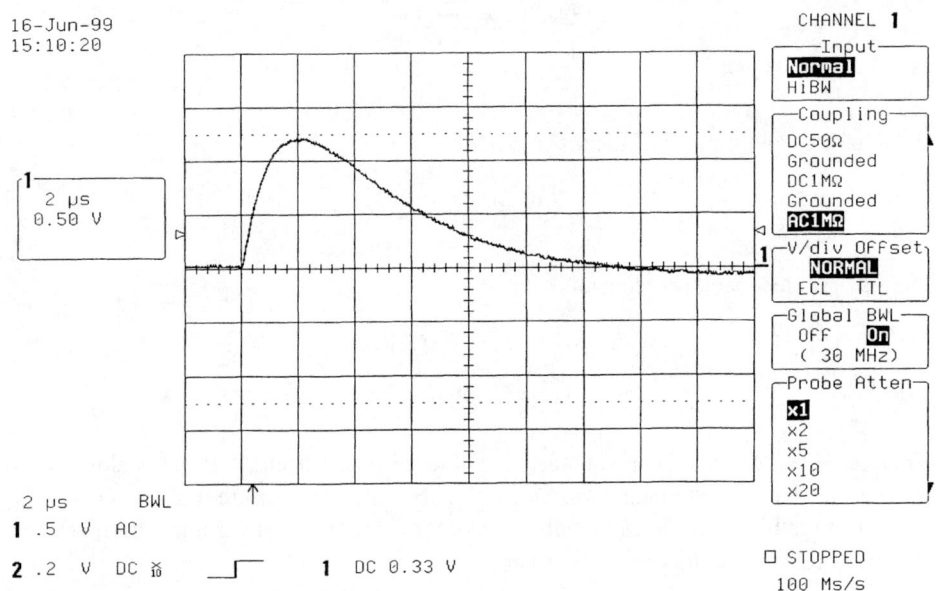

**FIGURE 7.** Output from a single channel

## Study of noise performance of the Viking readout circuit

An essential part in operating silicon sensors is a low noise amplifier circuit. The energy needed to create an electron-hole pair at room temperature in silicon is around 3.6 eV. A minimum ionising particle traversing 300 µm of silicon will on average create 25000 electron-hole pairs. For low energy X-ray applications the requirements for low noise is even more demanding. A 10 keV X-ray will only produce 2800 electron-hole pairs in silicon.

Without going into too many details on electronics design, I want to mention the main noise contributions from silicon sensors degrading the performance of the system. The main sources of noise for silicon detector assemblies are:

The input FET channel noise.

$$ENC_{FET} = \frac{Ce}{q}\sqrt{\frac{4kT}{3g_m \tau}}$$

The detector (and input FET) leakage noise

$$ENC_I = \frac{e}{q}\sqrt{\frac{qI_l \tau}{4}}$$

The bias and feedback resistor noise

$$ENC_R = \frac{e}{q}\sqrt{\frac{\tau kT}{2R}}$$

Where $C$ is the total load capacitance, $I_l$ is the leakage current, $\tau$ is the peaking time of the amplifier, $k$ Boltzmann constant, $T$ the absolute temperature in Kelvin and $R$ is parallel of the bias resistor and feedback resistor. The total noise contribution is the squared sum of the components listed above

$$ENC_{total} = \sqrt{(ENC_{FET}^2 + ENC_I^2 + ENC_R^2)}$$

For the Viking chip the noise is typically 70 e⁻ + 12 e⁻/pF. It is obvious that we want to keep the leakage small by choosing a sensor which has low leakage current at a voltage that fully depletes the sensor. We will study the contribution to the noise from the detector capacitance in this laboratory.

## *Measuring noise versus load capacitance*

In order to measure the noise the setup has to be calibrated. This is done by applying a known charge pulse $Q_{cal}$ on the input of the Viking chip and measuring the output from the chip $V_{outl}$. The input charge is generated by applying a known voltage pulse $V_{cal}$ over a known calibration capacitor $C_{cal}$.

$$Q_{cal} = \frac{C_{cal} V_{cal}}{e}$$

where $e$ is here the magnitude of the electron charge ($e$ = 1.6 x $10^{-19}$ C). The calibration capacitance $C_{cal}$ = 1.8 pF in our case

The RMS of the noise $V_{noise}$ from the Viking chip without test pulse in mV can be calculated using a modern digital oscilloscope. Since the calibration is known the measured value in mV can be converted to ENC by the relation

$$ENC = \frac{V_{noise}}{V_{out}} Q_{cal} \quad RMSe$$

Six channels have been pre-wired with different load capacitances to the input pads of the Viking chip.
1. Select a channel not connected to a capacitor in order to get the baseline noise of the amplifier, note down the RMS noise value.
2. Switch on the test pule, note down the amplitude and rise time of the test pulse
3. Choose a channel with load capacitor. Switch off the test pulse and note down the RMS noise value.
4. Switch on the test pulse, note down the amplitude og the output signal
5. Repeat 3 and 4 for a few different capacitances.

| Capacitance pF | Amplitude V | Noise(RMSe-) |
|---|---|---|
| 1 | 2 | 53 |
| 4 | 2 | 69 |
| 14 | 1.95 | 145 |
| 28 | 1.9 | 232 |
| 37 | 1.8 | 297 |
| 51 | 1.7 | 382 |

Giving the following Graph.

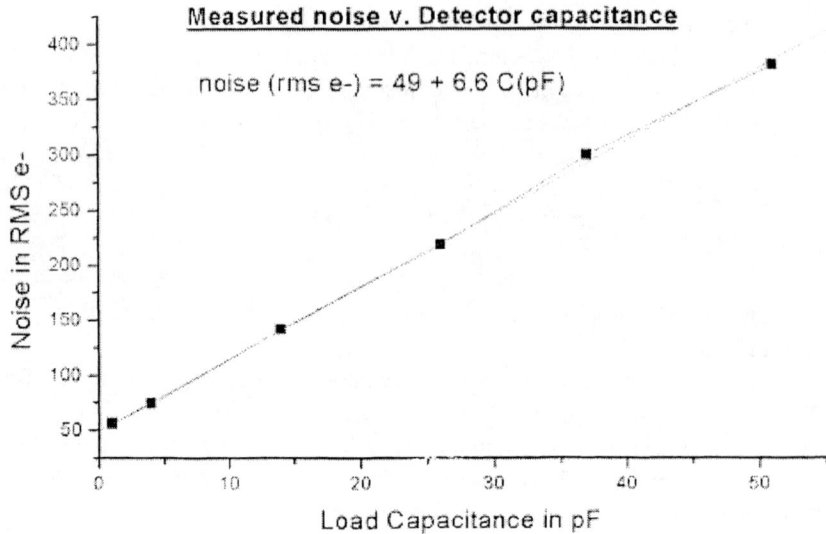

## 3. STUDY IF OF THE POSITION RESOLUTION OF A SILICON MICROSTRIP SENSOR

### Introduction

Silicon microstrip sensors are the most commonly used device for high resolution tracking in particle physics. The strip design allows a large sensitive area with relatively few readout channels. The basic strip detector is read out on one side giving information of the track position only in one dimension. Various solutions to measure track position in two dimensions exist. The simplest solution is to glue two single sided sensors back-to-back but a more demanding design is to process strips on both sides of the sensor. In this experiment we will study a single sided sensor which is illuminated by a pulsed laser source on the strip side (because the back of the sensor is fully aluminised).

# The laboratory setup

In this experiment we use the following equipment:

- Timing unit for readout circuit, VIKING TIMING
- Pulse generator for trigger and laser, PULSE GEN
- Current limited supply for readout circuit
- Laser diode driver, LASER DRIVER
- NIM crate
- Oscilloscope
- Battery pack for biasing the sensor

The silicon microstrip sensor is wire bonded to the Viking readout circuit, which has been placed on a readout PCB (Printed Circuit Board). The assembly has been mounted on a slide which can be precisely moved such that the translation direction is orthogonal to the strips. An optical fibre has been mounted about 100 μm away from the silicon surface. Figure 1 shows a schematic drawing of the assembly.

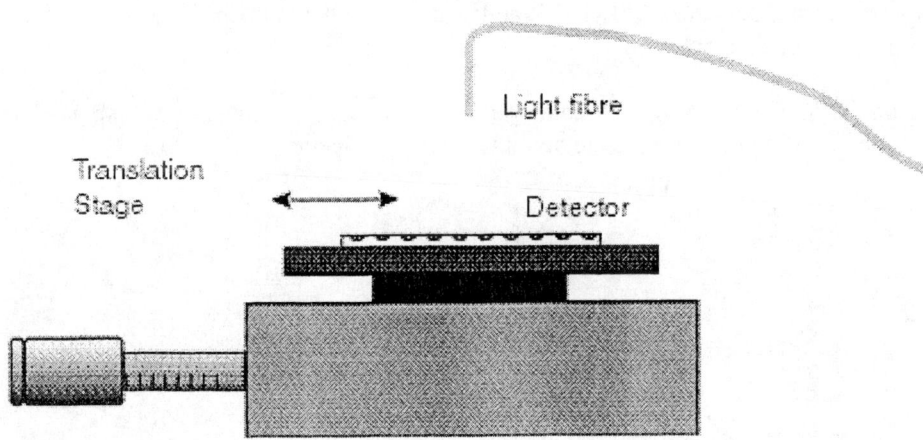

**FIGURE 1.** A schematic drawing of the detector assembly

# Measurement of position resolution

We will now try to determine the position resolution of our silicon microstrip sensor. The peak of the light source is a few strips wide. Optimal would be two or three strips wide clusters but since the light comes from a cut fibre without focusing lens we cannot achieve this small size. By moving the fibre closer to the silicon the laser spot size will reduce but on the other hand we risk mechanically damaging the sensor.

4. Move the slide by turning the micrometer screw on the right hand side of the box (seen from the repeater electronics board). You will now see the signal from the light source move from one strip to another. You may also notice that this translation is not very smooth. The reason for this is the aluminum on top of the implanted strips which reflect a fraction of the light and reduces the signal locally. This effect would be larger if we had a better focused light spot.
5. Place the sensor in a region with a nicely distributed signal.
6. Determine and write down in table I the amplitudes of the channels in the peak by moving the cursor.
7. Move the slide by *10 µm* and repeat 3.
8. Repeat 4. and 3.
9. Disconnect the optical fibre in the optofibre coupler between fibre coming from the LASER DRIVER and setup. Write down the pedestal value of the corresponding channels.

Figures 2-5 show the oscilloscope output with and without a laser spot. One horizontal division corresponds to about two readout channels.

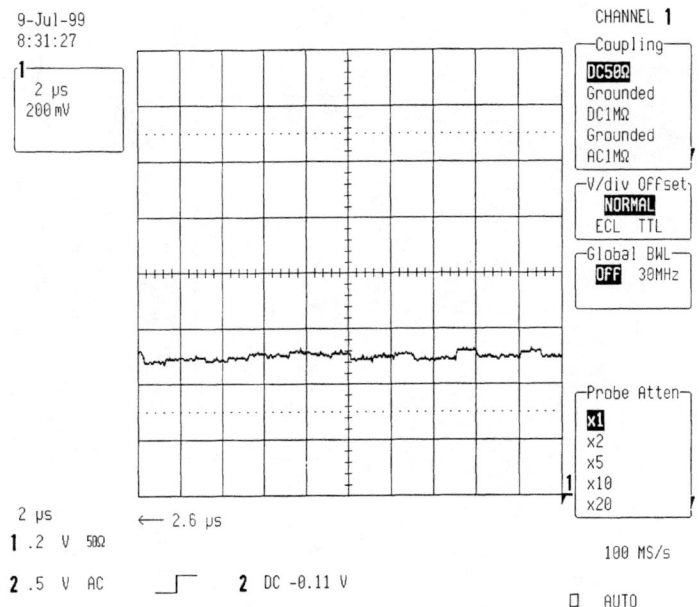

**FIGURE 2.** The pedestal amplitudes without laserspot

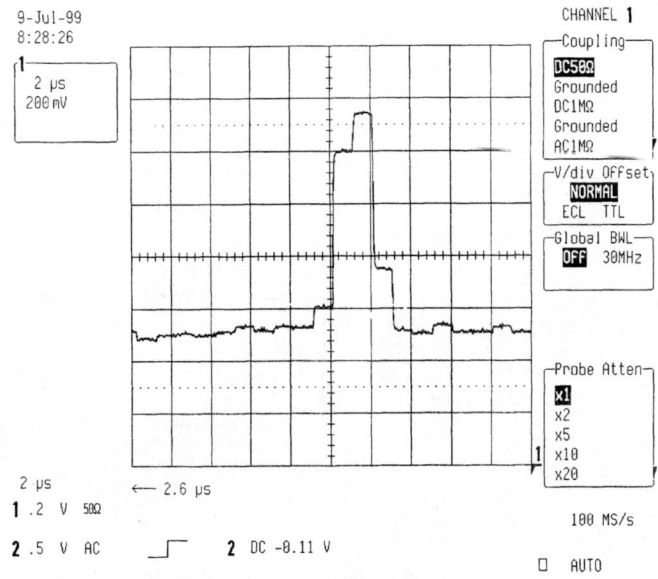

**FIGURE 3.** The amplitude of the signal at 0 μm

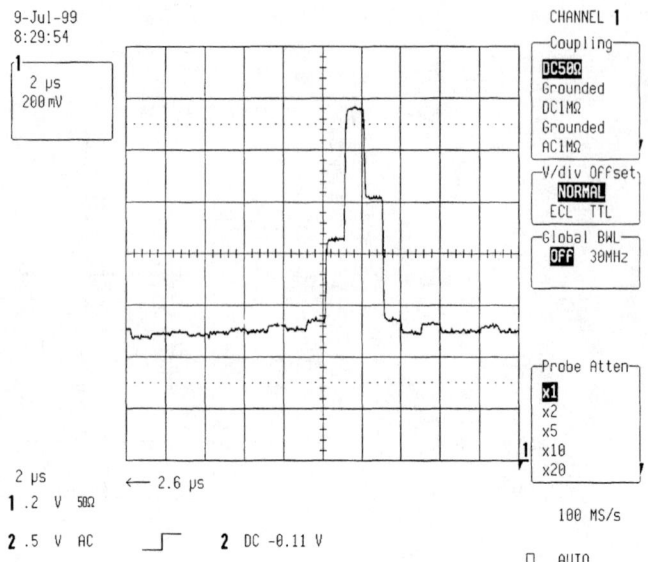

**FIGURE 4.** The amplitude of the signal at 10 μm

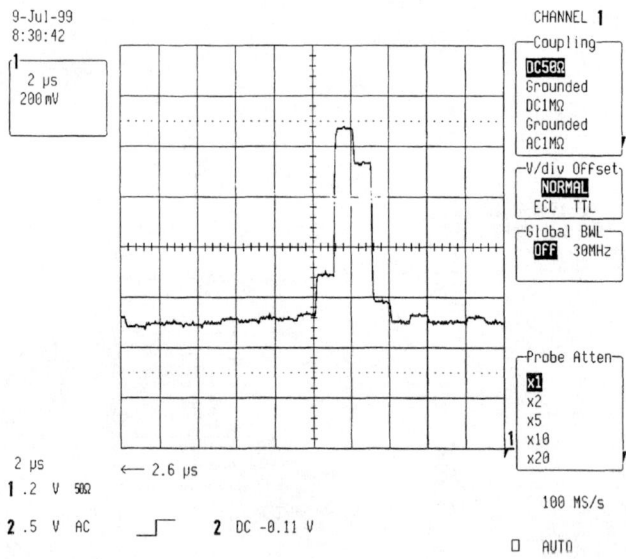

**FIGURE 5.** The amplitude of the signal at 20 μm

Table 1 shows the raw data and table 2 shows the pedestals subtracted from the data with laser spot.

**TABLE 1.** Raw data from the setup

|              | Pedestal [mV] | Signal 0 μm [mV] | Signal 10 μm [mV] | Signal 20 μm [mV] |
|--------------|---------------|------------------|-------------------|-------------------|
| Strip n-2    | 110           | 205              | 160               | 140               |
| Strip n-1    | 90            | 800              | 560               | 455               |
| Strip n      | 95            | 960              | 960               | 960               |
| Strip n+1    | 100           | 360              | 440               | 620               |
| Strip n+2    | 90            | 110              | 110               | 130               |

**TABLE 2.** Corrected data, signal-pedestal

|              | Signal 0 μm [mV] | Signal 10 μm [mV] | Signal 20 μm [mV] |
|--------------|------------------|-------------------|-------------------|
| Strip n-2    | 95               | 50                | 30                |
| Strip n-1    | 710              | 470               | 365               |
| Strip n      | 865              | 865               | 865               |
| Strip n+1    | 260              | 340               | 520               |
| Strip n+2    | 20               | 20                | 40                |

Since the laser spot is wider than two strips only the two edge strips of the cluster carry some information of the position. We now use a simple formula to calculate the position of the light spot using a strip pitch of 50 μm. Try two clustering methods:

1. Take the outermost strip on both sides of the peak, i.e. strips n-2 and n+2

2. Take the two outermost strips on both sides of the peak, i.e. strips n-2, n-1, n+1 and n+2. Add the left ones together forming PH(left) and the right ones for PH(right).

$$position = \frac{PH(left)}{PH(right) + PH(left)} \cdot pitch$$

The result is shown in table 3. In this exercise the sum of two strips gives a better result. The laser spot was not centred on top of a strip and therefore one should only compare the relative movement of the laser spot and not the absolute position.

**TABLE 3.** The calculated strip positions

|  | Signal 0 μm [μm] | Signal 10 μm [μm] | Signal 20 μm [μm] |
|---|---|---|---|
| 1 strip | 8,7 | 14,29 | 28,57 |
| Sum of two strips | 12,9 | 20,45 | 29,32 |

# 4. STUDY OF CHARGE TRANSPORT IN SILICON WITH A FAST AMPLIFIER

## Introduction

In gaseous detectors the mobility for electrons is several orders of magnitude higher than for positive ions. In semiconductors the mobility for holes is only slightly lower than for electrons. In general the signal propagation in semiconductors is a few nanoseconds while the signal propagation in gaseous detectors typically varies from microseconds to milliseconds. In order to study the drifting of electron and holes in silicon very fast electronics is required.

The drift velocity for electron and holes in silicon at low electric field strength is given by

$$v_e = \mu_e E$$

$$v_h = \mu_h E$$

where $\mu_e$ and $\mu_h$ are the mobilities for electrons and holes respectively and $E$ is the electric field. The mobility in silicon at room temperature is *1350 cm²/Vs* for electrons and *480 cm²/Vs* for holes. At high fields the velocity saturates with velocities of the order of *10⁷ cm/s*.

When the sensor is fully depleted the signals from holes and electrons will arrive almost at the same time. We can try to study the slower propagation of holes by shining a short laser pulse on the backside of a non-depleted n-type silicon diode. By choosing a wavelength which does not penetrate far in the silicon the charge can be generated close to the surface of the silicon. The holes have now to drift to the other side of the sensor while the electrons are formed at the interface. If the electrical field is low and the detector has a reasonably large depleted region we will now see a difference in the time between the signal arising from the electrons and the holes.

By starting with a full depletion voltage the signal from the setup will have the same shape as the light pulse from the laser which is shown in *figure 1*.

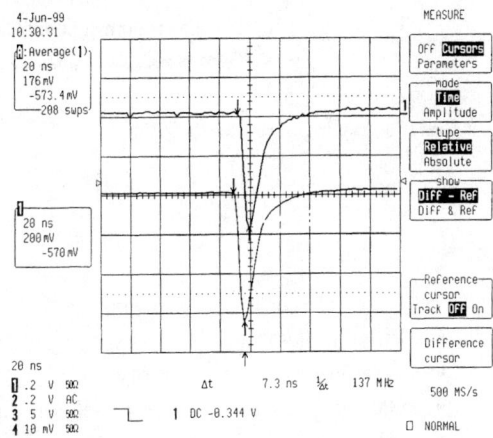

**FIGURE 1.** The amplifier output from the laser pulse with a fully depleted silicon diode.

At a voltage between 20 V and 40 V we expect to see a difference in pulse shapes between signals generated by electrons and holes.

## The setup

In this experiment we use a fast amplifier chain connected to a diode of n-type silicon with the back plane not covered with aluminium. A schematic drawing of the sensor setup is shown in figure 2.

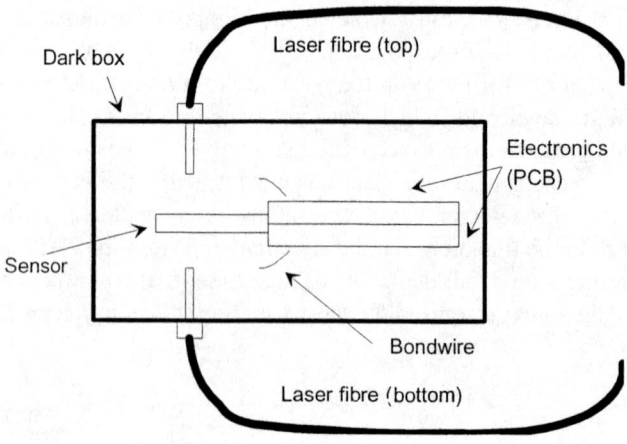

**FIGURE2.** A schematic drawing of the sensor setup

Other equipment used are:

- Oscilloscope
- Laser Driver
- Pulse generator
- Low voltage power supply
- Battery pack for biasing the silicon diode
- NIM crate for the units

## A. Study signals when shining laser on the junction side (p-side)

1. Connect optical fibre labelled p-side to the Laser Driver unit. You get the picture on oscilloscope shown in figure 3.

2. Turn on the sensor bias and ramp slowly up the voltage. The amplitude of the signal gets larger but the shape stays approximately unchanged. The sensor is depleted from the p-side and the n-side is conducting transporting the electrons to the amplifier.

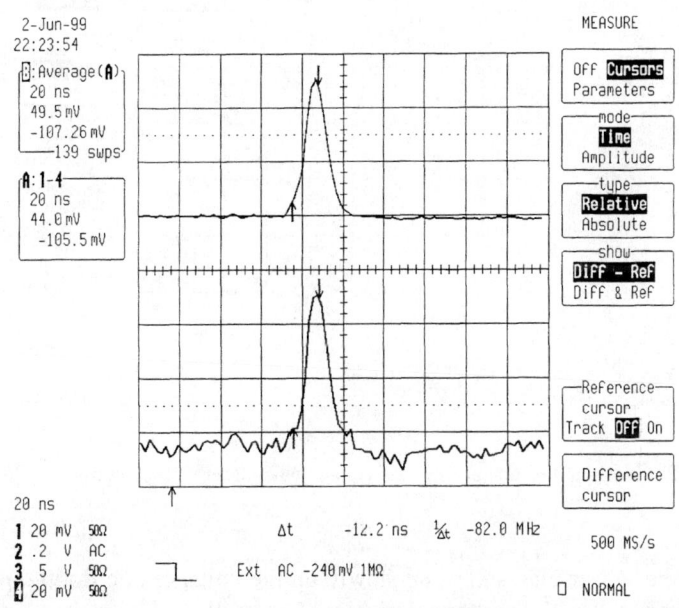

**FIGURE 3.** The signal from a unbiased sensor when laser shined on the p-side

## B. Study signals when shining laser on back plane (n-side)

1. Connect optical fibre labelled n-side to the Laser Driver unit. Very tiny signal. The holes are trapped close to the backplane.

At low bias voltage a double peak structure is visible with a contribution coming from fast arriving electron signals and a contribution from holes with a lover mobility which have to travel through the depleted region of the sensor

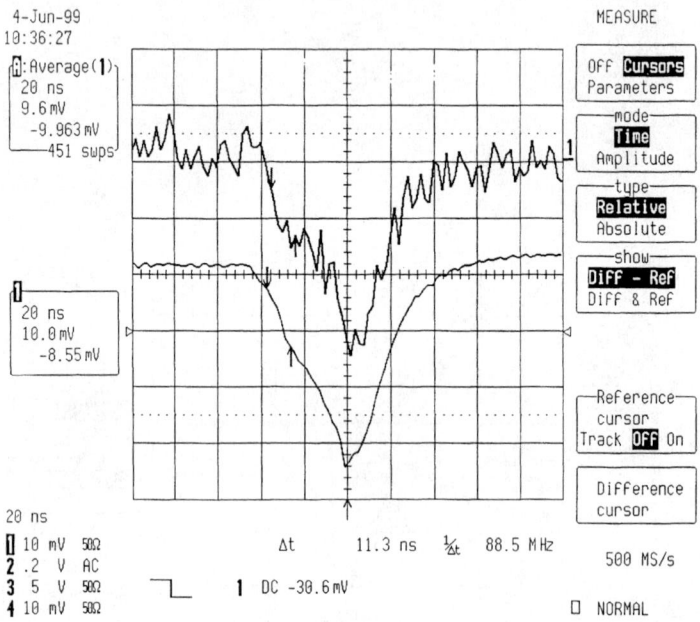

**FIGURE 4.** The signal amplitude at low bias voltage when laser shined on n-side of the silicon sensor.

Turn on the sensor bias and ramp slowly up the voltage. The double peak structure disappears when the bias voltage gets higher. At full depletion the amplitude of the signal, shown in figure 5, is comparable when laser shined on p- and n-sides.

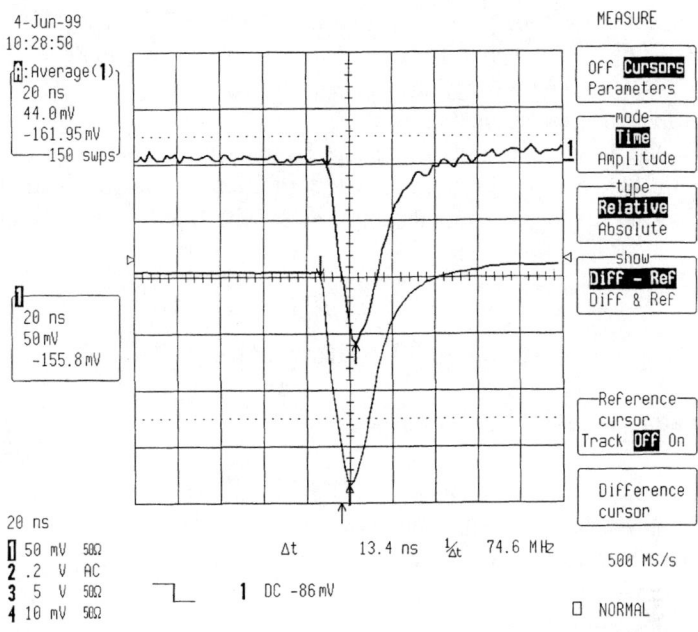

**FIGURE 5.** Signal amplitude at full depletion when laser is shined on n-side

# 5. ANALYSING PULSE HEIGHT SPECTRUM FROM SILICON DETECTORS

## Introduction

In a Pulse Height Analyser (PHA) the amplitude of each pulse from a silicon detector is recorded and histogrammed. With the appropriate calibration, the histogram can be interpreted as an energy spectrum. The PHA is a useful tool for studying performance of a radiation detector and its electronics.

The shape of the spectrum will show how much energy is deposited in the sensor. At the low end of the spectrum the efficiency will fall because the oxide surface layer prevents the penetration of low energy particles. For gamma radiation the efficiency will drop also at high energies because the photoabsorption cross section decreases. For high energy particles (1GeV) the interaction probability is 100%, and the particles traverse the silicon depositing an amount of energy which is approximately constant.

These particles are commonly called Minimum Ionizing Particles (MIP). In 300 μm silicon the most probable energy deposition from a MIP is around 88keV, which gives a signal of 25000 e$^-$. The energy distribution for MIP's has a long tail towards high energies because of Landau fluctuations.

A PHA can also be used to measure noise. By measuring the width of the peak of monoenergetic X-rays from a source the width of the peak is dominated by the sensor and electronics noise. Alternatively the a test pulse with fixed amplitude can be used.

## The laboratory setup

In this laboratory we use the following equipment:

- Timing unit for readout circuit, VIKING TIMING
- Pulse generator for trigger and laser, PULSE GEN
- Current limited supply for readout circuit
- NIM crate
- (Oscilloscope)
- Battery pack for biasing the sensor
- PC with Nucleus PCA, Pulse Height Analyser card

## A. Pulse height spectrum from Ru β source

The Viking setup has been studied in previous laboratory sessions. We will now only concentrate on the PHA part. The electrons from a ruthenium source have a most probable energy close to that of minimum ionising particles. Such particles give an energy deposition in silicon which follows the characteristic Landau curve shown in figure 1.

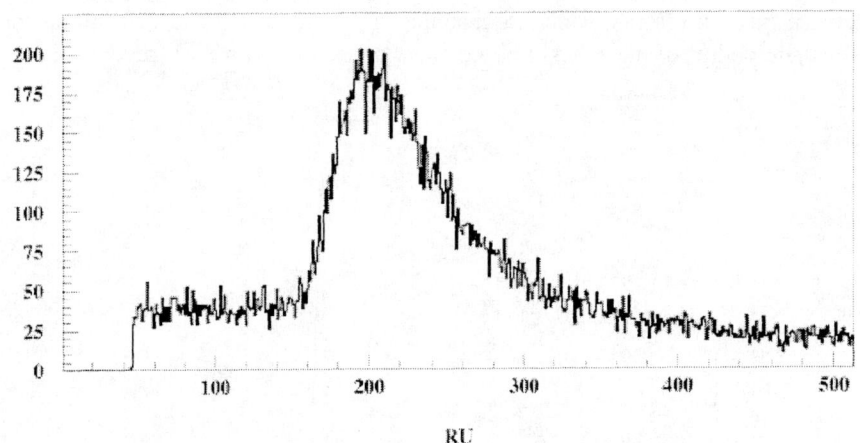

**FIGURE 1.** A beta spectrum from a Ru source (horizontal axis is pulse height, arb. Units; vertical axis frequency)

## B. Noise measurement with PHA

Irradiate the sensor with an $^{241}$Am γ source. The source has peaks at following energies:

| | |
|---|---|
| Am | 59.95 keV |
| Np ($L_{\alpha 1}$) | 13.95 keV |
| Np ($L_{\beta 1}$) | 17.74 keV |

The sensor is supported by a printed circuit board with copper and the contact to the backplane of the sensor is done with silver loaded glue. We will see additional peaks from copper and silver.

| | |
|---|---|
| Cu ($K_\alpha$) | 8.03 keV |
| Cu ($K_\beta$) | 8.90 keV |
| Ag ($K_\alpha$) | 21.99 keV |
| Ag ($K_\beta$) | 24.94 keV |

Acquire a spectrum from a Americium source and measure in the spectrum the peak positions and widths of the peaks. Fully deplete the sensor with 80 V reverse bias.

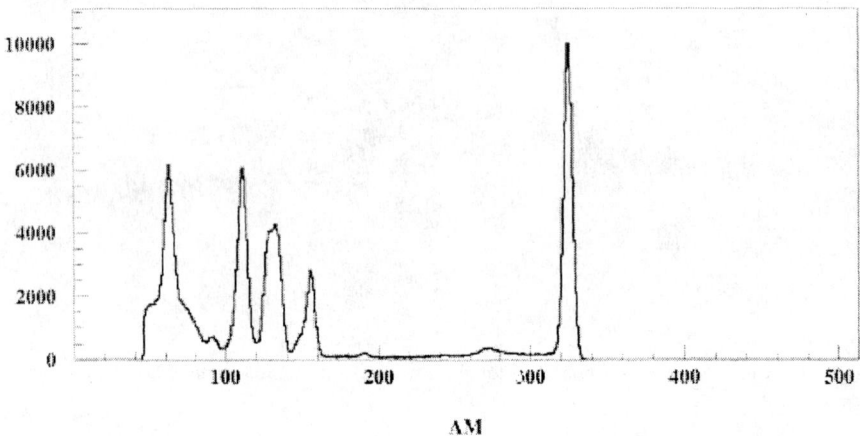

**FIGURE 2.** A X-ray spectrum from an Am source

| Energy | Channel | FWHM |
|--------|---------|------|
| 8,03   | 61      | 6,8  |
| 13,95  | 91      |      |
| 17,49  | 109     | 6,98 |
| 21,99  | 129     | 11,55|
| 24,94  | 153     | 6,55 |
| 59,95  | 323     | 6,49 |

The channel number versus energy with data points fitted with a linear fit is shown in figure 3.

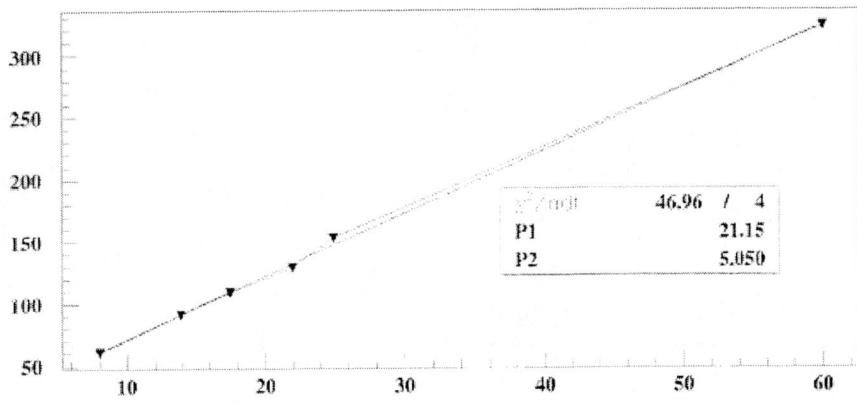

**FIGURE 3.** Energy plotted versus channel number

By knowing the energy or the size of the test pulse the pulse height spectrum can be calibrated the width of the peak can be converted to RMS electrons or eV. Calculated for the Am-peak at 60 keV.

$$FWHM(Am) = FWHM[ch] * calibration = 6.49\ ch \bullet (1\ keV/5.05\ ch) = 1.28\ keV$$

The RMS of a Gaussian distribution is related to the full width at half maximum by the factor 2.33, giving:

$$RMS(Am) = FWHM(Am)/2.33 = 1.28\ keV/2.33 = 0.54\ keV$$

## A few more spectra obtained during the school with the setup

The spectrum from a Cd source with peaks at 22.98 keV and 26.10 keV is shown in figure 4.

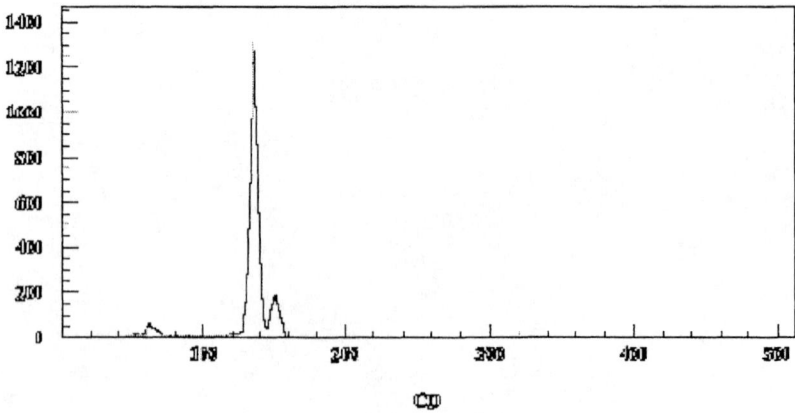

**FIGURE 4.** An X-ray spectrum froma Cd source

The low energetic X-ray lines of Mn created be a Fe-source are shown in figure 5. The two main K-lines at 5.9 keV and 6.5 keV are not separated.

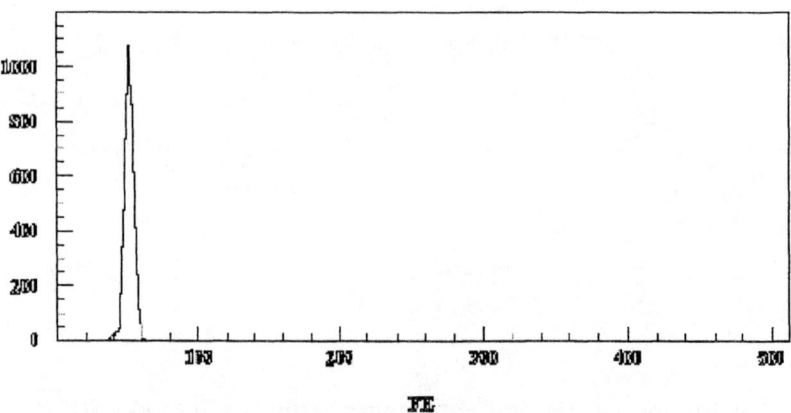

**FIGURE 5.** A X-ray spectrum from a Fe source

# GaAs Pixel Detector with Integrated Readout for Digital Mammography

S.R.Amendolia[a,c], M.G.Bisogni[a,b], U.Bottigli[a,b], P.Delogu[a,b], G. Dipasquale[a,b], M.E.Fantacci[a,b], V.Marzulli[a,b], P.Maestro[a,b], E.Pernigotti[a], V.Rosso[a,b], A.Stefanini[a,b], S.Stumbo[a,c], O.Venier[a,b]

[a] INFN (Istituto Nazionale di Fisica Nucleare), Italy
[b] Dept. of Physics, Univ. of Pisa, Italy
[c] Dept. of Mathematics and Physics, Univ. of Sassari, Italy

**Abstract.** Aim of this paper is to give a general idea of the characteristic of a system for medical digital imaging based on a GaAs pixel detector bump-bonded to a VLSI electronic chip. After a discussion on the characteristics of the GaAs detector, and its detection and charge collection efficiency properties, it follows a description of the high-density channel VLSI chip named Photon Counting Chip (PCC). Then, a description of the apparatus used to characterize the detector and to perform electrical measurements is made. Finally are discussed irradiation tests and the laboratory setup used at the laboratory course of the ICFA 99 school.

## GALLIUM ARSENIDE DETECTORS

Gallium Arsenide (GaAs) is an III-V compound semiconductor with atomic numbers 31/33 and a density of 5.32 g/cm$^3$. The band gap energy is 1.42 eV at room temperature and the average energy required to produce an electron-hole pair is approximately 4.2 eV. The Fano factor is close to 0.18, indicating that a good energy resolution can be achieved. The electron mobility at low electric field can exceed 8000 cm$^2$/Vs. At fields above $3 \times 10^3$ V/cm, the electron mobility decreases and the velocity saturation value is $10^7$ cm/s. The hole mobility is near 400 cm$^2$/Vs and for electric fields above $4 \times 10^4$ V/cm the holes velocity reaches the same value for the electrons. The intrinsic carrier concentration of GaAs is $1.79 \times 10^6$ cm$^{-3}$ which corresponds to a resistivity of $10^8$ Ohm cm, indicating that thermally produced current is low at room temperature. Typical carrier lifetimes are of the order of few nanoseconds. GaAs detectors have been studied over the past 30 years mostly for gamma and X spectroscopy at room temperature. Vapor Phase Epitaxy (VPE) or Liquid Phase Epitaxy (LPE) GaAs detectors can achieve high-energy resolution (5% for 59.5 keV gamma rays from $^{241}$Am source at room temperature is reported). Nevertheless, epitaxial detectors are limited to a thickness up to 100 μm and are very difficult to reproduce. Bulk GaAs material was inferior in quality until early '80. Since then,

GaAs has undergone extensive research because of its use as material for VLSI applications as well as substrate for several high-speed devices. Due to demand for high quality GaAs, high purity bulk is now quite readily available. Detectors built with Semi-Insulating (SI) Liquid Encapsulated Chochralski (LEC) materials are now very appealing as room temperature gamma ray and X spectrometers, as dark matter detectors and in particular for medical imaging applications[1-5].

## Detection properties

### Detection Efficiency

Due to its high atomic number and density, GaAs shows very high detection efficiency for X-rays in the diagnostic energy range (10-100 keV). Figure 1 shows the detection efficiency of a 300 μm thick silicon detector and of GaAs detectors of different thickness as a function of the photon energy. Points are experimental results, lines Monte Carlo simulations.

At 59.5 keV, the efficiency of a 600 μm thick GaAs detector is about 50% while for a 300 μm thick silicon detector it is only 1%.

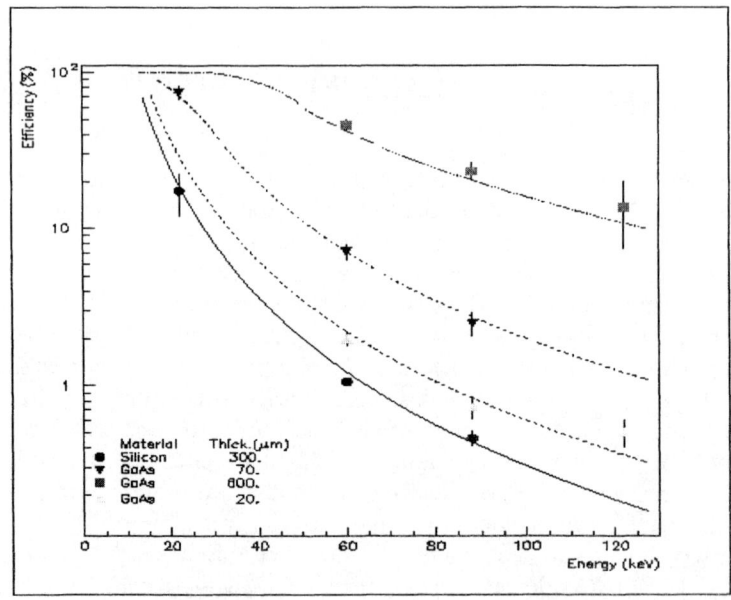

**Figure 1** Efficiency vs. energy for a 300 μm thick silicon detector and for GaAs detectors of different thickness.

Most important is to compare the performance of the GaAs with respect to the conventional radiographic film. At the typical average energy used for mammography examinations, 20 keV, the detection efficiency of a 200 μm thick GaAs crystal is about 98%. This value should be compared with the efficiency of a screen-film system for mammography, which is made up by a fluorescent screen coupled to a photographic film (the screen is needed to enhance the overall detection efficiency). This stack can reach a maximum efficiency of 60% at 20 keV. The GaAs higher efficiency permits to reduce the patient dose (typical 2 mGy), allowing to extend mass screening programs to women younger than 50. Nevertheless, for the same amount of dose a GaAs detector is able to detect very low contrast details (of order of 1%) inaccessible to a screen-film stack due to the microscopic structure of the photographic emulsion. The low contrast detection is of great relevance in the early diagnosis of the breast cancer, detected in a pre-calcification stage.

In the past ten years, an important research work has been conducted to understand structural and electrical characteristics of the Semi Insulating GaAs [6-13].

*Electric field distribution*

The study of electric field distribution inside the crystal is one of the most interesting arguments in this research field. Semi Insulating materials are obtained by compensation with the residual shallow acceptor impurities (usually Carbon) of the native deep donor defects. The detector active region width is determined by the concentration of deep donors ($10^{16}$ cm$^{-3}$) and shallow acceptor ($10^{15}$ cm$^{-3}$).

Schottky barrier SI GaAs diodes reverse biased experimentally show a linear behavior of the active zone as a function of the bias voltage, instead of the classic square root dependence. The active layer W follows the relation:

$$W = W_0 + b \times V_a \qquad (1)$$

where $W_0 = 30$ μm and $b = 0.7$ μm/V.

According to theoretical models, this behavior accounts for the presence of a "quasi-neutral" region, which determines a constant field inside the crystal. The quasi-neutral region width increases linearly with the detector bias voltage.

*Charge Collection Efficiency*

The trapping centers also affect the collection of the charge produced by the interaction of the photons within the crystal. The charge collection efficiency (CCE) is defined as:

$$CCE = \frac{Q}{Q_0} \qquad (2)$$

where Q is the charge collected at the read-out electrode and $Q_0$ is the total charge produced in the detector. Incomplete charge collection can be a serious problem especially in low energy applications as mammography. It can become very hard to detect photon signals due to the read-out electronic noise. Thanks to the improvement in the material manufacturing and in the contact technology, now SI GaAs detectors with CCE higher than 80 % are available with a significant thickness (> 100 μm).

Figure 2 shows the emission spectrum of a $^{241}$Am source (59.5 keV characteristic emission peak) recorded with a 200 μm thick GaAs detector. The detector has been produced by Alenia S.p.A. from a SI LEC <100> oriented undoped 200 μm thick GaAs crystal. A square Schottky contact (200 μm) has been realized on one side of the bulk with an Au/Pt/Ti alloy. On the other side an ohmic contact, able to sustain bias voltages up to 450 V, has been realized. The vertical line marked in the graph represents the position where the 59.5 keV peak should be in case of full charge collection. As can be seen, the peak recorded by the GaAs diode is shifted respect to this position due to the incomplete charge collection.

Figure 3 shows the CCE of this detector as a function of the bias voltage: above 250 V the CCE value saturates at the 89% level.

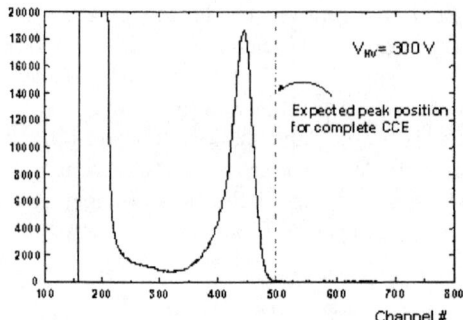

**Figure 2** Emission spectrum of a $^{241}$Am source recorded with a 200 μm thick GaAs detector.

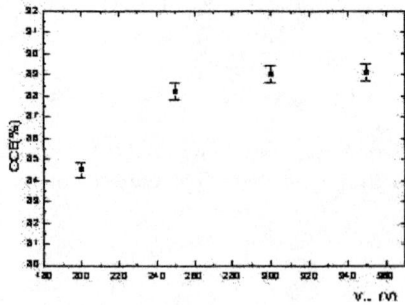

**Figure 3** CCE of the same GaAs detector as a function of the bias voltage.

# GaAs Pixel Detectors

A square electrode matrix realized on the surface of the semiconductor makes out a pixel detector crystal. The charge produced by the interaction of a photon within the detector and collected by a pixel provides the coordinates of the impact point on the detector surface.

Due to the intrinsic direct mapping of the detector structure to the image, pixel architecture is the most natural solution for radiological applications. In pixel detectors, every sensitive cell is connected to an electronics read-out channel for signal processing.

In the single photon counting mode a digital acquisition system allows to record in a digital memory the number of events detected by each pixel in a fixed time span. The information of the energy release in the detector is not considered. This feature is very important when a polychromatic spectrum has to be detected: with respect to an integrating charge read-out system the single photon counting mode allows to detect lower contrast details. The MEDIM collaboration of the Italian Istituto Nazionale of Fisica Nucleare realized an X-ray imaging system prototype based on a small GaAs pixel detector [14-16].

Alenia S.p.A. has built the detector on SI LEC <100> oriented 200 μm thick substrate. Its pixel architecture features a matrix of 36 200 μm wide square pixels with 20 μm spacing between adjacent pixels. At 400 V the average CCE is 90 % [17]. The read-out electronics realized by Laben S.p.A. operates in single photon counting mode. Each pixel is connected to the corresponding electronic channel by means of wire bonding. With this system images of a phantom with low contrast details (down to 1.2%) have been acquired in standard mammographic conditions. This performance overcomes the contrast obtainable with a standard mammographic film at the full standard dose.

The main constraint for a pixel detector of significant area has always been the difficulty of the planar arrangement of a large number of channels and the relevant problem of electrical connections. For the 36 pixel detector, the sensitive area was only 1.7 $mm^2$.

A big step toward the solution of this problem has been offered by the bump-bonding technique. A front-end VLSI integrated circuit is designed with the read-out cells, which have the same shape and dimensions of the detector pixels. Each electronics cell is connected to the corresponding pixel by means of a metal bump (typically In or Sn) of few tens micron in diameter (see figure 4) [18].

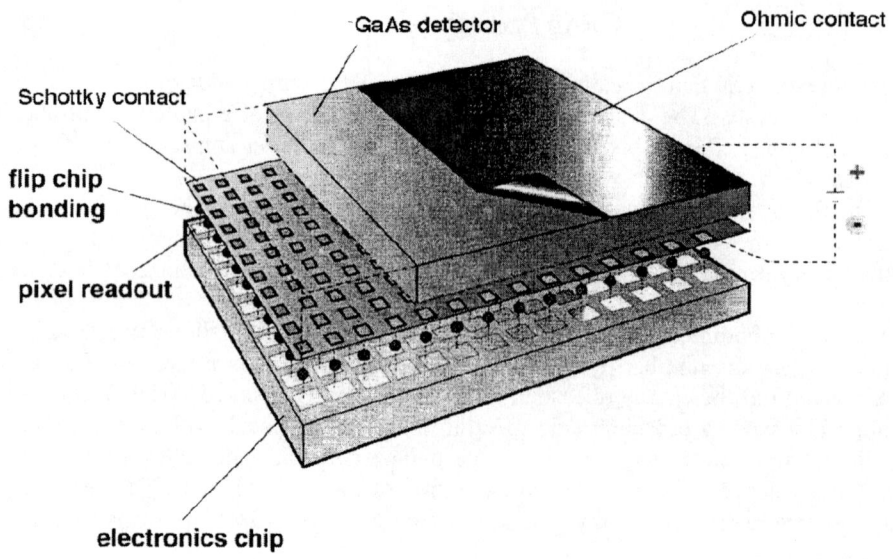

**Figure 4** Bump-bonding technique

Taking advantage of this technique, the MED46 experiment of the Italian Istituto Nazionale di Fisica Nucleare and the Microelectronics group of CERN realized the X-ray digital imaging prototype presented in this laboratory session.

The pixel detector is a matrix of 64 x 64 square Schottky contacts realized on one side of a 200 μm thick SI LEC GaAs (Alenia S.p.A.) bulk. The pixel size is 150 x 150 μm$^2$ with a 20 μm gap between adjacent pixels. The total active area is 1.18 cm$^2$. On the other side, an ohmic contact covers the whole detector surface. The operating bias voltage is 300 V. A VLSI front-end chip called Photon Counting Chip (PCC) has been realized by the Microelectronics group of CERN. It is composed of a 64 x 64 matrix of square cells whose size is 170 μm. The detector and chip cells are connected via Sn bump bonding. A digital system called MEDIPIX READ-OUT SYSTEM (MRS) (Laben S.p.A.) performs the data acquisition and provides bias and controls for the PCC [19].

## THE VLSI READOUT CHIP

### Description of the chip

The Photon Counting Chip (PCC - also referred to as "Medipix" in this text) is a 64 x 64 matrix of identical square pixels whose side measures 170 μm.

The chip is designed in the SACMOS1 technology of FASELEC (Zurich). The total area of the chip is about 1.7 cm² with a sensitive area of 1.18 cm². Each cell contains a full readout chain comprising preamplifier, discriminator, counter and shift register and can be bump bonded to an identically segmented semiconductor detector.

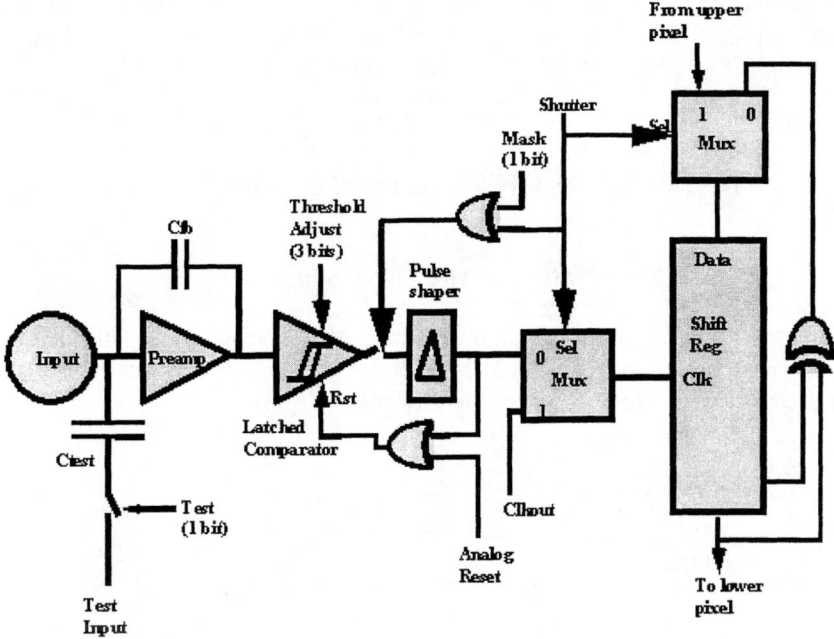

**Figure 5** Block diagram of the pixel cell.

The block diagram of the pixel cell is shown in Figure 1. A charge sensitive preamplifier with leakage current compensation, able to compensate up to 10nA/pixel, can receive signals from the detector via the bump-bonding or test signals from an external pulse generator via a test capacitance, implemented in each pixel, for electrical characterization. The leakage current is sensed by an additional row of dummy cells and subtracted from the inputs of the other sensitive cells. The output of the preamplifier is sent to a latched comparator with an externally controllable common threshold, which can be locally adjusted via 3 fully static flip-flops. There are 2 additional static flip-flops to mask noisy pixels and to enable electrical testing. These flip-flops along with the threshold adjust flip-flops are organized as a configuration shift register. The signal produced by the comparator is connected to a short delay line producing a pulse which is used to reset the comparator itself and to act as a clock for the counter when the shutter is low, that means during the data acquisition. The counter is a shift register whose 14$^{th}$ and 15$^{th}$ bits are fed back via an exclusive-or gate to the input. This is in fact a pseudo-random counter. When the shutter is high, the feedback loop of the counter is broken and the 15 bits are connected in series to the upper and lower pixels of the same column in order to form a 15 × 64-bit shift register. As the flip-flops used for the counter are fully static there is no need for refresh signals and data may be accumulated slowly or stored on the chip for long periods prior to readout. An external clock can be used to shift out the contents of the counter in a serial fashion avoiding the need for a space consuming bus. The 64 columns of the chip are connected in groups of four and multiplexed onto the 16-bit input/output bus. The same bus is also used to load the 5 x 64-bit configuration register. Both loading and readout operations can be performed at a frequency of 10 MHz, one complete frame is then read out in 384 µm [20-21].

## DETECTORS CHARACTERIZATION

### Description of the apparatus

The characterization of GaAs detectors in terms of charge collection efficiency (c.c.e.) and energy resolution (R) starts from the analysis of the spectra acquired irradiating detectors with photons from an X-ray source.

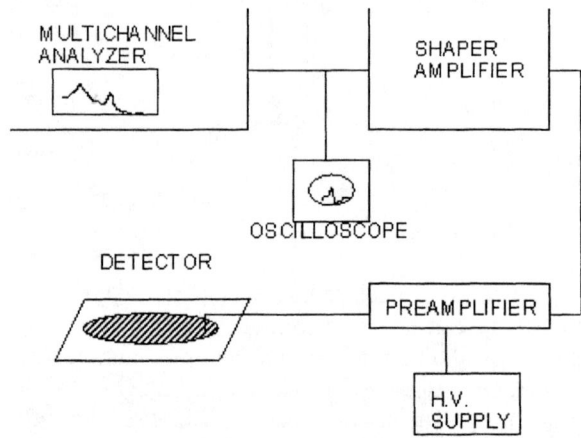

**Figure 6** electronics chain to readout the detector.

The electric charge (Q) created by a photon interacting in the detector is collected at the electrodes by the action of an external reverse-bias voltage (HV), sent via a preamplifier. The charge-sensitive preamplifier integrates the weak signal from the detector, so that its output voltage is proportional to the charge carried by the incoming pulse, i.e.

$$V_{out} = -\frac{Q}{C_f} \qquad (3)$$

where $C_f$ is the feedback capacitor of the preamp circuit.
Moreover the charge Q is proportional to the energy $E_g$ of the detected photon according to the equation

$$Q = -\frac{E_r}{w} \qquad (4)$$

where w is the average energy needed to create an electron-hole pair in semiconductor detectors.

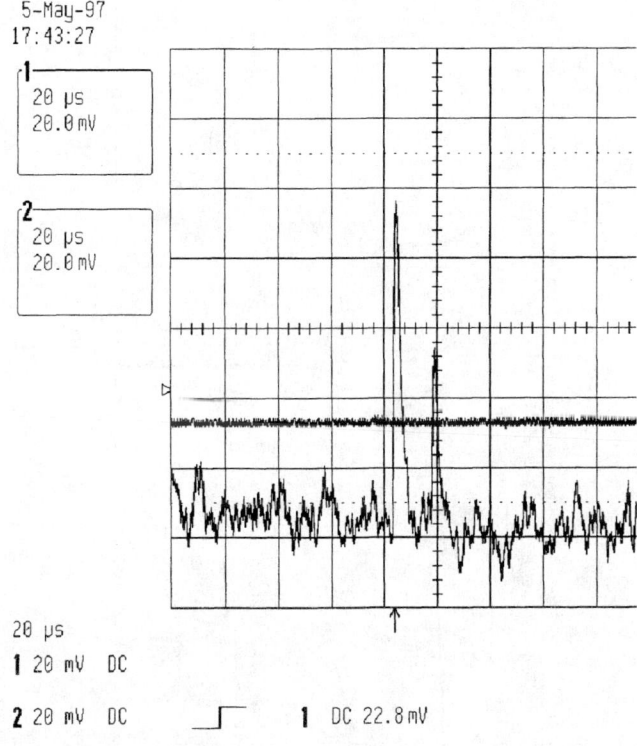

**Figure 7**: amplifier output signals, corresponding to photons from $^{241}$Am of 60 and 26 keV respectively.

So in GaAs (w equal to 4.3 eV) a 60 keV photon produces in the detector a charge of about $1.4 \times 10^4$ electrons i.e. a preamp output of the order of some millivolts ($C_f$ has a typical values of few pF). This signal is again amplified with a gain of $10^3$ or more (shaper amplifier), and shaped to a semi-gaussian form (fig. 2) for further processing by the multichannel analyzer (MCA).

The MCA is a device that operates analog to digital conversion of the pulse height from the shaper amplifier. The number of the equal height pulses is recorded in a memory channel of the MCA, whose address is proportional to the digitized value of the pulse height itself. In this way the information contained in each channel is the number of pulses with a given height, i.e. the number of detected photons with energy proportional to that height pulse. Displaying the contents of each channel on a screen the energy spectrum of the radioactive source can be obtained (fig 3).

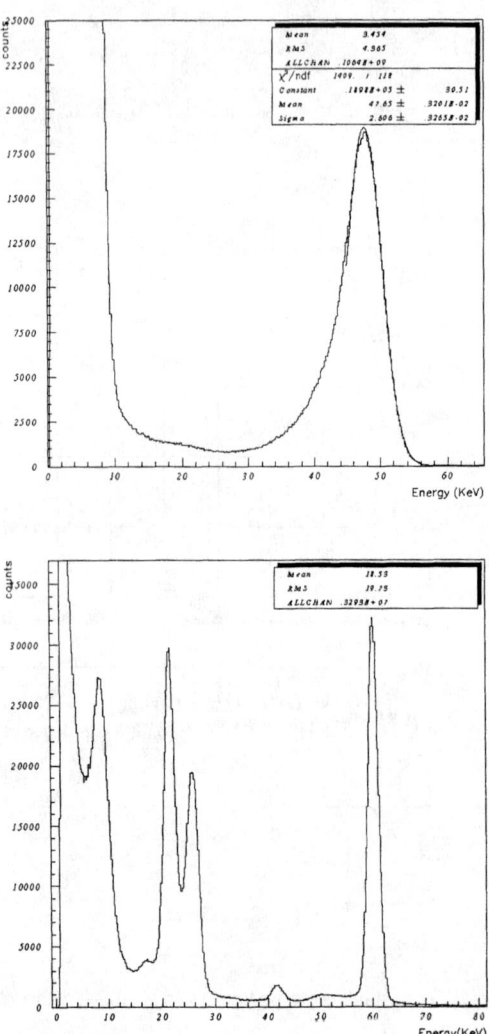

**Figure 8**: spectrum of $^{241}$Am from a Si detector (up) and a GaAs detector (down).

In figure 8 we can observe the difference between the spectrum of a $^{241}$Am source as acquired by a GaAs detector and as acquired by a standard silicon detector. The decay scheme of $^{241}$Am is reported in the following table.

Table 1

| $E_g$ (keV) 241Am | Emission probability (%) |
|---|---|
| 26 | 2-5 |
| 33 | 0.1 |
| 43 | 0.1 |
| 59.5 | 35.3 |
| Np L X-rays (12-22) | 40 |

While the spectrum by Si reproduces correctly the peak energies of the source with a good energy resolution, the spectrum by GaAs indicates a lower peak energy for the 59.5 keV peak emission and does not show the other decays. That is due to the incomplete charge collection of the GaAs detector; the charge produced by the incoming photon is partially lost through phenomena of trapping and recombination of the e--h+ pairs in the detector volume, giving a lower input signal in the preamp. This effect also decreases the energy-resolution of the device.

It's possible to enhance the c.c.e. with higher external voltage applied to the detector; the resulting electric field increases the electron drift-velocity and then diminishes the probability of recombination and trapping. The limit of applicable voltages is given by the detector breakdown, at which the device begins to conduct.

## GENERAL DESCRIPTION OF THE ELECTRICAL MEASUREMENTS

These measurements are aimed at characterizing electrically the chip, which is to know the distributions of the output data for a known input signal. This is possible by sending input pulses of a known amplitude through the already mentioned input capacitance, the value of which is known only approximately from the design, due to processing inaccuracies, but has been calibrated accurately after assembly to a detector and tests with radioactive sources (Cinput= 25 fF). Given an input charge, the same for all the enabled pixels, the output we get for each pixel, after is has been processed by the Medipix Readout System, is a number: 1 if the pulse was higher than the threshold previously set, 0 if it was smaller. If a train of pulses is sent, the single pixel output will be a number representing the number of pulses that passed the threshold. The chip output will then be a 64 x 64 matrix containing this information for each pixel.

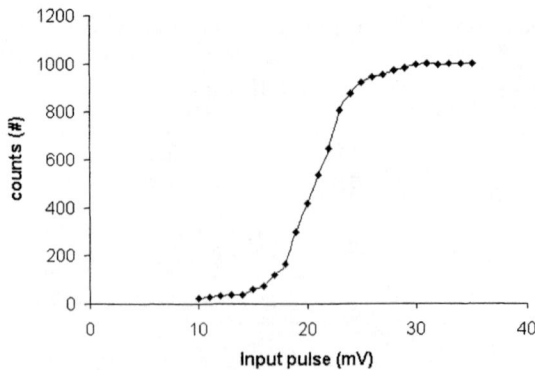

**Figure 9**: pixel response to an increasing input pulse

Varying the input charge (pulse) for a previously set common threshold will then allow to state the exact value of the threshold that will be represented by the value of the input pulse for which we get 50% of the total number of counts.

The threshold values of the 4096 pixels will then be distributed on a gaussian, whose mean value will give the common threshold set for the whole matrix, and the *rms* will give the dispersion of the threshold on the matrix.

We mentioned before the possibility to adjust the threshold pixel by pixel by 3 externally controllable fully static flip-flops. This means that after a common threshold is set, each pixel can add a 3-bit correction to it. This feature reduces the threshold non-uniformity all over the matrix, and an electrical test similar to the previous one will give a much smaller rms on the threshold distribution. Fig.10 show a comparison between the curves obtained with and without threshold adjustment [20-21].

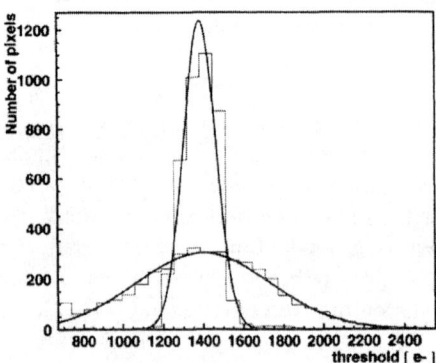

**Figure 10**: threshold distribution on the matrix, before and after threshold adjustment

Noise measurements can also be performed. The noise of the electronics of the single pixel will be derived by the input pulse range for which each pixel passes from

2% to 98% of the total number of counts when increasing the input pulse for a given threshold (see distribution in Fig.9).

The minimum threshold allowed for each matrix will be the minimum $V_{th}$ value for which no output is seen when no input is given. If this value is low enough to see the charge deposited by a single photon in 300 μm GaAs detector (typically ~4500 e-) then the chip is suitable for photon counting operation, that is the aim of the whole system!

## IRRADIATION TESTS

To perform these tests, the detector will be irradiated with a $^{241}$Am source. The aim of the irradiation tests is to evaluate the detector performance in terms of detection efficiency and imaging capability.

### Efficiency Calibration

As mentioned previously, the detection efficiency of the GaAs is one of the most important features that makes this material appealing for medical imaging applications. It is convenient to subdivide counting efficiency in two classes: absolute and intrinsic. The absolute efficiency can be defined as:

$$\varepsilon_{abs} = \frac{number\ of\ recorded\ pulses}{number\ of\ photons\ emitted\ by\ the\ source} \quad (5)$$

This quantity depends not only on the detector properties but also on the details of the geometry (primarily on the source-detector distance).
The intrinsic efficiency is defined as:

$$\varepsilon_{int} = \frac{number\ of\ recorded\ pulses}{number\ of\ photons\ incident\ on\ the\ det ector} \quad (6)$$

and it does not include the solid angle subtended by the detector as an implicit factor. These two efficiencies can be related, in case of isotropic sources, by:

$$\varepsilon_{int} = \varepsilon_{abs}\left(\frac{4\pi}{\Omega}\right) \quad (7)$$

where $\Omega$ is the solid angle of the detector seen from the actual source position.

The intrinsic efficiency can also be written as:

$$\varepsilon_{int} = \frac{N 4\pi}{S \Omega} \quad (8)$$

where N is the sum of the counts of all 4096 pixels recorded in a fixed time interval, S is the number of photons emitted by the source over the same measurement period and $\Omega$ represent the solid angle (in steradians) subtended by the detector at the source position.
In our case the solid angle is given by:

$$\Omega = 4 \arctan \frac{a^2}{4d\sqrt{d^2 + \frac{a^2}{2}}} \quad (9)$$

where d is the source-detector distance and a is the detector side (a = 1.09 cm).
For a full depleted (HV = 300 V) 200 μm thick GaAs detector:

$$(59.5 \text{ keV}) = 18 \%$$

## Imaging capability

In digital radiology, two parameters, the "contrast" and the "signal-to noise" ratio quantify the image quality.
These two quantities depend on the physical properties of the radiographed object but also on the imaging system characteristics.
The contrast C is defined as:

$$C = \left| \frac{n - n'}{n} \right| \quad (9)$$

where n' is the average count in the target object and n is the average count in the background region (see figure 1). Both regions must have the same area.

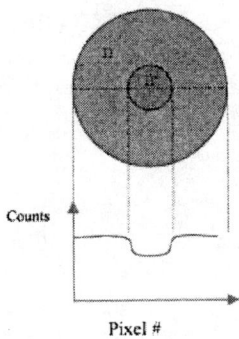

**Figure 11** Radiographic contrast definition.

For details larger than the pixel area, the contrast is independent from the detected photon flux. It depends only on the radiographed materials and on the photon energy.
If one considers the quantum noise as the only source of noise in the image, the Signal-to-Noise ratio SNR is defined as:

$$SNR = \left|\frac{n-n'}{\sqrt{n}}\right| \qquad (10)$$

In order to distinguish the signal from the background, SNR must overcome a threshold K. For a fixed K value, the minimum number of photons to be recorded on the background to detect a contrast $C_K$ is given by:

$$n = \frac{K^2}{C_K^2} \qquad (11)$$

Typical K values range from 2 to 5.
The aim of the imaging test is to make images of different details irradiated with the $^{241}$Am source and to evaluate the contrast and the SNR of the radiographed objects.

## Spatial resolution

To evaluate the spatial resolution of the imaging system, a test object called "bar pattern" will be used. A bar pattern is made out of 50 µm thick equally spaced Pb bars and grouped in series. Each series is characterized by a fixed Pb bar width, which is equal to the pitch between two adjacent bars. In this way the spatial frequency of the bars is defined as the number of line pairs per millimeter (f = lp/mm). The spatial frequency in the bar pattern ranges from 0.25 to 6 lp/mm.

The spatial resolution is related to the maximum number of line pair per millimeter that the imaging system is able to distinguish.
For a pixel size a, fmax is given by:

$$f_{max} = \frac{1}{2a} \qquad (12)$$

The aim of this work is to measure the contrast of the bar pattern images for the different bar series and evaluate the Contrast Transfer Function (CTF) as a function of the spatial frequency.
The CTF is defined as:

$$CTF(f) = \frac{C_{meas}(f)}{C_{th}} \qquad (13)$$

where Cmeas(f) is the experimental contrast measured for different spatial frequencies and Cth is the contrast between a 50 μm thick Pb sheet and air at 59.5 keV. Cth is 24.7%.

## ICFA LABORATORY SETUP

### Instruments used during imaging with $^{241}$Am source

The MRS acquisition system to control and setting the MEDIPIX chip, consist (as shown in figure 12) of a PC, a VME crate, the National Instrument VXI-PCI8015 kit, the Laben kit MRS (VME, MOTHER and CHIP board), three voltage generator, a GPIB interface board (NI 1014p) a HV generator and a $^{241}$Am source. The PC hosts the Medisoft software (realized in C language) by means of which it is possible to control the totality of the settings that are necessary to the configuration of the Medipix chip.
The control via PC of the bus VME is realized by means of the National Instrument interface kit VXI-PCI8015. The VME crate hosts the GPIB National Instrument board (mod. 1014p) and one of the Laben MRS kit of board (the VME one).

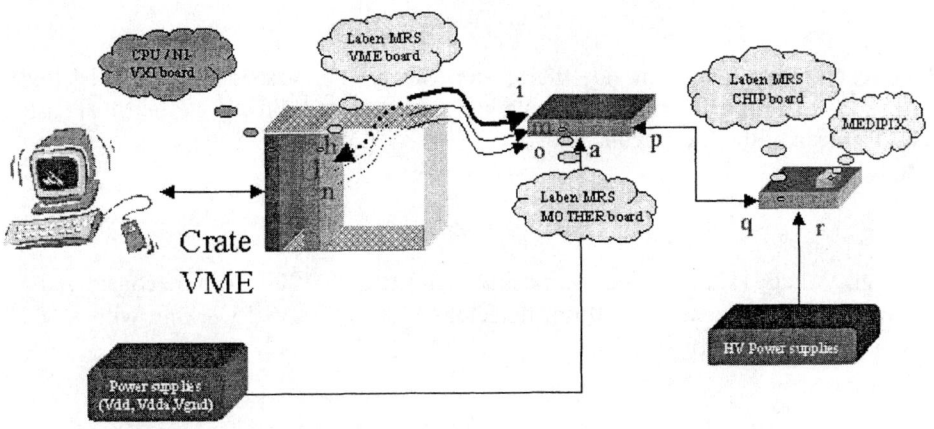

**Figure 12** Schematic picture of the setup used during acquisition with $^{241}$Am source.

*The PC and the Medisoft software*

The PC is based on the Pentium (or Pentium II) processor and Windows NT/95 operating system. Medisoft has been realized in the C language using the National Instrument "Labwindows CVI" as platform of development. The software is designed to have a great modularity so that different user can integrate it with additional and particular features without changing anything of the original core. There are library files in which are defined the variables used and low level function for read/write operation on the memory registry are implemented. There are source files where every kind of function and routine used are defined, there are files which contains all the default values used by the Medipix chip and a look-up table file for correspondence between the number sequence generated by the pseudo random generator counter of Medipix and a sequential numbering order.

*National Instrument VXI-PCI8015 kit*

The VXI-PCI8015 interface kit links any PCI-based computer directly to the VXI/VME bus using the high speed Multisystem eXtension Interface bus (MXI).
The system makes the computer perform as if it were plugged directly into the VXI/VME back plane, giving the external computer the capability of an embedded computer. The VXI-PCI8150 kit includes one half size PCI plug-in board (the PCI-MXI-2), which is installed directly in an available PCI slot in the computer; one B-size (or C-size) VXI-MXI-2 Slot 0 module that plugs directly into the VXI/VME mainframe; a flexible MXI-2 cable; the NI-VXI/VISA VXI bus interface software. The Labwindows CVI platform is the natural platform of development for software in order to take the full and maximum advantages by routines that have to deal with the hardware we came to describe.

*Power supplies*

Three external generators provide the tension called Vcc, Vss,Vdd, Vdda and Vgnd respectively. These three generators voltages are connected directly to the MOTHER board by mean of the "Aux" connector.

*High Voltage generator*

The High Voltage (HV) applied is a positive voltage and +300V can be considered a safe value if working with a 200 µm detector (+400V/+500V if working with a 600 µm detector).

# REFERENCES

1. W. Bencivelli et al., "Evaluation of elemental and compound semiconductors for X-ray digital radiography", Nucl. Instr. and Meth. A310, 1991, 210-214
2. W. Bencivelli et al., "Experimental Results from Gamma-Irradiated GaAs Crystals Grown with Different Techniques", Mat. Res. Soc. Proc. Vol 302, 1993, 381-387
3. W. Bencivelli et al., "Comparison of different GaAs detectors for X-ray digital radiography", Nucl. Instr. and Meth. A338, 1994, 549-555
4. W. Bencivelli et al., "Electrical characterization and detector performances of a LPE GaAs detector for X-ray digital radiography", Nucl. Instr. and Meth. A346, 1994, 372-378
5. W. Bencivelli et al., "Electrical characterization and detector performances of various semi-insulating GaAs crystals for low energy X-rays", IEEE Trans Nucl Sci NS42(4), 1995, 1522-1525.
6. V. Marzulli, "Semiconductor Detectors in Medical Physics", Nuovo Cimento Vol. 112 A, N. 1-2, 1999, 159
7. E. Bertolucci, "Digital Medical Imaging with GaAs Detectors", Proc. of Third International Wokshop on Gallium Arsenide and Related Compounds, S. Miniato World Scientific, 1995, 211-216.
8. M.E. Fantacci,"Pixel GaAs detectors for Digital Radiography", Workshop on Gallium Arsenide and Related Coumpounds, S. Miniato, 1995, 527-530
9. W. Bencivelli et al., "Some new results on semi-insulating GaAs detectors for low energy X-rays", Nucl. Instr. and Meth. A355, 1995, 425-427"
10. E. Bertolucci et al.,,"X-Ray Imaging using a Pixel GaAs detector", Nucl. Instr. and Meth. A362, 1995, 547-550.
11. S.R. Amendolia et al.., "Experimental Study of LEC GaAs Detectors for X-Ray Digital Radiography", Nucl. Instr. and Meth. A380, 1996, 410-413.
12. S.R. Amendolia et al.., "Digital Mammography based on GaAs Pixel Detector and on a Self-Triggering Acquisition System", IEEE 1995 Medical Imaging Conference, S. Francisco.
13. E. Bertolucci et al.,"Electrical characterization and detection performances of various semi- insulating GaAs crystals for low energy x-rays", IEEE Transactions on Nuclear Science in press
14. S.R. Amendolia et. al., "A project of digital mammography based on a GaAs pixel detector and on a self-triggering single photon counting acquisition system", Physica Medica, 13 (4), 1997, 157.
15. S.R. Amendolia et. al., "Imaging Performance of a GaAs Pixel Detector", Nuovo Cimento Vol. 112 A, N. 1-2, 1999, 167
16. S. Stumbo, "Experimental Results in Mammography", Physica Medica, 14 (suppl. 2) 1998

17. S.R. Amendolia et. al.,"Charge Collection Properties of a SI-GaAs Detectors for Digital Radiography ",Physica Medica, 14 (suppl. 2) 1998,
18. S.R. Amendolia et. al., " MEDIPIX: a VLSI chip for a GaAs pixel detector for digital radiology ", 1998 Symposium on Radiation Measurements, Ann Arbor 1998, Nucl Instr Meth A, A422, 1999, 247.
19. E. Bertolucci et. al.,"GaAs pixel radiation detectors as an autoradiography tool for genetic studies ",1998 Symposium on Radiation Measurements, Ann Arbor 1998, Nucl Instr Meth A, A422, 1999, 242.
20. M.G. Bisogni et. al., "Performance of a 4096 pixel photon counting chip ", 43nd SPIE International Symposium on Optical Science, Engineering and Instrumentation, San Diego, CA, USA 1998, SPIE Proceedings, 1998, 3446.
21. M. Campbell et al., "A readout chip for a 64 x 64 pixel matrix with 15-bit single photon counting", IEEE Trans. Nucl. Sci. NS-45 (1998) 751.

# LIST OF PARTICIPANTS

## Lecturers

| | |
|---|---|
| **Physics of Particle Detection** | C. Grupen (Siegen, Germany) |
| **Gaseous Detectors** | A. Sharma (CERN, Switzerland) |
| **Calorimetry** | D. Green (FNAL, USA) |
| **Silicon Detectors** | P. Giubellino (Torino, Italy) |
| **RICH Detectors** | E. Nappi (Bari, Italy&CERN, Switzerland) |
| **Particle Identification Techniques** | M. Sheaff (Wisconsin, USA/CINVESTAV, Mexico) |
| **High Rate Data Acquisition** | S. Erhan (UCLA, USA) |
| **Analog and Digital Circuits** | W. Dabrowski (Cracow, Poland) |
| **Future Accelerators** | Yokoya Kaoru (SDK & KEK, Japan) |
| **Detector Systems for Future HEP Experiments** | A.G. Clark (Geneva, Switzerland) |

## Review Talks

| | |
|---|---|
| **Review on Neutrino Detectors** | E. Arik (Bogazici, Turkey) |
| **Perspectives in HEP** | C. Quigg (FNAL, USA) |
| **Review on HERA Experiments** | G. Herrera (DESY, Germany & CINVESTAV, Mexico) |
| **Confronting New Technological Challenges in HEP** | A. Savoy-Navarro (Paris 6-7, France) |
| **Web Tools for Physicists** | P. Palazzi (CERN, Switzerland) |
| **Dark Matter Detection** | R.L. Dixon (FNAL, USA) |
| **Photodetectors for Tracking, Medical Imaging and Astronomy** | M. Atac (FNAL, USA) |
| **Biological and Industrial Applications** | A.H. Walenta (Siegen, Germany) |

## Laboratory Instructors

| | |
|---|---|
| **Lifetime of Cosmic Ray Muons** | J. Anderson, M. Atac, S. Cihangir (FNAL, USA), M.N. Erduran and S. Kartal (Univ. of Istanbul, Turkey), S. Erturk (Nigde Univ., Turkey) |
| **Muon Lifetime Measurement** | T. Conka, S. Cetin and A. Mailov (Bogazici, Turkey) |

| Test of a Position Sensitive P/M and Measurements of Diffraction Pattern by Single Gamma Counting | S.Korpar, P.Krizan, A.Stanovnik (Josef Stafan Inst., Slovenia) |

| Scintillation Fibers and Advanced Photodetectors | M. Atac, S. Cihangir (FNAL, USA), M.N. Erduran and S. Kartal (Univ. of Istanbul, Turkey), S. Erturk (Nigde Univ., Turkey) |

| Drift Chambers | A.H. Walenta (Siegen, Germany) |

| Microstrip Gaseous Detectors | A. Cattai (CERN, Switzerland) |

| Silicon Detectors | R. Brender (Uppsala, Sweden) P. Weilhammer (CERN, Switzerland) |

| High Badwidth Data Acquisition | M.Sheaff (Wisconsin/ CINVESTAV) |

| LABVIEW for DAQ, Monitoring and Controls | D.Anderson & S.W.L. Kwan (FNAL,USA) |

| Synchrotron Radiation Simulation | A.H. Walenta (Siegen, Germany) |

| New GaAs Pixel Detector with Integrated Readout for Mammography Imaging | R. Amendolia (PISA-INFN, Italy) |

## Students

| NAME | SURNAME | E-MAIL | COUNTRY |
|---|---|---|---|
| Fazil | Aliyev | Rauf@lan.ab.az | Azerbaijan |
| Azad | Bayramov | azhep@lan.ab.az | Azerbaijan |
| Yashar G. | Guseinaliev | Semic@lan.ab.az | Azerbaijan |
| Gilles de | Lentdecker | Gilles.delentdecker@hep.iihe.ac.be | Belgium |
| Germano | Guedes | Germano@cbpf.br | Brazil |
| Andre | Massafferri | Masaferi@lafex.cbpf.br | Brazil |
| Miroslav | Danchev | Miroslav@phys.uni-sofia.bg | Bulgaria |
| Nikolai | Durmenov | Durmenov@heph.phys.uni-sofia.bg | Bulgaria |
| Kamil | Sedlak | Ksedlak@fzu.cz | Czech Rep. |
| Markus | Hauser | Hauser@lhep.unibe.ch | Germany |
| Heinz | Tilsner | Tilsner@physi.uni-heidelberg.de | Germany |
| Michael K. | Vowotor | Lafoc@ncs.com.gh | Ghana |
| Socrates | Kiourkos | Socrates.kiourkos@cern.ch | Greece |
| Anand | Dubey | Akdubey@iopb.res.in | India |
| Mehrdad | Atabak | Mardanik@physic.sharif.ac.ir | Irania |
| Gohar | Rastegarzadeh | Rastegaa@physics.sharif.ac.ir | Irania |
| Antonio | Lagatta | Antonio.Lagatta@cern.ch | Italy |
| Andrea | Manara | Andrea.manara@cern.ch | Italy |
| Emad Al | Mahmoud | Enidal@yu.edu.jo | Jordan |
| Woon-Seng | Choong | Wschoong@lbl.gov | Malaysia |

| | | | |
|---|---|---|---|
| Mario Ivan Martinez | Hernández | Mim@fis.cinvestav.mx | Mexica |
| Claudia | Moreno | Yaya@fis.cinvestav.mx | Mexica |
| Serghei | Malkov | Malkova@lises.asm.md | Moldovia |
| Prem Raj | Dhugel | Glocom@mos.com.np | Nepal |
| Hari prasad | Lamichhane | Glocom@mos.com.np | Nepal |
| Neeraj | Nepal | Glocom@mos.com.np | Nepal |
| Durga | Paudyal | Glocom@mos.com.np | Nepal |
| Ramesh | Paudyal | Glocom@mos.com.np | Nepal |
| Yamal Chandary | Rajbhandary | Yamalr@usa.net | Nepal |
| Ramesh Raj | Subedi | Glocom@mos.com.np | Nepal |
| Pawel | Majewski | Pawel.Majewski@ires.in2p3.fr | Poland |
| Keith | Mathieson | K.Mathieson@physics.gla.ac.uk | Scottland |
| Craig | Whitehill | c.whitehill@physics.gla.ac.uk | Scottland |
| Assane | Ndiaye | Mmfaye@telecomplus.sn | Senegale |
| Peter | Mkhabela | Franklyn@aec.co.za | South Africa |
| Elsie | Monale | Elsie106@hotmail.com | South Africa |
| Gabriela | Llosa | Gllosa@mailcity.com | Spain |
| Patrik | Ekström | Ekstrom@physto.se | Sweden |
| John | Watt | Jwatt@physics.gla.ac.uk | UK |
| Dmytro | Pugachov | Pugachov@mppmu.mpg.de | Ukraine |
| Stephen | Bailey | Bailey@physics.harvard.edu | USA |
| Suzanne | Nicol | Suzanne.Nicol@cern.ch | USA |
| Naubet | Bisenov | Nissan@suninp.tashkent.su | Uzbekistan |
| Azmi | Barut | Azmibarut@hotmail.com | Turkish |
| Ilkay | Cakir | Iturk@science.ankara.edu.tr | Turkish |
| Orhan | Cakir | Ocakir@science.ankara.edu.tr | Turkish |
| Ozlem | Celik | | Turkish |
| Cuneyt | Celiktas | Celiktas@fenfak.ege.edu.tr | Turkish |
| Yesim | Cetin | Yoktem@hotmail.com | Turkish |
| Isa | Dumanoglu | Asim@pamuk.cc.cu.edu.tr | Turkish |
| Sefa | Erturk | Esefa@hotmail.com | Turkish |
| Gulhan | Gurdal | Gulhangurdal@hotmail.com | Turkish |
| Metin | Kantar | Kantar@fndaub.fnal.gov | Turkish |
| Zerrin | Kirca | | Turkish |
| Nil | Koc | | Turkish |
| Ozgul | Kurtulus | Ozgul@ocean.phys.boun.edu.tr | Turkish |
| Yildirim | Mutaf | mutafy@gursey.gov.tr | Turkish |
| Erdal | Recepoglu | Recepogl@science.ankara.edu.tr | Turkish |
| Iskender | Reyhancan | Reyhanca@cnaem.nukleer.gov.tr | Turkish |
| Saim | Selvi | | Turkish |
| Bukem | Tanoren | Bukem@ocean.phys.boun.edu.tr | Turkish |
| Oktay | Ureten | Ureten@ieee.org | Turkish |
| Mehtap | Yalcinkaya | Yalcin@nucleus.istanbul.edu.tr | Turkish |

## AUTHOR INDEX

### A

Amendolia, S. R., 381
Anderson, J., 291
Ataç, M., 227, 240, 291, 294

### B

Bisogni, M. G., 381
Bottigli, U., 381
Brenner, R., 349

### C

Cattai, A., 324
Çetin, S. A., 301
Cihangir, S., 291, 294
Clark, A. G., 115
Çonka-Nurdan, T., 301
Corral, G. Herrera, 267
Crescio, E., 35

### D

Delogu, P., 381
Dipasquale, G., 381
Dixon, R. L., 253

### E

Erduran, N., 291, 294
Erturk, S., 291, 294

### F

Fantacci, M. E., 381

### G

Giubellino, P., 35
Gorišek, A., 340

Grupen, C., 3

### H

Hernandez, R., 35

### I

Idzik, M., 35

### J

Johnson, M., 329

### K

Kartal, S., 291, 294
Korpar, S., 340
Križan, P., 340

### M

Maestro, P., 381
Mailov, A., 301
Malina, R., 324
Marzulli, V., 381

### N

Nappi, E., 60
Nouais, D., 35

### P

Pernigotti, E., 381

### Q

Quigg, C., 165

## R

Rivetti, A., 35
Roe, S., 349
Rosso, V., 381
Rudge, A., 349

## S

Savoy-Navarro, A., 192
Sheaff, M., 87, 329
Stanovnik, A., 340

Stefanini, A., 381
Stumbo, S., 381

## V

Venier, O., 381